Advances in Theoretical and Experimental Research of High Temperature Cuprate Superconductivity

Peking University–World Scientific Advanced Physics Series

ISSN: 2382-5960

Series Editors: Enge Wang *(Peking University, China)*
Jian-Bai Xia *(Chinese Academy of Sciences, China)*

Vol. 10 *Advances in Theoretical and Experimental Research of High Temperature Cuprate Superconductivity*
edited by Rushan Han

Vol. 9 *Electron–Phonon Interaction and Lattice Dynamics in High T_c Superconductors*
edited by Han Zhang

Vol. 8 *Physics on Ultracold Quantum Gases*
by Yongjian Han, Wei Yi and Wei Zhang

Vol. 7 *Endless Quests: Theory, Experiments and Applications of Frontiers of Superconductivity*
edited by Jiang-Di Fan

Vol. 6 *Superconductivity Centennial*
edited by Rushan Han

Vol. 5 *Quasi-One-Dimensional Organic Superconductors*
by Wei Zhang and Carlos A. R. Sá de Melo

Vol. 4 *Applied Symbolic Dynamics and Chaos (Second Edition)*
by Bailin Hao and Weimou Zheng

Vol. 3 *Computer Simulations of Molecules and Condensed Matter: From Electronic Structures to Molecular Dynamics*
by Xin-Zheng Li and En-Ge Wang

Vol. 2 *Geometric Methods in Elastic Theory of Membranes in Liquid Crystal Phases (Second Edition)*
by Zhanchun Tu, Zhongcan Ou-Yang, Jixing Liu and Yuzhang Xie

Vol. 1 *Dark Energy*
by Miao Li, Xiao-Dong Li, Shuang Wang and Yi Wang

Peking University-World Scientific Advanced Physics Series

Advances in Theoretical and Experimental Research of High Temperature Cuprate Superconductivity

Editor

Ru-shan Han
Peking University, China

Published by

World Scientific Publishing Co. Pte. Ltd.
5 Toh Tuck Link, Singapore 596224
USA office: 27 Warren Street, Suite 401-402, Hackensack, NJ 07601
UK office: 57 Shelton Street, Covent Garden, London WC2H 9HE

British Library Cataloguing-in-Publication Data
A catalogue record for this book is available from the British Library.

B&R Book Program

Peking University-World Scientific Advanced Physics Series — Vol. 10
ADVANCES IN THEORETICAL AND EXPERIMENTAL RESEARCH OF
HIGH TEMPERATURE CUPRATE SUPERCONDUCTIVITY

Copyright © 2020 by Han Rushan

This Work is originally published by Peking University Press in 2014.
This edition is published by World Scientific Publishing Company Pte Ltd by arrangement with Peking University Press, Beijing, China.

All rights reserved. No reproduction and distribution without permission.

ISBN 978-981-3271-16-6 (hardcover)
ISBN 978-981-3271-17-3 (ebook for institutions)
ISBN 978-981-3271-18-0 (ebook for individuals)

For any available supplementary material, please visit
https://www.worldscientific.com/worldscibooks/10.1142/11011#t=suppl

Desk Editor: Nur Syarfeena Binte Mohd Fauzi

Typeset by Stallion Press
Email: enquiries@stallionpress.com

Contents

Preface		xix
1.	The Electronic State Phase Diagram of Copper Oxide High-Temperature Superconductors	1
	Jian-Lin Luo	
	1.1 The Region of the Undoped "Parent" and the Low Carrier Doping Compounds with Long-Range Anti-ferromagnetic Order	5
	1.2 Superconducting State	6
	1.2.1 d-Wave Pairing Symmetry	7
	1.2.2 Very High Upper Critical Field (H_{C2}) and Short Coherent Length	9
	1.2.3 The Very Strong Superconducting Fluctuations	10
	1.3 Normal State Pseudogap	11
	1.4 Studies on HTSC's Phase Diagram Using High-Resolution Differential Specific Heat Measurement Technique	20
	1.5 Summary of the Phase Diagram	25
	1.6 Unsolved Problems	25
	Acknowledgments	26
	References	26

2.	The Complement of Phase Diagram		27

Ru-Shan Han

- 2.1 The Characteristic of the Full Phase Diagram's Transition is the Basic One of HTSC, Including Five Transitions and Five Revelations 29
  - 2.1.1 Transition One 29
  - 2.1.2 Transition Two 34
  - 2.1.3 Transition Three 36
  - 2.1.4 Transition Four 37
  - 2.1.5 Transition Five 39
- 2.2 Main Features of the Superconducting State 41
  - 2.2.1 Inhomogeneity of the Superconducting State 41
  - 2.2.2 Indicate that the Local Pair in Real Space and Coherence Condensation Do Not Occur Simultaneously 42
- 2.3 Phase Diagram 44
  - 2.3.1 Alloul Phase Diagram 44
  - 2.3.2 Phase Diagram Under Strong Magnetic Field 45
  - 2.3.3 Guo-Qing Zheng's Important Work in NMR Study........................ 46
- References 49

3.	Infrared and Optical Response of High-Temperature Superconductors		51

Nan-Lin Wang

- 3.1 Introduction 51
- 3.2 An Introduction about Optical Properties of Solids 53
- 3.3 Optical Properties of High-T_C Superconductors..... 60
- 3.4 Whether the Superconducting Energy Gap Can be Observed in the In-Plane Infrared Measurement 65
- References 77

4.	NMR Study of High-Temperature Superconductors		79
	Guo-Qing Zheng		
	4.1	Introduction	79
	4.2	Local Hole Distribution	80
	4.3	Superconducting State Properties	83
		4.3.1 Energy Gap and the Nuclear Spin-Lattice Relaxation Rate	83
		4.3.2 Cooper Pair Symmetry	85
		4.3.3 Impurity Effect	89
	4.4	Electron Correlations and the Ground State of the Pseudogap	91
	4.5	Concluding Remarks	96
	References		97
5.	Angle-Resolved Photoemission Spectroscopy Studies of Many-Body Effects in High-T_C Superconductors		99
	Xing-Jiang Zhou		
	5.1	Introduction to Angle-Resolved Photoemission Spectroscopy	99
	5.2	Angle-Resolved Photoemission Spectroscopy Studies of Many-Body Effects in High-T_C Superconductors	102
		5.2.1 The First Observation of Many-Body Effects in High-T_C Superconductors — Nodal Direction	102
		5.2.2 The Origin of the 70 meV "Kink" Structure	104
		5.2.3 Non-Traditional Electron–Phonon Coupling	107
		5.2.4 Extracting the Boson Spectral Function from the ARPES Experimental Results	108
		5.2.5 Many-Body Effects in Anti-Nodal Direction	110
		5.2.6 The Possible and New Electronic Coupling Modes in High-T_C Superconductors	113
	References		116

6. ARPES Study on the High-Temperature
   Superconductors — Energy Gap, Pseudogap
   and Time-Reversal Symmetry Breaking   117
   Shan-Cai Wang

   6.1 The Application of ARPES in the High-Temperature
       Superconductivity 117
   6.2 Superconducting Gap, Pseudogap and Two-Gap
       Problems 120
   6.3 Pseudogap 123
   6.4 Time-Reversal Symmetry Breaking 132
   References 140

7. Progress in the Scanning Tunneling Microscopy Study
   of High-Temperature Superconductors   141
   Qiang-Hua Wang

   7.1 Introduction to Scanning Tunneling Microscopy 141
   7.2 Impurity Resonance States 142
   7.3 Vortex Core States 143
   7.4 Gap Inhomogeneity 144
   7.5 Quasi-particle Scattering Interference 144
   7.6 Density Waves 145
   7.7 Gap Closing Versus Temperature 146
   7.8 Bosonic Modes 147
   References 148

8. Intrinsic Tunneling Spectroscopy of $Bi_2Sr_2CaCu_2O_{8+\delta}$
   Cuprate Superconductors   149
   Shi-Ping Zhao

   8.1 Tunneling Studies of BCS Superconductors 150
       8.1.1 Characteristics of Electron Tunneling
             Spectroscopy 150
       8.1.2 Strong-coupling Theory and Eliashberg
             Equations 151
       8.1.3 Other Effects Beyond BCS 154
   8.2 Tunneling Studies of Bi2212 Cuprate
       Superconductors 156

		8.2.1	STM Experiments: Precursor Pairing Scenario of the Pseudogap	157

        8.2.2 BJ Experiments: Superconducting Pairing Glue Mechanism 159

        8.2.3 IJJ Experiments: Superconducting Gap and Pseudogap from Different Origins 160

  8.3 Intrinsic Tunneling Studies of Submicron Bi2212 IJJs 162

        8.3.1 Sample Fabrication 162

        8.3.2 Surface Layer Characterization and Control 163

        8.3.3 Junction Size Versus Self-heating 163

        8.3.4 Temperature- and Magnetic-field-dependent Tunneling Spectra 166

        8.3.5 Tunneling and Transport Properties Along the C Direction 167

        8.3.6 Tunneling Spectra at Low Temperatures 169

        8.3.7 Tunneling Spectra at High Temperatures 171

  8.4 Conclusion 175

References 175

9. The Transport Properties of High-Temperature Cuprate Superconductors   177

  Xian-Hui Chen

  9.1 Magnetic Structure and Phase Diagram 177
  9.2 Resistivity and Hall Coefficient 179
  9.3 Charge Confinement and *c*-Axis Transport 188
  9.4 Pseudogap and Resistivity 189
  9.5 Transport Properties under High Magnetic Field 192
  9.6 The Transport Properties of the Underdoped Samples 197

References 205

10. Nernst Effect and Phase Fluctuation Picture of High-T_C Superconductors   207

  Zhu-An Xu

  10.1 Introduction to Nernst Effect 207

	10.2	High-T_C Superconductor Phase Diagram and Pseudogap State	210
		10.2.1 Electronic Phase Diagram	210
		10.2.2 Pseudogap	212
	10.3	Nernst Effect of Hole-Doped High-T_C Superconductor	215
		10.3.1 Vortex Nernst Signal Above T_C	215
		10.3.2 Upper Critical Field H_{C2}	220
		10.3.3 2D Nature of Vortex Excitations Above T_C	221
		10.3.4 Superconducting Phase Fluctuations Picture	223
		10.3.4.1 Conventional fluctuations theory	225
		10.3.4.2 Strong phase fluctuation theory (preformed pair theory)	226
		10.3.4.3 Resonance valence bond (RVB) theory	227
	10.4	Nernst Effect in Electron-Doped High-T_C Superconductors	228
	10.5	Further Experimental Evidence of Superconducting Phase Fluctuation Picture: Magnetic Susceptibility Measured by Torque Technique	231
	10.6	Conclusions	234
		Acknowledgments	235
		References	235
11.	Very Low-Temperature Heat Transport Properties of High-Temperature Superconductors		237

Xue-Feng Sun

	11.1	Introduction of Heat Transport	237
	11.2	Universal Thermal Conductivity	242
	11.3	Very Low-Temperature Thermal Conductivity and Metal–Insulator Crossover	257
	11.4	Wiedemann–Franz Law	268
	11.5	Acquirement and Analysis of the Intrinsic Low-Temperature Thermal Conductivity Data	271
	11.6	Conclusions	278
		References	278

12. A Brief Overview of Raman Scattering in Cuprate
    Superconductors   281

    Mei-Jia Wang, An-Min Zhang and Qing-Min Zhang

    12.1 Introduction   281
    12.2 Electron Raman Scattering in High-Temperature
         Superconductors   283
         12.2.1 Phenomenological and Microscopic Theoretical
                Pictures   283
         12.2.2 Superconducting State   287
                12.2.2.1 Superconducting gap and anisotropic
                         pairing   287
                12.2.2.2 Two gaps   291
                12.2.2.3 Impurity effect   292
         12.2.3 Normal State   293
                12.2.3.1 Pseudogap   293
                12.2.3.2 Quantum critical point   295
                12.2.3.3 Electronic stripe phase   296
    12.3 Raman Scattering of Two-Magnon Process   298
    12.4 Overview   300
    References   301

13. Quasi-particle Excitations in High-T_C Cuprate
    Superconductors Probed by Specific Heat: Implications
    on the Superconducting Condensation   303

    Yue Wang and Hai-Hu Wen

    13.1 Introduction   304
    13.2 Low-Temperature Specific Heat of a D-Wave
         Superconductor: Theoretical Background   306
         13.2.1 Nodal Excitation Spectrum of a D-Wave
                Superconductor   306
         13.2.2 The Effect of a Magnetic Field: The Volovik
                Effect and Scaling Relations   307
         13.2.3 Effect of Impurity Scattering 309
         13.2.4 Specific Heat Anomaly at T_C   310

	13.3	Experimental Background	311
		13.3.1 Experimental Techniques: Principle of the Relaxation Method	311
		13.3.2 Data Analysis: Separation of Different Contributions	312
		13.3.3 Overview of Previous Specific Heat Studies in High-T_C Cuprates	314
		13.3.3.1 Low-temperature specific heat: Evidence for d-wave symmetry of the superconducting gap	314
		13.3.3.2 Specific heat anomaly at T_C	316
	13.4	Low-Temperature Specific Heat Results in $La_{2-x}Sr_xCuO_4$ (LSCO) and $Bi_2\ Sr_{2-y}La_yCuO_{6+\delta}$ (La-Bi2201)	319
		13.4.1 Evidence for a d-wave Pairing State Throughout the Phase Diagram	320
		13.4.2 Bulk Measure of the Superconducting Gap and its Anomalous Doping Dependence	322
		13.4.3 Evidence for an Intimate Relation Between Pseudogap and High-T_C Superconductivity . . .	325
		13.4.4 Residual Linear-T Specific Heat in Zero-Field: Implications on Phase Separation	327
		13.4.5 Implications on the Superconducting Condensation	329
		13.4.5.1 Evidence for a "Fermi arc" ground state in the underdoped regime	329
		13.4.5.2 Indication of a second superconducting energy scale	334
		13.4.5.3 Doping dependence of T_C, the superconducting condensation energy E_C, and the upper critical field H_{C2}	336
		13.4.5.4 Additional remarks	341
	13.5	Summary	343
	References		343

14. Recent Results on the 2D Hubbard, $t-J$ and Gossamer
    Models and Relevance to High-Temperature
    Superconductivity   347

    Gang Su

    14.1 Introduction 347
    14.2 Two-Dimensional Hubbard Model 350
    14.3 The 2D $t-J$ Model 356
    14.4 Gossamer Superconducting Model 361
    14.5 Concluding Remarks 364
    References 366

15. The High-Temperature Superconducting Cuprates
    Physics: The "Plain Vanilla" Version of RVB   369

    Yue Yu and Ru-Shan Han

    15.1 Introduction 369
    15.2 The Method 373
    15.3 Overview 378
    15.4 Conclusion 380
    15.5 Postscript 381
    References 381

16. The Two-Band Model of Electron-Doped High-T_C
    Superconductors   383

    Hong-Gang Luo and Tao Xiang

    16.1 Introduction 383
    16.2 Experimental Evidence for the Two-Band
         Model 387
         16.2.1 ARPES [16] 387
         16.2.2 Transport Measurements [20–23] 387
         16.2.3 Penetration Depth Experiments
                [5, 6, 24] 389
    16.3 The Two-Band Model 390
    References 391

17. Theoretical Investigations on the Spin Dynamics in High-T_C Cuprates 393

Jian-Xin Li

  17.1 Experiments of Spin Dynamics — Nuclear Magnetic Resonance and Neutron Scattering Experiment 393
      17.1.1 Nuclear Magnetic Resonance 393
      17.1.2 Neutron Scattering Experiment 394
      17.1.3 Commensurate and Incommensurate Spin Response 394
      17.1.4 Universality in the Spin Dynamics — The Hourglass Dispersion 395
      17.1.5 Comparison with the Dispersion in the Parent Compounds La_2CuO_4 and YBCO 396
      17.1.6 Doping Dependence of the Incommensurability 398
      17.1.7 Frequency Dependence of the Dynamic Spin Susceptibility–Spin Resonance 399
  17.2 Two Theoretical Pictures for the Explanation of the Neutron Scattering Data 401
      17.2.1 Striped-phase Scenario 402
            17.2.1.1 Striped-phase model and Yamada plot 402
            17.2.1.2 Metallic stripe phase 403
      17.2.2 The Weak-coupling Theory Based on the Fermiology and D-wave Superconductivity . . . 407
            17.2.2.1 Incommensurate peaks along $(\pi, \pi \pm \delta\pi)$ direction in the momentum space 409
            17.2.2.2 Spin resonance at the anti-ferromagnetic wave vector (π, π) — collective spin excitation mode 410
            17.2.2.3 Spin excitation spectrum 411
            17.2.2.4 Yamada plot — δ (incommensurability) $= 2x$ (doping concentration) 413
            17.2.2.5 Temperature and doping dependences of the spin-resonance mode energy 414

		17.2.2.6	Anisotropy of the spin incommensurate structure in detwinned YBaCuO$_{6.6}$	414
		17.2.2.7	Spin dynamics in electron-doped high-T_C cuprates	418
	17.3	Comparison between the Striped-Phase Scenario and the Weak-Coupling Theory		420
		17.3.1	Universal Dispersion	420
		17.3.2	Yamada Plot and Anisotropic Spin Fluctuations	421
		17.3.3	Origin of the Spin Resonance	421
		17.3.4	Incommensurate Peaks along Diagonal Directions at Very Low Dopings (0.02−0.05)	421
	17.4	Concluding Remarks		421
	Acknowledgments			422
	References			422

18. **Slave-Boson Effective-Field Theory of RVB State and Its Application for the Mechanism of High-Temperature Superconductivity**   **423**

   Tao Li

	18.1	Background		423
	18.2	Doped Mott Insulator and RVB State		425
	18.3	Slave-Boson Representation and the Mean-Field Theory of $t-J$ Model		427
		18.3.1	Half-filled System	428
		18.3.2	The Physical Meaning of the Gauge Field . . .	430
		18.3.3	The Staggered-Flux Phase and the SU(2) Gauge Structure	432
		18.3.4	Finite Doping	434
	18.4	Gauge Fluctuation and Its Physical Consequences . . .		436
		18.4.1	U(1) Gauge Fluctuation in the Strange Metal Phase	436
		18.4.2	Z_2 Gauge Fluctuation in the Pseudogap Phase	439
		18.4.3	The Ioffe–Larkin Composition Rule	440
		18.4.4	Flux Quantization and Vortex Core State . . .	441
		18.4.5	Quasi-particle Properties	444

	18.4.6 Spin Dynamics	446
	18.4.7 The Issue of Confinement	446
18.5	Variational Study of the RVB Theory: Gutzwiller Projected Wave Function	447
	18.5.1 Ground State	448
	18.5.2 Quasi-particle Properties	449
	18.5.3 Spin Dynamics	451
18.6	Conclusions	454
	Acknowledgments	455
	References	455

19. Superconductivity of Cuprates — A Phenomenon of Strong Correlation of Electrons 457

Wei Guo and Ru-Shan Han

19.1	Electronic Structure of the Cuprate and Magnetism	458
19.2	The Magnetic Origin of Superconductivity in Cuprates	460
19.3	Effect of the Local Spin Polarization	462
19.4	The Resonating Valence Bond State in the Superconducting Cuprates	464
19.5	Summary	467
	References	468

20. Magnetic Excitations in High-Temperature Superconductors: Search for Universal Features in Different Classes of Copper Oxides 469

Peng-Cheng Dai and Shi-Liang Li

20.1	Introduction	470
20.2	The Evolution of Static Anti-ferromagnetic Order when Hole- or Electron-Doped to the Copper–Oxygen Plane	472
20.3	The Evolution of Spontaneous Excitation as a Function of Hole Concentration	474
20.4	The Evolution of the Spin Excitation in the Electron-Type High-T_C Superconductors	479
20.5	Summary and Conclusions	483

Acknowledgments 485
References 485

A Letter to the High-Temperature Superconducting Colleagues 487

Conference Summary 489

Preface

"The Assessment Seminar on High-Temperature Superconducting Mechanism Research" was jointly held by China Center of Advanced Science and Technology (CCAST); Institute of Physics, Chinese Academy of Sciences(CAS); National Lab for Superconductivity, Institute of Physics, CAS; School of Physics, Peking University; and Center of Advanced Study Tsinghua University on March 1-5 2008. Experts in experimental and theoretical aspects on the first front were invited to give reviews and attempt to conclude the main consensuses, differences, and solvable and unsolvable aspects by analyzing some influential theoretical models in the last twenty years in superconductivity from a global perspective, to provide the striving direction and inspire enthusiasm and to motivate high-temperature superconductivity (HTSC) research further. This required the scientist to introduce the raw experimental data and the problems objectively, to point out the trend of the theoretical study impartially and to discuss the real situation and provided recommendations more deeply.

The conference was a great success, and basically achieved its purpose. The scientists made careful preparation, and provided concise and insightful data. Participants' enthusiastic involvement and serious discussion made the conference very fruitful.

At the time of the arrangement, having a meeting and preparing for the corpus, i.e. 2008, a new class of superconductors — FeNi-based oxygen-chalcogenide layered superconductors — occupied the central position in the superconducting field, and Chinese scientists made outstanding contributions in this regard. After about a year of exploration, the scientists have basically sorted out the general picture and could summarize that only the Fe-based small family has superconductivity among the hundreds of materials with ZrCuSiAs class structure. From the point of experimentation, they are on the border between BCS superconductors and cuprate oxide superconductors, but closer to the BCS superconductors with only traces of cuprate oxide superconductors. Its characteristics are as follows: strong e-p interaction with distorted s-wave symmetry; while above the T_C, the appearance of e-e interaction turning out–the resistivity $\rho \alpha T^2$; the parent compound is bad metal, with neither nearly free electron nor insulator; not k-space pairing and not completely real space pairing, and its H_C2 is between the BCS and cuprate oxide superconductors, about 50–60 Tesla, (the low T_C samples showed a higher H_C2); the angle-resolved photoemission spectroscopy (ARPES) data is in good agreement with the band structures calculation, indicating that they are weakly correlated multi-band system; the penetration depth is approximately 190 nm; direct exchange in intra-layer and ionization interaction in interlayer in this layered compounds; the superconducting area and magnetic area are close neighbors, but it is unclear whether they co-exist so further study is needed; including the additional data for the magnetic ordering with the doping evolution, it is one of the controversial issues whether the pseudogap exists or not. In short, the preparation of good single crystals still presents a major challenge, and the use of these will provide conclusive results with regard to some experimental properties. It will help to accurately conclude the similarity and sample differences and promote the study of the superconducting mechanism.

I think it is certain that the mechanism of superconductivity in cuprates is a more imperative task. Finding out that they are the extreme cases can promote the thinking of this new family situation between the BCS and cuprate oxide superconductors. As an important control system of the high-T_C cuprate oxide superconductor, Fe-based superconductors will also promote the study of the mechanism of the cuprate oxide superconducting system. Fe-based superconductors work by direct exchange interaction, while cuprates work by super-exchange interaction. So, exchange interaction should be given an extremely important position and considered as the starting point to build the framework of the theory of high-temperature superconducting cuprate mechanism. At this point, it seems that most of superconducting researchers have reached a consensus. Outwardly, the difference lies in the model, and in fact there are very fundamental differences. In my personal view, the following points should be emphasized:

1. The super-exchange concept was proposed by H. A. Kramers seventy years ago. The difference is that the transition metal wave function does not directly overlap. Owing to the presence of the oxygen ion O^{2-} and the

transition metal ions such as Cu ions, the electron wave functions overlap. However, the lack of oxygen ions O^{2-} will affect the strength of the super-exchange. The phase diagram shows that hole injection at the oxygen sites makes the antiferromagnetic ordering gradually disappear and convert to a conventional Fermi liquid, during the emergence of high-temperature superconductivity in cuprate oxide superconductors. Thus, oxygen ions O^{2-} should be seriously discussed. The simplified models, ignoring the oxygen ions, may be successful in other antiferromagnetic materials, but can be considered unsuccessful when studying high-temperature superconducting mechanism. For such cases, the three-band model or two-component model should be taken as the starting point. Unfortunately, the Zhang–Rice model has led the problem in a wrong direction, back to the single-band model from the three-band model. This is the theoretical reason why some people still insist on using, a single-band model.

2. Using a pure two-dimensional (2D) model on layered compounds is successful in many systems. But the research of high-temperature superconductivity mechanism in the high-temperature superconducting couprate oxide is not very successful. A quasi-2D (actual three-dimensional (3D)) model should be used, which will greatly expand the imagination and processing space.

3. The RVB picture may be the correct picture, especially for the bad metal system. In the three-band or two-component model, there should be short-range pairing of oxygen holes (or electrons). It is still hotly debated whether the glue exists or not.

4. In recent years, there is a debate on the pseudogap state, i.e. below the T^*, the density of states near Fermi surface decreases, and Fermi surface decreases a lot, up to T_C, the superconducting gap appears, and Fermi surface disappears — strictly speaking leaving only some points (nodes). There are two opposing views about the pseudogap state, one is pre-pairing, also known as single energy gap; the other is two-gap ordering competition. In fact, please do not ignore the correlated order of Cu^{2+}, which also plays a role in the decrease of density of state near Fermi energy. At zero doping, the system is an insulator, completely eating the density of state near Fermi energy. This process is gradual, and the localization of Cu^{2+} co-exists with the superconducting state.

This book, named "Advances in Theoretical and Experimental Research of High-Temperature Cuprate Superconductivity", is a collection of reports presented at the seminar of high-T_C superconductors and is the work of 20 years of research. As the name of the seminar suggests "The Assessment Seminar on High-Temperature Superconducting Mechanism Research", the meeting is a clearing inventory. After two decades of research, the mysteries of high-temperature superconductivity, despite not reaching a consensus on the mechanism, for the existing problems from the global perspective, it is a clean-up to guide future work. There are some people, who do the same work internationally, such as J. S. Brooks and J. R. Schrieffer, editors of "Handbooks of High-Temperature Superconductivity" (Springer, 2007). I hope this collection will play a role in promoting high-temperature superconductivity mechanism research. I thank all those who have contributed to the publication of this book.

Ru-Shan Han
During the Beijing Olympic Games in 2008

1
The Electronic State Phase Diagram of Copper Oxide High-Temperature Superconductors

Jian-Lin Luo

Institute of Physics, Chinese Academy of Sciences, Beijing 100190, China

High-temperature superconductors were discovered by J. G. Bendnorz and K. A. Müller in 1986. Before the discovery of high-temperature superconductivity in copper oxide perovskites, the transition temperature had increased from 4 K to 22–23 K when Kammerlingh Onnes first observed superconductivity in mercury in 1911. But since the Ba–La–Cu–O system at 40 K, discovered by Bednorz and Müller, the transition temperature has increased to 163 K in the Hg–Ba–Ca–Cu–O system under pressure in a very short time, which is several times higher than that considered before. Recently, an iron-based superconductor with T_C higher than 50 K has become the superconductor with the highest transition temperature, except for cuprates. The superconducting transition temperature and the discovery time are shown in Fig. 1.1.

As the superconductors' transition temperature T_C is very high, higher than the temperature of liquid nitrogen, it makes them have a wide range of applications, and in turn promotes the high-temperature superconductors' application research. In addition, the high-temperature superconductor is a strongly correlated electron system, and its physical properties are very complex and rich; so, it is very important to study the mechanism of the superconductivity. In condensed matter physics, there are no other systems

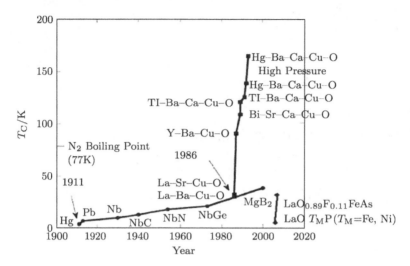

Fig. 1.1. The superconducting transition temperature T_C and the discovery time.

like the high-temperature superconductors, and so this is of much concern to physicists; thus, there are many different experimental methods to study.

Crystal structures of the cuprates: the high-temperature superconductors adopt a perovskite structure, which mainly include the weakly coupled copper-oxide layers and other charge-providers components (see Fig. 1.2). According to the type of carrier, the cuprates can be divided into hole- and electron-type superconductors; the hole-type superconductors, based on the number of CuO_2 layers, can be divided into single-, two-, three- and infinite-layer superconductors. The main characteristics of cuprates are dominated by the CuO_2 planes. The copper-oxide planes are checkerboard lattices with squares of O^{2-} ions and a Cu^{2+} ion at the centre of each square (see Fig. 1.3). Its structure is as follows: the Cu^{2+} ion has a $3d^9$ configuration, and the anti-ferromagnetic spin correlation between copper ions is the super- exchange interaction by the medium of the oxygen 2p electrons.

Electron- and hole-doped electronic state phase diagram: with the change in doping concentration, the properties of high-temperature superconductors have changed a lot. Figure 1.4 shows the phase diagram of high-temperature superconductors. At the left is the electron-doped region, while at the right is the hole-doped region. The undoped "parent" and the low carrier doping compounds are Mott insulators with 3D long-range anti-ferromagnetic ordering. However, as the doping increases, it slowly

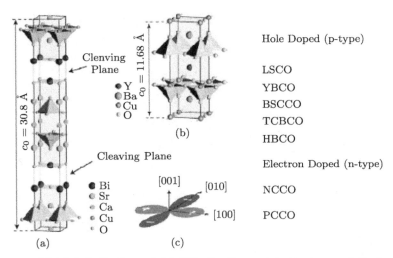

Fig. 1.2. (a) $Bi_2SrCaCu_2O_8$ and (b) $YBa_2Cu_3O_7$ crystal structure and the type of high-temperature superconductors.

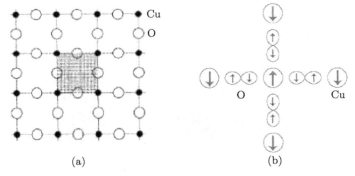

Fig. 1.3. The structure of CuO_2; spin correlation between copper ions is due to the super-exchange interaction by the medium of the oxygen 2p electron.

becomes conductive; then, there is a relatively large region of doping, called the pseudogap region, which means there is a gap found in the normal state. With the carrier concentration increasing, it becomes a superconductor, and then as the carrier concentration continues to increase, it will basically become a Fermi liquid, i.e. normal metallic state. In cuprate superconductors, the T_C as a function of doping concentration of the different systems can basically be represented by the universal parabola, the

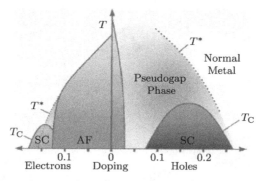

Fig. 1.4. The electronic state phase diagram of the high-T_C superconductor (AF: antiferromagnetic region; SC: superconducting region).

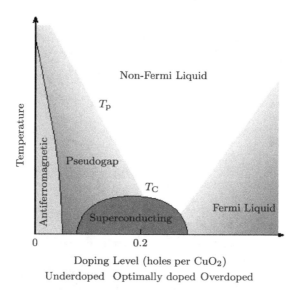

Fig. 1.5. The electronic state phase diagram of the hole-type high-T_C superconductors.

empirical formula. This chapter discusses the electronic state phase diagram of hole-type superconductors.

A more comprehensive electronic state phase diagram of the hole-type cuprate superconductors is shown in Fig. 1.5. The phase diagram shows that it is basically a non-Fermi liquid state in a large region, but as the doping concentration increases it takes on the behavior of a Fermi liquid. Why does its properties undergo such a big change, as a function of doping concentration? Why does the system finally reach the Fermi liquid state of normal metal from a strongly correlated 3D anti-ferromagnetic ordering system? How to explain these phenomena in the phase diagram of cuprate superconductors is very important.

The following is a discussion about the basic physical properties of the various parts of the phase diagram. Finally, I will introduce my work about using the high-resolution differential specific heat measurement technique to study the phase diagram.

1.1. The Region of the Undoped "Parent" and the Low Carrier Doping Compounds with Long-Range Anti-ferromagnetic Order

In Fig. 1.6, in the CuO_2 plane, the copper and oxygen ions are shown, and the anti-ferromagnetic order is the result of a super-exchange interaction between the copper and oxygen ions.

Table 1.1 lists the neutron scattering experimental results of different parent materials. Its anti-ferromagnetic ordering temperature differs for different systems: for single layer it is $\sim$300 K, bilayer it is $\sim$400 K and

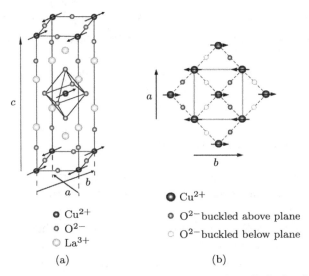

Fig. 1.6. The lattice structure and magnetic structure of La_2CuO_4 [2, 3]. Taken from Ref. [3].

Table 1.1. The neutron scattering experimental results of different parent materials [4].

Compound	T_N (K)	m_{Cu} (μ_B)	J (meV)	Layers/cell
La_2CuO_4	325	0.60	146	1
$Sr_2CuO_2Cl_2$	256	0.34	125	1
$Ca_2CuO_2Cl_2$	247	0.25		1
Nd_2CuO_4	276	0.46	155	1
Pr_2CuO_4	284	0.40	130	1
$YBa_2Cu_3O_{6.1}$	410	0.55	106	2
$TlBa_2YCu_2O_7$	>350	0.52		2
$Ca_{0.85}Sr_{0.15}CuO_2$	537	0.51		∞

infinite layer it is ∼500 K. But the energy scale of the super-exchange coupling J between copper and copper spins is far higher than that of kT_N, where T_N is the AF ordering temperature. For example, the J in La_2CuO_4 is about 146 meV, i.e. almost 1500 K.

So, why is the magnetic ordering temperature much lower than the coupling J? This is because it is a 3D long-order, whose intra-plane coupling J is very large, while the spin coupling between planes is small; so, the magnetic ordering temperature T_N is dependent on the inter-plane coupling. In fact, when the temperature is much higher than T_N, the 2D spin correlation is not destroyed completely, it instead becomes a short-range correlation, rather than a long-range magnetic order. Figure 1.7(a) shows the neutron scattering experiment results, which indicate that the 2D spin correlation peak is still there, even if the temperature reaches as high as 625 K. Figure 1.7(b) shows the relationship between the inverse magnetic correlation length as a function of temperature. We can see that the correlation length is indeed in decline, when the temperature increases, and it can almost extend to the magnitude of J.

1.2. Superconducting State

The properties of the superconducting state are very vast. Here, we will list some simple differences between the superconducting state of high-T_C cuprates and that of the conventional superconductors.

The copper oxide superconductors have very high superconducting transition temperature. Under pressure, its T_C can reach higher than 160 K, and so far no other systems can achieve this temperature. The superconducting paring symmetry is d-wave, which is the most significant result obtained after more than 20 years of research in the field of

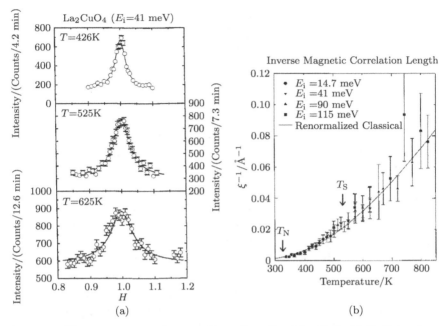

Fig. 1.7. The neutron scattering experimental results of La_2CuO_4 [5, 6]. Taken from Ref. [5].

high-temperature superconductivity, and it has been recognized by most people, although there may be some different opinions. There is a very high upper critical field. The upper critical field in YBCO can be more than 100 T. The coherence length is very short, the in-plane coherence length is about 10–12 Å and the inter-plane is about several Ångströms.

The superconducting fluctuations are very strong near T_C.

There is a formula which can express the T_C as a function of the concentration of carriers of different systems, i.e. $T_C/T_C^{max} = 1 - 82.6(p - 0.16)^2$.

Electron–phonon interaction is likely not the medium for the high-T_C pairing mechanism to occur.

The following will highlight some of the related aspects.

1.2.1. d-Wave Pairing Symmetry

According to the BCS theory, superconductivity is derived from electron pairing, and the conventional superconductors are s-wave paired. However, for the high-temperature superconductors, there is much evidence to prove

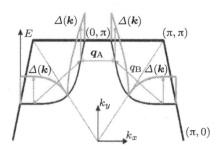

Fig. 1.8. The superconducting gap map of $d_{x^2-y^2}$ pairing in the $\mathbf{k}$ space.

that it follows the d-wave pairing symmetry. The main experimental evidences are of two types: one is the result of phase-sensitive experiments; the other is to study the pairing properties of high-temperature superconductors by studying the low-energy excitation. If it follows the s-wave pairing, the energy gap should be the same in all $\mathbf{k}$ directions on the Fermi surface; however, if it is $d_{x^2-y^2}$ pairing, as shown in Fig. 1.8, the superconducting gap reaches a maximum at $(\pi, 0)$, while the gap is zero along (π, π) (line nodes). Thus, if observing the physical quantities related to the density of states, the relationship between these quantities and energy should be a power law, rather than an exponential change like the conventional s-wave pairing superconductors. Therefore, we can confirm if there are nodes or not by measuring different physical quantities. Experimental examples include angle-resolved photoemission spectroscopy (ARPES), scanning electron microscope (STM) measuring tunnel spectrum, and measurements of penetration depth, nuclear magnetic resonance (NMR), optical conductivity, specific heat, thermal conductivity, Raman scattering, etc. Of course, NMR measurement can also provide some additional information, such as the relationship between the Knight shift and temperature and can be used to study whether the spin pairing is singlet or triplet.

In the following, we will present a brief summary of the tricrystal phase-sensitive experiments on pairing symmetry of HTSC. Tsuei *et al.* have made a π-tricrystal in YBCO film [8], and the three cuprate crystals involved are controlled to have different orientations, as shown in Fig. 1.9. If the $d_{x^2-y^2}$ pairing is the universal nature in cuprate superconductors, a spontaneously generated half-flux quantization will be observed only in the three-junction ring but not in the grain boundaries and the internal. The flux is directly measured with a high-resolution scanning SQUID microscope

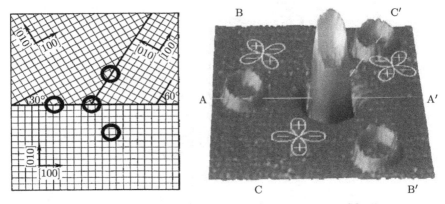

Fig. 1.9. The phase-sensitive experiments in YBCO film: a scanning SQUID microscope image of a thin-film YBCO tricrystal ring sample [7, 8]. At low temperature and in nominally zero magnetic field, the central three-junction ring has half of a superconducting quantum of flux spontaneously generated in it. Taken from Ref. [7].

(SSM). A detailed analysis of the SSM data shows that the flux threading through the center ring is indeed $\Phi_0/2$, where Φ_0 is magnetic flux quantum, which provides convincing evidence for a $d_{x^2-y^2}$ pairing symmetry.

Furthermore, the low-energy excitation experiments also indicate the difference between the high-T_C superconductors with line nodes in gap function and conventional s-wave superconductors. The tunneling spectroscopy experiment results are shown in Fig. 1.10, which show that the differential conductance is proportional to the density of states of a single quasi-particle. At very low temperature, when the bias is smaller than the superconducting gap, there is no density of states observed; then, there is a sudden increase in density of states where the bias is almost equal to the superconducting gap of ∼1 meV, showing the standard s-wave superconductor behavior for the conventional superconductor Nb (Fig. 1.10(a)). For the different systems of superconductors, the density of states with energy shows a linear relationship at very low temperature in the low energy part, indicating the existence of line nodes in the superconducting gap function.

1.2.2. Very High Upper Critical Field (H_{C2}) and Short Coherent Length

Besides the d-wave pairing, the high-T_C superconductors have another characteristic, a very high H_{C2}. Figure 1.11 shows the value of H_{C2} that was measured early on in 1989 using the pulsed-field in YBCO. The result shows

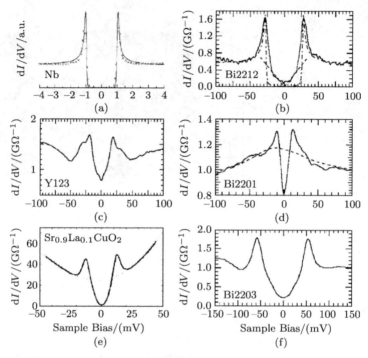

Fig. 1.10. C-axis SIN vacuum tunneling conductance spectra of conventional and high-T_C superconductors [1]. Taken from Ref. [1].

that even when the magnetic field increases up to 100 T along the ab plane, the superconductivity cannot be completely suppressed. So, its upper critical field is higher than 100 T, about 110 T. When the magnetic field is applied perpendicular to ab plane, the upper critical field is about 40 T, and the coherent length is about $\xi_{ab} = 25$ Å and $\xi_c = 8$ Å. Of course, these were very early works, and now we can achieve higher H_{C2} using better quality samples.

1.2.3. *The Very Strong Superconducting Fluctuations*

The superconducting state in HTSC has another characteristic the superconducting fluctuations are very strong. How can we judge if the superconducting fluctuations are strong or weak? We can do this by comparing the two values. One value is the result of condensation energy multiplied by coherent volume $(F_n - F_s)\Omega$, which indicates the total condensation energy

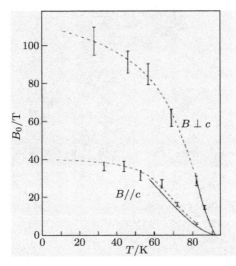

Fig. 1.11. The upper critical field as a function of temperature in a YBCO single crystal. Taken from Ref. [9].

of the related volume; the other is thermal fluctuations $k_B T_C$. The coherent volume is the square of its in-plane coherent length multiplied by its interplane coherent length, $\Omega = \xi_{ab}^2 \xi_c$. So, we can judge the fluctuations by comparing the size of condensation-energy and thermal fluctuations $k_B T_C$.

Because the coherent lengths are very small, the coherent volume is very small as well, but the transition temperature is rather high, so the superconducting fluctuation is very large. The value of $k_B T_C/(F_n - F_s)\Omega$ is 3~4 orders of magnitude larger than that of a conventional superconductor. Thus, the high-T_C superconductors have very strong superconducting fluctuations. For example (see Fig. 1.12), for the conventional superconductors, the specific heat curve has a very steep change near T_C, which agrees with the BCS theory very well. But for the high-T_C superconductors, the superconducting fluctuations occur in a large temperature range above T_C. For Ca-doped YBCO and Bi2212, this region is probably extended to 20–30 K.

1.3. Normal State Pseudogap

The pseudogap in YBCO was first observed using NMR measurement, which also enabled the obtainment of the relationship of $1/T_1 T$ as a function

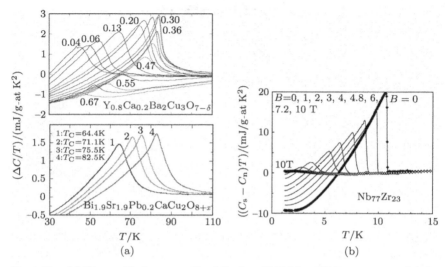

Fig. 1.12. (a) The specific heat near T_C of Ca-doped YBCO and Bi2212 systems; (b) the specific heat of conventional superconductor $Nb_{77}Zr_{23}$.

of temperature, where T_1 is the spin-lattice relaxation rate. $1/T_1T$ is a constant in a Fermi liquid, i.e. Korringer law. Also, $1/T_1T$ is directly proportional to the integral of the average dynamic susceptibility. Then for the normal states of the under-doped high-T_C superconductors, when the temperature is far higher than T_C, the $1/T_1T$ falls with an increase in temperature, indicating that the spin susceptibility also decreased. In addition, because the Knight shift is directly proportional to the spin susceptibility, the results from it can be seen more clearly. Figure 1.13 shows the sudden decrease in Knight shift in the vicinity of T_C of optimally doped and under-doped samples, because it is spin singlet pairing.

For under-doped samples, the Knight shift falls with an increase in temperature, indicating that the spin susceptibility also decreases. So, there is a spin gap, and it was thought the normal state pseudogap was a spin gap at the beginning. Then, further experiments were done, including thermodynamics experiments, transport and ARPES, which also demonstrated this gap; thus, it was called the normal-state pseudogap. The gap not only opens in the spin degree of freedom but also in the charge degree of freedom. Figure 1.14 shows, in the measurement of transport properties, that the resistivity at the temperature of pseudogap opening deviates from the linear resistivity behavior, which is observed at high temperatures. We can

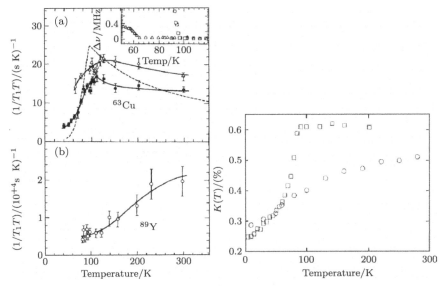

Fig. 1.13. The NMR experimental results for ^{63}Cu and ^{89}Y in YBCO$_{6.7}$ [11]. The left one shows $(TT_1)^{-1}$ as a function of temperature, the right one shows the Knight shift as a function of temperature ($\square$ represents the optimally doping samples; $\bigcirc$ represents the under-doped samples). Taken from Ref. [11].

see this more clearly from the specific heat measurement. The electron-specific heat is almost independent of the temperature in the over-doped region above T_C. The electron heat is directly proportional to the density of states (DOS) above the Fermi energy, so the DOS in normal state also does not change with temperature in the over-doped region. But in the under-doped samples, C/T falls as temperature decreases, from a temperature far higher than the T_C, indicating the loss of density of states; thus, there is a normal-state pseudogap gap. The ARPES showed similiar results. There is a complete Fermi surface in the over-doped region; however, the gap opens up at the $(\pi, 0)$ direction in the under-doped region. Furthermore, the surface reduced to a point in the node direction in the very under-doped region.

There are two theoretical pictures about the origin of pseudogap, and the present experiments also cannot make sure which understanding is right.

The first one (Fig. 1.15(a)) is that the pseudogap is derivated from the superconducting pre-pairing. It means the electrons first form the

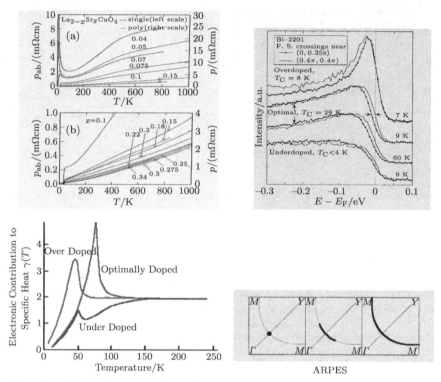

Fig. 1.14. The performances of pseudogap in normal states in transport [12], specific heat, and ARPES [14]. Taken from Refs. [12, 14].

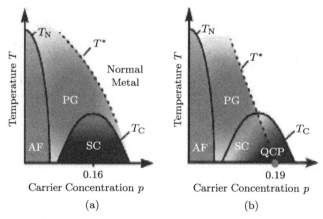

Fig. 1.15. Two pictures for pseudogap in the normal states as a function of doping concentration.

Cooper pairs at high temperatures, and open a gap which is called the pseudogap. Then as the temperature decreases to T_C, the Cooper pairs begin coherent and form superconductivity. It actually divides the process of the superconductivity into two steps. First, the Cooper pairs are formed. Because of the large phase fluctuations, the Cooper pairs cannot be coherent at higher temperature, and only start being coherent at a lower temperature T_C. In this setting, the pseudogap is actually the same as the superconducting gap.

The second setting (Fig. 1.15(b)) is one in which the pseudogap is not related to superconducting pairing, i.e. it has another origin. But it is not clear what the origin is — charge density wave (CDW), spin density wave (SDW) or other possible mechanisms. However, because the state in the pseudogap phase is competing with superconductivity, the pseudogap phase can extend into the superconducting phase. Many experiments show that the carrier concentration $p = 0.19$ may be a critical point, and the pseudogap can survive below this point, and vice versa.

The first view is supported by STM and the Nernst effect experiments (see Zhu-an Xu's paper; page 220).

In the STM experiment, Fig. 1.16(a), the result of an under-doped sample Bi2201 whose T_C is about 10 K can be seen. In the STM spectrum, the superconducting gap can be seen clearly below T_C, and the gap still exists above T_C, only the coherence peak is not obvious at the gap position, and the signal of the energy gap can be seen even at very high temperatures above the T_C. The energy gap has no sudden change and almost remains the same as the temperature changes. This experiment seems to imply that the superconducting energy gap and the normal state are the same. Figure 1.16(b) shows the different systems' results, such as over-doped and under-doped samples of Bi2201 and Bi2212, with the curves plotted together below and above T_C. It is found that there is indeed a gap that exists above T_C, and there are no significant changes comparing with values of that below T_C. The temperature at which the gap opens can be scaled in different systems. As shown in Fig. 1.16(c), T^*/T_C is almost linear independent with $2\Delta/k_B T_C$, where T^* is the temperature at which the gap opens.

Of course, STM can measure the local density of states (LDOS) within the nanometer scale, so we can use it to study LDOS at the vortex core. The work by Renner et al. [15, 16] measures the pseudogap changes in the scale of 0–4.2 nm (see Fig. 1.17), at $Y = 0$, which is in the location of the vortex core, while $Y = 4.2$ nm is the slow crossover to the superconducting region.

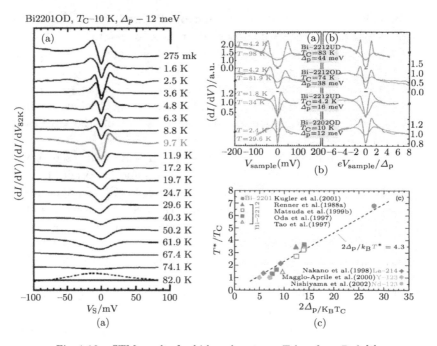

Fig. 1.16. STM results for bi-based systems. Taken from Ref. [1].

The experimental condition is at 4.2 K, 6 T. In the superconducting state, we can see the superconducting coherent peak clearly and can also observe an energy gap (i.e. pseudogap) at the vertex core.

Figure 1.17(b) shows the change process of the coherent peak in the superconducting state, as the temperature changes, which basically agrees with the results shown in Fig. 1.16. Figure 1.17(c) shows the tunneling conductance of the vortex core and the superconducting region. The characteristic properties are almost the same whether above or below T_C. Thus, it is thought that the pseudogap and the superconducting gap are the same gap. The magnitudes of both the superconducting gap and the pseudogap are almost the same and do not vary with temperature, and the temperature of the pseudogap opening increases monotonically with decreasing doping in the under-doped region. In addition, the pseudogap also exists in the over-doped region.

Below is an overview of the results of these measurements:

(1) The superconducting gap and the pseudogap are independent of temperature.

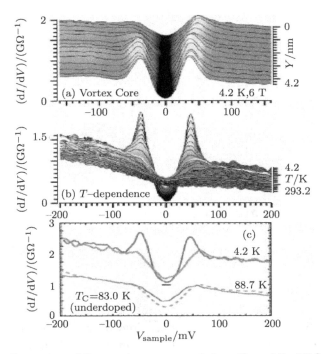

Fig. 1.17. Comparison of the pseudogap measured above T_C and the LDOS inside the vortex core of Bi2212 (OD). (a) Spatial evolution of the LDOS when moving the tip from the center ($Y = 0$) to the periphery of the vortex core ($Y = 4.2$ nm) at 4.2 K and 6 T[15] (over-doped Bi2212). (b) T dependence of the DOS[16] (under-doped Bi2212). (c) Top: superconducting gap measured between vortices at 4.2 K and 6 T compared to a vortex-core spectrum [15] (under-doped Bi2212). Taken from Ref. [1].

(2) They have the same amplitude.
(3) For under-doped samples, with the reduction in the amount of doping, the pseudogap temperature T^* increases monotonically.
(4) The pseudogap also exists in over-doped samples.

The STM results are the strong support for the setting shown in Fig. 1.15(a).

However, many experimental results demonstrate that the picture shown in Fig. 1.15(b) could be correct, i.e. the superconducting gap and the pseudogap are different gaps. Tallon and Loram wrote a good article analyzing of a variety of different experimental methods including heat capacity, NMR Knight shift and spin-lattice relaxation rate, in-plane resistivity, inter-plane resistivity, Raman scattering, ARPES, tunneling spectroscopy,

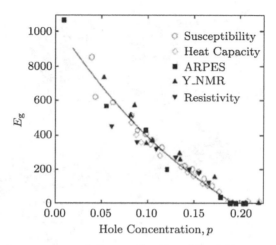

Fig. 1.18. Pseudogap energy variation as a function of the doping concentration, p, and E_g is determined by the magnetic susceptibility, specific heat, ARPES, ^{89}Y-NMR and resistivity. Taken from Ref. [17].

Zn-doped results, and so on. All these results indicate that the pseudogap extends into the superconducting state, and disappears when the carrier concentration is almost equal to 0.19. They obtained such a conclusion about the pseudogap for different systems, and using different methods, that a line could express the relationship between the pseudogap and doping. At zero doping, the magnitude of the pseudogap can be extended to the super-exchange energy scale J, which strongly suggests that the origin of the pseudogap may be related to the anti-ferromagnetic correlations of copper spins (see Fig. 1.18).

Zheng et al. carried out some interesting work on the NMR spectrum of Bi2201 system under a strong field [18]. Because the Bi2201's T_C is relatively low, a magnetic field of certain strength can completely destroy the superconductivity. Thus, the pseudogap behavior can be observed once the superconductivity is suppressed by the applied magnetic field. They studied the NMR spectrum in a large doping concentration from the over-doped region to the under-doped region. In the over-doped region, a magnetic field of 28.5 T can stop the superconductivity completely.

Looking at over-doped region in Fig. 1.19, when the hole concentration was probably 0.21, $1/T_1T$ showed a marked decrease at T_C under zero magnetic field, indicating the superconducting transition. However, $1/T_1T$ is nearly unchanged as temperature changes when the applied magnetic field destroys the superconductivity completely, in accordance with Korringer

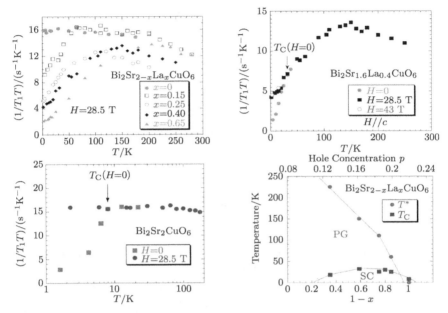

Fig. 1.19. The NMR experimental results of $Bi_2Sr_{2-x}CuO_6$ under strong magnetic field. Taken from Ref. [18].

law, which states that the over-doped region of $p > 0.2$ has a Fermi liquid behavior.

For other doping concentrations, $1/T_1T$ is obviously dependent on the temperature, indicating the existence of the pseudogap. Actually, the behavior of the spin-lattice relaxation rate falling with a decrease in temperature does not change even when the magnetic field destroys the superconductivity. When the magnetic field increases to over 40 T, halting the superconductivity completely, the spin-lattice relaxation rate behavior does not change. This is direct evidence for the scenario shown in Fig. 1.15(b).

The ARPES and some other experimental methods also support the two gaps picture, i.e. the pseudogap and the superconducting gap are not the same. Further details regarding the two gaps picture can be found in Refs. [21–23], or in the parts of this book that deal with ARPES.

With regard to the physical properties around the critical concentration 0.19, Uchida et al. performed some interesting research work on this topic [24] (see Fig. 1.20). His results indicate that the real carrier concentration increases with the increasing of hole-doping concentration x below 0.19, while it changes as $(1-x)$ above 0.19. This may be related to a crossover

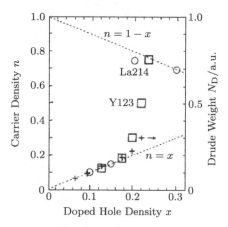

Fig. 1.20. The carrier density as a function of doping concentration comes from the relationship between the residual resistance of Zn-doped samples and doping concentration. Also, the Drude spectral weight is drawn, except for the infrared part, as measured by the photoconductivity. Taken from Ref. [24].

from a small Fermi surface below 0.19 to a large surface above 0.19. The conclusion is still controversial, but if it was to be confirmed, the existence of two Fermi surfaces, a small one and a large one, would be proved. Above 0.19, many experiments demonstrate that it is a Fermi liquid, and the change of $\Delta_0/k_\mathrm{B}T$ with doping concentration is in accordance with the BCS theory, indicating that it is in a kind of normal metallic state in the over-doped region.

1.4. Studies on HTSC's Phase Diagram Using High-Resolution Differential Specific Heat Measurement Technique

The specific heat can be used to study the high-T_C superconductors both in the normal state and the superconducting state. However, in HTSC the T_C is very high, and the phonon contribution to specific heat is rather high. Therefore, it is essentially impossible to use the common measurement methods to obtain an accurate electronic specific heat in a wide temperature range. The differential specific heat technique is a useful method, i.e. measuring the differences in specific heat between two samples. It involves choosing an appropriate reference sample, which has a similar phonon contribution to the measured sample, so that we can back off most of the

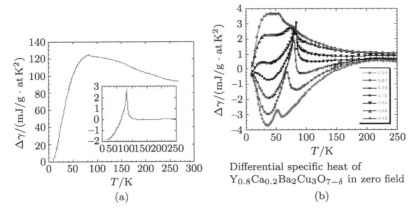

Fig. 1.21. (a) The total specific heat; the specific heat jump at $T_C \sim 80$ K accounts for just only 2–3% of its total specific heat. Inset shows the differential specific heat after deducting the phonon and the normal state electronic specific heat. (b) It shows the raw data of $Y_{0.8}Ca_{0.2}Ba_2Cu_3O_{7-\delta}$ using the differential specific heat technique.

phonon specific heat by this technique. Figure 1.21(a) shows the total specific heat of YBCO, and the inset shows specific heat jump around T_c, but the specific heat jump associated with the superconducting transition accounts for just 2–3% of its total specific heat. Figure 1.21(b) shows the raw data of $Y_{0.8}Ca_{0.2}Ba_2Cu_3O_{7-\delta}$ at different oxygen contents, obtained by using the differential specific heat technique. It can be seen that most of the specific heat is from the electronic contribution, while the phonon contribution only accounts for a minor fraction.

We use the differential heat capacity study of several systems' electronic specific heat as a function of the doping concentration, magnetic field and temperature, including the $Y_{0.8}Ca_{0.2}Ba_2Cu_3O_{7-\delta}$ and Bi2212 systems [25–29]. Figure 1.22 shows, with different doping concentration in the $Y_{0.8}Ca_{0.2}Ba_2Cu_3O_{7-\delta}$ system, the variations of the electronic specific heat coefficient ($\gamma \equiv C_e/T$), electronic entropy S and S/T with temperature, and (a) denotes the under-doped region and (b) the over-doped region. In $Y_{0.8}Ca_{0.2}Ba_2Cu_3O_{7-\delta}$, $\delta = 0.3$ is the optically doped concentration, $\delta > 0.3$ is the under-doped and $\delta < 0.3$ is the over-doped region. In the over-doped region, its behavior has no evident difference compared with that of conventional superconductors. The electronic specific heat coefficient above T_C does not vary with temperature, showing the Fermi liquid behavior. There is a sharp jump near T_C, indicating the opening of the superconducting gap. Only at very low temperatures does the low-energy

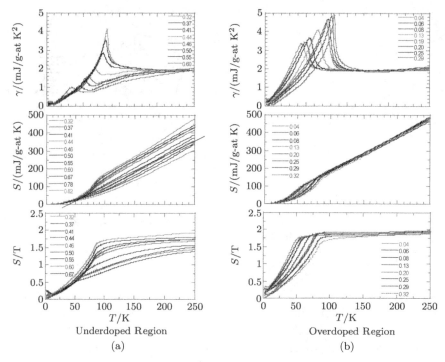

Fig. 1.22. Electronic specific heat coefficient ($\gamma \equiv C_e/T$), electronic entropy S, and S/T in $Y_{0.8}Ca_{0.2}Ba_2Cu_3O_{7-\delta}$ with different doping concentration. (a) It shows the under-doped region, and (b) shows the over-doped region.

excitation behave as a power law, which indeed reflects the d-wave pairing behavior. However, in the under-doped region, its behavior has two major differences compared to that of the over-doped region: first, at very high temperatures, the normal-state electronic specific heat coefficient falls with a decrease in temperature; second, although T_C is still high, the magnitude of the specific heat jump at T_C falls very quickly with a decrease in doping concentration. A similar behavior also exists in the Bi2212 system (see Fig. 1.23).

Looking at the doping and temperature dependences of electronic entropy, we can see that there exists the same behavior between the overdoped region of HTSC and traditional superconductors. However, in the under-doped region, the entropy could linearly extend to a negative value at zero temperature. Entropy is a direct measure of the number of thermally excited states $S = k_B \ln \Omega$, and a negative value of entropy indicates

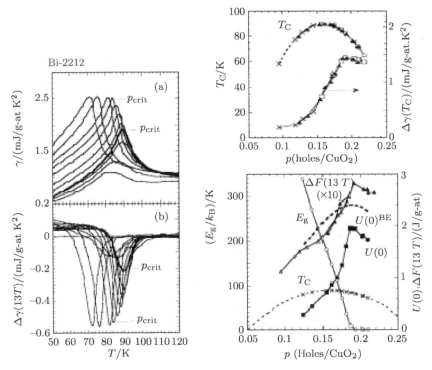

Fig. 1.23. The electronic specific heat coefficient and pseudogap in Bi2212 system with different doping concentration.

that some density of states that exist at high temperatures are lost at low temperatures, i.e. the pseudogap opens in the normal states. S/T reflects the average density of states on the energy scale from zero to several $k_B T$, which also clearly shows the existence of the pseudogap. In the under-doped region, the magnitude of specific heat jump falls with a decrease in doping because of the loss of the normal states' entropy in the pseudogap phase.

In addition, in other systems of HTSC, such as Bi2212 system, the behavior of specific heat is in accordance with the previous conclusion (see Fig. 1.23). The electronic specific heat coefficient jump $\Delta\gamma$ at T_C is almost unchanged with doping concentration above 0.19; but $\Delta\gamma$ drops very quickly at T_C below 0.19. It indicates the 0.19 is indeed a critical point, and there exists a pseudogap below 0.19, but no pseudogap above 0.19.

If there is an energy gap in an electronic system, one can get the total size of the gap by measuring the electronic heat capacity and S/T

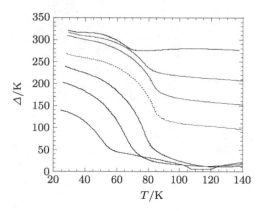

Fig. 1.24. The relationship between the total energy gap (different doping concentration in $Y_{0.8}Ca_{0.2}Ba_2Cu_3O_{7-\delta}$) and temperature, from bottom to top, $\delta = 0.06, 0.13, 0.21, 0.30, 0.36, 0.44$ and 0.52.

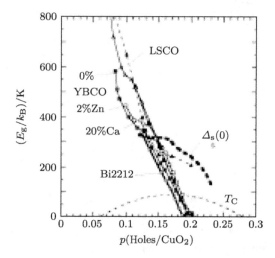

Fig. 1.25. Pseudogap as a function of hole-doping concentration.

as a function of temperature [20]. Figure 1.24 shows the energy gap of $Y_{0.8}Ca_{0.2}Ba_2Cu_3O_{7-\delta}$ obtained from electronic specific heat. The gap has already existed in the normal state of the under-doped and optimally doped HTSC, which does not change with temperature. Also, the superconducting gap opens up only below T_C.

If we put the results of different systems together, as shown in Fig. 1.25, observing the relationship of the normal-state gaps with the doping concentration, we can see the curve, which has extended to almost 0.19 when the normal-state gap vanishes. The pseudogap and the doping concentration have the following relationship: $E_g \sim J(1 - p/p_{\text{crit}})$, where J is about 1200–1800 K, which is similar to the magnitude of the antiferromagnetic coupling constant and $p_{\text{crit}} = 0.19$. The result seems to imply that the pseudogap has a magnetic origin.

1.5. Summary of the Phase Diagram

First, the parent compounds are the strongly correlated 3D antiferromagnetic ordered insulators, and by doping they reach a metallic Fermi liquid ground state with carrier concentration $n = 10^{20}$–10^{21}cm^{-3}.

Second, there exists a critical doping concentration where the superconductivity occurs. The T_C increases with an increase in doping concentration p in the under-doped region, and then attained the maximum value at the optimally doped point. The T_C will decrease when doping increases in the over-doped region. The superconducting state of HTSC has many unique properties, such as d-wave pairing, extremely high upper critical field, very short coherence length and very strong superconducting fluctuations.

Third, the pseudogap exists in the under-doped region, and the origin of the pseudogap is still under debate.

Fourth, near the optimally doped point, the resistivity in the normal state is linear temperature dependent.

Fifth, there exists a magnetic resonance mode in the superconducting state, as observed by neutron scattering experiments.

1.6. Unsolved Problems

First, do the pseudogap and the superconducting gap have the same origin? If not, what is the origin of the pseudogap? CDW, SDW, QCP or competitive results of multi-band?

Second, if the magnetic field has completely destroyed the superconductivity, what is the ground state in the under-doped region? Fermi arc, Fermi pocket or Fermi point?

Third, is there a real Fermi liquid state in the over-doped region?

Fourth, what is the high-T_C superconducting mechanism?

Acknowledgments

I would like to thank Professor Ru-shan Han for his help and support in the preparation of this chapter.

References

[1] O. Fischer et al., Rev. Mod. Phys. **79**, 353 (2007).
[2] D. Vaknin et al., Phys. Rev. Lett. **58**, 2802 (1987).
[3] Y. S. Lee, Phys. Rev. B **60**, 3643 (1999).
[4] J. M. Tranquada, *Handbook of High-Temperature Superconductivity: Theory and Experiment* (Springer, New York, 2007).
[5] R. J. Bergeneau et al., Phys. Rev. B **59**, 13788 (1999).
[6] S. Chakravarty et al., Phys. Rev. B **39**, 2344 (1989).
[7] C. C. Tsuei, J. R. Kirtley, Rev. Mod. Phys. **72**, 969 (2000).
[8] C. C. Tsuei, J. R. Kirtley, Physica C **341–348**, 1625 (2000).
[9] K. Nakao et al., Phys. Rev. Lett. **63**, 97 (1989).
[10] A. Jonud et al., Physica C **317–318**, 333 (1999).
[11] W. Warren et al., Phys. Rev. Lett. **62**, 1193 (1989).
[12] H. Takagi et al., Phys. Rev. Lett. **69**, 2975 (1992).
[13] J. W. Loram et al., Phys. Rev. Lett. **71**, 1740(1993).
[14] J. M. Harris et al., Phys. Rev. Lett. **79**, 143 (1997).
[15] C. Renner et al., Phys. Rev. Lett. **80**, 3606 (1998).
[16] C. Renner et al., Phys. Rev. Lett. **80**, 149 (1998).
[17] J. L. Tallon, J. W. Loram, Physica C **349**, 53 (2001).
[18] G. Q. Zheng et al., Phys. Rev. Lett. **94**, 47006 (2005).
[19] W. S. Lee et al., Nature **450**, 81 (2007).
[20] Kiyohisa Tanaka et al., Science **314**, 1910 (2006).
[21] M. Le Tacon et al., Nature Phys. **2**, 537 (2006).
[22] M. C. Boyer et al., Nature Phys. **3**, 802 (2007).
[23] T. Kondo et al., Phys. Rev. Lett. **98**, 267004 (2007).
[24] S. Uchida et al., Physica C **282**, 12 (1997).
[25] J. L. Luo et al., Physica C **341**, 1837 (2000).
[26] J. W. Loram, J. L. Luo et al., J. Phys. Chem. Solid **62**, 59 (2001).
[27] J. L. Luo et al., Physica B **284**, 1045 (2000).
[28] J. W. Loram, J. L. Luo et al., Physica C **341**, 831 (2000).
[29] J. L. Luo et al., arXiv: 0112065, unpublished.

2
The Complement of Phase Diagram

Ru-Shan Han

School of Physics, Peking University, Beijing 100871, China
China Center of Advanced Science and Technology, Beijing 100080, China

A high-T_C superconducting phase diagram should always be at the center of studies on the superconducting mechanism. In this chapter, a complement of phase diagram will be created and a discussion about the abnormal knowledge which phase diagram tells us.

Figure 2.1 shows the n-type and p-type superconducting phase, taking $Nd_{2-x}Ce_xCuO_4$ and $La_{2-x}Sr_xCuO_4$ as representatives, and shows that the phase includes the superconducting, anti-ferromagnetic, pseudogap and normal metal areas. From the anti-ferromagnetic insulators, by doping the electron or hole into the CuO_2 plane, the compound can pass through the superconducting region and finally into the non-superconducting normal Fermi liquid (FL) region (see Fig. 2.2). A detailed study of the phase diagram reveals that in most of the regions, using the conventional FL picture and superconducting BCS theories are inappropriate. The high-temperature superconducting (HTSC) material attracts significant attention because of its various applications and the significance of fundamental science. This also highlights the crisis in the quantum theory of solids. Not only do the FL and BCS theories, which were successful in the past, fail but since the discovery of HTSC, many concepts and theories have also been proposed, used, tested and combined. Some of the theoretical concepts and methods, which were used, are quite advanced, but looking at the whole phase diagram and its detailed contents, they tend to show their weakness and cannot show basic facts. Experiments are constantly resolving new facts, eliminating a large number of theories. This is science — experiments, always come first!

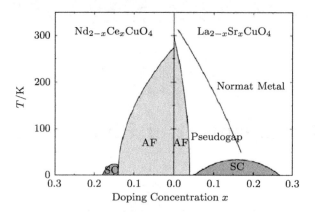

Fig. 2.1. N-type and p-type superconducting phase.

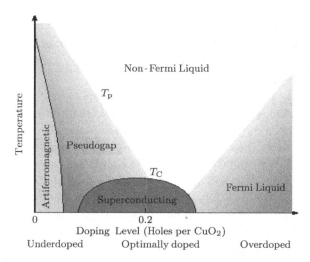

Fig. 2.2. Phase diagram of hole-doped type superconductor.

The following will focus on analysis of the phase diagram of the hole-doped type, which will provide us a complete, variable, whole area, some parts of which were not familiar to us in the past. We knew the extreme left and right regions very well: the extreme right is the FL region, in other words, the region often discussed by the energy band theory and nearly free electron approximation; the extreme left part is the anti-ferromagnetic order region, which was studied for decades in condensed matter physics.

Here, the anti-ferromagnetic order is accompanied by strong correlation and insulating state. Actually, it is the transitional region in the phase diagram that which holds the unusual information and is still unfamiliar. Some considerable points are as follows:

(1) From the FL (non-superconducting, weak correlation) area, the compounds make the transition to a strong correlated system by decreasing the hole concentration, and the key point is the itinerant holes on Cu^{2+} transform into localized holes. The lack of contribution for the Fermi surface makes the large U appear, which stabilizes the Cu^{2+} ions. (No positive trivalent ions.)
(2) The anti-ferromagnetic super-exchange interaction makes the Cu ions spin in an anti-ferromagnetic arrangement. The process is that 2D gradually changes into 3D (from right to left), i.e. the 2D anti-ferromagnetic region extends to 3D as the hole concentration decreases. Holes on the oxygen play an active part, which can weaken the anti-ferromagnetic super-exchange interaction. 2D anti-ferromagnetism and superconductivity could co-exist. When the holes on oxygen decrease, the 3D anti-ferromagnetism recovers.

First, the following discusses the five transitions, then the two curves and some other complements.

2.1. The Characteristic of the Full Phase Diagram's Transition is the Basic One of HTSC, Including Five Transitions and Five Revelations

2.1.1. *Transition One*

Transiting from anti-ferromagnetic order to disorder, this system is not one that is conventional or familiar to us. We are familiar with magnetic ordering materials from the past, mostly appearing as magnetic ordered alloy, for example, from the anti-ferromagnetic order to the disordered system, which is called the temperature effect, or impurity (vacancy) effect, and this is a transition from the anti-ferromagnetic order to the paramagnetic. We base discussion on the super-exchange interaction. The following will give a brief and significant description of the super-exchange, and then point out the unconventional magnetic model that must be using here for the following reasons:

(1) A high-temperature superconductor is only one of the O-electron-deficient super-exchange system and the first one with the full implementation and verification system of Kramers–Anderson super-exchange model. The HTSC system is a special one in which the oxygen (O) ions are in spontaneous anti-ferromagnetic order owing to the presence of two copper (Cu) ions in the medium, and this is a state of electron deficiency, due to vacancies in the oxygen (O) ions.
(2) In the system, the 3D anti-ferromagnetic order transits to paramagnetic order via the 2D short-range anti-ferromagnetic order, which is unlike the conventional one that transits from the 3D anti-ferromagnetic order to paramagnetic order directly. There are many people who mechanically applied the anti-magnetic order model in the past in discussions about the HTSC transiting from anti-ferromagnetic order to paramagnetic order directly, but one of the biggest features of HTSC is that the 2D anti-ferromagnetic order plays a very important part. In the superconducting region of the phase diagram, the 2D anti-ferromagnetic order and superconductivity co-exist, and the former plays an important part in the superconducting mechanism.

Revelation One

The presence of a vacancy in oxygen, called electron deficiency in oxygen, plays a very active part, so the model which lacks vacancy is inappropriate.

Some of the experiments and concepts involved will be discussed here. First is the concept of super-exchange [1–3].

According to Kramers–Anderson's analysis, we will take the example of a pair of Cu ions for analysis (see Figs. 2.3 and 2.4).

What needs to be stressed is that the super-exchange between the Cu ions depends on the O^{2-} ions. Because of the oxygen ions, the interaction energy between the two copper ions is very strong. The easiest, and earliest, discussion is on the direct exchange interaction in the hydrogen molecules. The two identical electrons in hydrogen molecule are indistinguishable, so there is an exchange interaction. Because of the electronic wave functions overlapping, when the two electrons are in the anti-spin order, they possess an energy advantage. So, the energy of the paramagnetic order is much higher than the energy of the two separate hydrogen atoms, which is the reason for the stability of the hydrogen molecule.

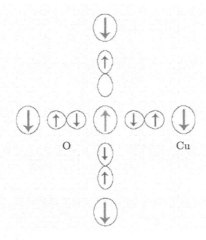

Fig. 2.3. Schematic diagram of vacancy in oxygen.

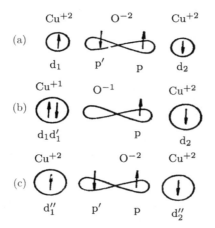

Fig. 2.4. A pair of Cu ions indicates the concept of super-exchange.

Because the distance between the two ionic bands in Cu is larger, the direct exchange is very weak in HTSC. There is also an interaction which comes from the extension of the electron cloud overlaps by the intermediary of the oxygen ions (O^{2-}), often called super-exchange interaction. This is the Kramers–Aderson model. So, if there is electron deficiency in the oxygen ions, then the super-exchange is weakened, and the magnetic order is also weakened. This weakening, as can be seen from the phase diagram, is even stronger than the effect of temperature. In the transition from magnetic order to the disorder, doping holes weakens the super-exchange strength, which is an integral component, loading to average weakening of the super-exchange intensity. In understanding the HTSC mechanism, sufficient attention should be paid to thus, as it is the reason for the appearance of the local, nearly free, but not completely free Cu moments, and it is also the reason for the collapse of the anti-ferromagnetic background. With regard to the super-exchange interaction model, we can find the relevant material in the basic textbooks. The best is to look at the Kramers' or Anderson's original, and in the text Anderson gives a good explanation for Kramers' model in the first half, but the second half has been simplified. It is just the simplification used for the HTSC, but this causes the difficulties.

Another characteristic of magnetic order is the high anisotropy.

The most important characteristic of HTSC is that, in the CuO_2 plane, the 2D spin anti-ferromagnetic order exists in a certain limited area. From the 3D anti-ferromagnetic transformation point, when the temperature increases, it does not transform into the paramagnetic state, but into a 2D type of the anti-ferromagnetic order. Because of the severe anisotropy, the super-exchange parallel to the surface is much larger than the super-exchange normal to the surface, and the Neer temperature (T_N) are between them, i.e.:

$$J_{//} > T_N \gg J_\perp,$$
$$0.13 \text{ eV} > 0.03 \text{ eV} \gg 2 \times 10^{-6} \text{ eV}.$$

When the doping is zero and the temperature T is slightly above T_N, it can be observed that in Cu-plane a few hundred angstroms (taking La_2CuO_4 as an example, which has an anti-ferromagnetic coherence length of 650) of the anti-ferromagnetic order area remain, providing conditions for transition to 3D anti-ferromagnetic order: a large area of 2D order provides the conditions for the 3D order; although a couple of Cu ionic bonds are weak, whereas in 2D plane, a large number of Cu ions order provide a massive force between the surface. When heating up to T_N, interface decoupling happens, but when this is done parallel to the plane, considerable range ordered zones are seen. With regard to information on the 2D order, Neutron diffraction provides evidence of transition from 2D to 3D peak, and the 2D characteristic "Rod". This was determined early on by neutron diffraction, as can be seen Figs. 2.5 and 2.6 [4]. In CuO_2 plane, the spin magnetic order on the Cu atoms has easy magnetization direction, i.e. severe anisotropic direction [17].

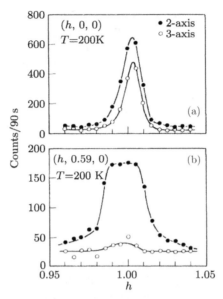

Fig. 2.5. Two-axis and three-axis scans across the 3D (100) magnetic peak (a) and 2D QSF ridge (rod) (b). Taken from Ref. [4].

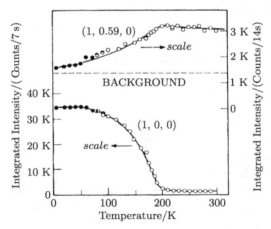

Fig. 2.6. Integrated intensities of the (100) 3D anti-ferromagnetic Bragg peak and the (1, 0.59, 0) 2D QSF ridge (rod). As the temperature increases, a decoupling between the CuO_2 planes occurs first. Taken from Ref. [4].

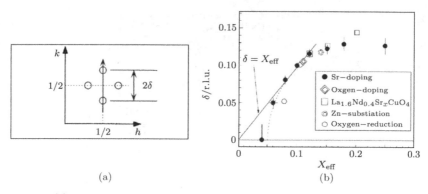

Fig. 2.7. (a) The incommensurate splitting; (b) doping dependence of the splitting. Taken from Ref. [5].

Doping is more efficient than the temperature effect, as can be seen from the phase diagram. Doping makes T_N decline rapidly. It is because the vacancy doping greatly affects the 3D coupling. In the under-doped region, the incommensurate splitting of magnetic peaks, based on data provided by the neutron experiments, confirms that the 2D anti-ferromagnetic order continues to exist, and is the evidence of its co-existence with superconductivity, although the 2D anti-ferromagnetic zone gradually reduces as the doping concentration increases (Fig. 2.7) [5].

To conclude what is described above, vacancy doping is more efficient in destroying the 3D anti-ferromagnetic order. The under-doped region does maintain the characteristic of short-range anti-ferromagnetic order, and anti-ferromagnetic order is related to the doping concentration, and the signal disappeared at around 0.19.

Summary: The concept of super-exchange interaction plays a major role in spontaneous magnetization. HTSC should be a new system of super-exchange concept owing to its high anisotropy, electron deficiency in the oxygen ions and its complete destruction of the magnetic order. The first transition is important — the transition from Kramers–Anderson super-exchange to no super-exchange — and it is here the electron-deficient oxygen ions play a key role.

2.1.2. *Transition Two*

As the doping concentration increases, the system transits from the strong correlation to the over-doped weak correlation: traversing the strong–weak coupling mixed zone, i.e. the potential energy U in Cu tends to zero ($\to 0$) when the doping concentration is very large (the electronic state accords with the band theory calculation). Meanwhile, the potential in the oxygen (O) ions remain zero.

The characteristic of strong correlation is that there is a very strong correlation characteristic in the left part of the phase diagram, which, of course, only impacts the performance on Cu, meaning the potential energy is very small in O. Some people may not be familiar with the so-called strong correlation concept. That is when each level is filled from the bottom up, until the highest level that can be filled, i.e. the FL. Also, filling rules are such that each level can be filled with two electrons, one spin up, one spin down, and the two electrons are degenerate. But the strong correlation tells us that when an electron orbit already has an electron and you want to add another electron with opposite spin direction, its energy will increase substantially. That is, as we have often heard, there is a difference energy U between the lower Hubbard band and upper Hubbard band, and $U \geq W$ (band width). This is because of the strong correlation; in our detection using low-energy experiments, the Cu^{3+} state so far has not been seen, i.e. there is one more hole in the Cu ions. Also, there is no experimental evidence for the Cu^{3+} state.

With regard to the role of U, usually people simply divide the band into two, the Cu's lower and upper Hubbard bands. There is one electron in the lower Hubbard band; if this is filled by another electron, the energy will increase by U (or equivalent to the two hole-states). But there is actually an O band in the system also. The role of the strong correlation is to stabilize the Cu spin, so that the spin states will be stable over a wide doping region. Finally, the strong correlation will disappear, and the band theory will work again at a doping concentration 0.3, and the same level can be filled with two electrons, which also shows that the holes are no longer strongly correlated in Cu.

The reverse transition from the weak correlation to strong correlation, from right to left, should be described by the vacancy or spin localization in Cu. Only the local electrons or holes can be strongly correlated. In this process, there exists a transition of the Cu hole state from the itinerant to the local state. This change corresponds to the change in energy U from almost 0 to 6–10 eV magnitude. That is why I say the energy U transits from weak to strong correlation or from strong to weak correlation. Any model with a fixed U cannot describe the whole phase diagram

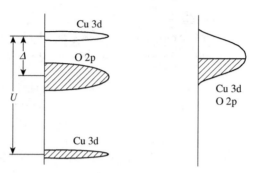

Fig. 2.8. The charge transfer insulator; the potential energy U of Cu stabilizes the antiferromagnetic order (no Cu^{3+}, hole doubly occupied). With an increase in doping, it transits to the normal metal band.

correctly; even making minor amendments to it may not be a good starting point.

There is one characteristic that needs to be re-emphasized: the system is actually a strong–weak mixed correlated system, and there is strong correlation in the vacancy of Cu, but weak correlation in that of O, with all finally transiting into weak correlation.

Summary: The role of U is to stabilize the Cu spin; also, note that there are no Cu^{3+} ions existing in the strong correlated region (see Fig. 2.8).

Revelation Two

The fixed U-component model is not a good one.

2.1.3. *Transition Three*

When doping increases, a transition from the charge transfer insulator to non-superconducting metal region occurs. It should be emphasized that the mother compound of the HTSC is the change transfer (CT) insulator, not the Mott insulator, because $\Delta \ll U$, where Δ is the gap between the O band and upper Hubbard band. As the doping increases, U and Δ tend to zero (meets the band theory calculation):

$$\Delta \sim 1.7\text{--}2.0 \text{ eV}, \quad U \sim 6\text{--}10 \text{ eV},$$
$$[x \to 0.34 \quad \Delta, U \to 0].$$

It is universally acknowledged in most of the theoretical models that the mother compound of HTSC is the Mott insulator. Mott insulators, as shown in Fig. 2.8, are actually the band structure after removal of the O band, considering only a single Cu-band model, and it is also called single-band model, or a single-component model. Actually it should have three bands, i.e. besides the two Cu bands there is an O band, thus forming the three-band model; it is also called the two-component model as it included the Cu and O components. The holes also include two parts: the Cu ions and O ions. There are holes in the Cu ions before doping, and in the O ions after. Therefore, even if the Zhang–Rice effective Hubbard model or other models adopt an overlarge U, it cannot reflect the feature of low energy excitation, $\Delta \sim 2\text{eV}$[6].

Revelation Three

The single-component Mott–Hubbard model can be handled with mathematics, but it is not a good physical starting point.

2.1.4. Transition Four

This section discusses the transition from the non-FL to FL. Note that the underdoped region, and even the optimally doped region, is still a bad metal.

About bad metal, there are complements as follows.

The meaning of the so-called bad metals involves the limit of metal proposed by Loffe-Regel. That is to say, the collision-free path l should essentially be larger than the Fermi wavelength λ_F. At this time, the quasi-particle picture, the classical particle picture, can also work. The metal that is close to the lower limit is a bad metal. When the formula ($l<\lambda_\text{F}$) becomes possible, we should be careful. k is no longer a good quantum number, although we can see the behavior of a single particle in projection in k-space. We must also pay attention to k, which is not a good quantum number any more, and even the concept of wave packet, particle collisions and Boltzman model have been challenged.

Some data provided are as follows:

The following is the data of normal metals Li, Na, Cu, Ag and Au, whose carrier is generally of the order of 10^{22}, and HTSC magnitude is about 10^{20}–10^{21}. Compared with normal metals, the HTSC material resistivity

Table 2.1.

	Normal metal					Bad metal	
	Li	Na	Cu	Ag	Au	$La_{1.85}Sr_{0.15}CuO_4$	$Y_{1236.5}$
$n/(10^{22}\ cm^{-3})$	4.70	2.65	8.47	5.86	5.90		$n \sim 10^{21} cm^{-3}$
$k_F/(10^8\ cm^{-1})$	1.11	0.92	1.36	1.20	1.21		
$\lambda_F/(10^{-8}\ cm)$	5.66	6.83	4.62	5.24	5.19		
$v_F/(10^8\ cm \cdot s^{-1})$	1.28	1.06	1.57	1.39	1.40		$8.4/(10^6 cm \cdot s^{-1})$
$\rho(273\ K)/(\mu\Omega \cdot cm)$	8.55	4.2	1.56	1.51	2.04	$0.21/(m\Omega \cdot cm)$	
$\tau(273\ K)/(10^{-14} s)$	0.88	3.2	2.7	4.0	3.0		
$l = v_F \tau /$ $(10^{-6}\ cm = 100\ Å)$	1.13	3.39	4.24	5.56	4.2		$1.6/(2\ K)$

Note: n is the carrier number, k_F is the Fermi momentum, λ_F is Fermi wavelength, v_F is Fermi velocity, ρ is the resistivity, τ is the relaxation time and l is the mean free path.

has a difference on the order of thousands of times, which makes it a poor conductive material; for further details, see Table 2.1.

It is common knowledge that the resistance of $La_{2-x}Sr_xCuO_4$ in relation to temperature shows a nearly linear behavior. For example, $La_{2-x}Sr_xCuO_4$, 0.17, $\rho_{ab}(900\ K) \sim 0.7\ \mu\Omega \cdot cm$, $\rho_{ab}(x \sim 0.06)$, which is larger, has a smooth temperature change, and should be a unified mechanism. The resistance corresponds to:

$$l < \lambda_F.$$

That is to say, the HTSC system we encounter, despite having metal behavior, and even a resistivity that is relatively low at low temperature, has a smooth transition to high temperature, caused by the same mechanism. In the case of the high-temperature side with low doping, it is a very bad metal. The concept of it being a very bad metal means that we should carefully apply some traditional, known theories, formulas or pictures. Because the so-called k is not a good quantum number, we should perform a careful ARPES measurement, Fermi surface measurement and measure the size of the Fermi surface. Also taking into consideration the Hall doping, the Hall coefficient is dependent on temperature, which in itself is a deviation from the normal behavior, thus we should explore other mechanisms, such as the historical Scew model. Discounting the temperature behavior and applying mechanically the original formula for the Hall conductance to discuss whether the small Fermi surface is hole-type or electronic-type, or how the hole–electron type transition happens etc. may artificially increase the confusion.

The carrier is a poor itinerant, which is just the condition at which the RVB state occurs (*note*: at $x = 0$, it is not the RVB state, but the anti-ferromagnetic state; it will be the RVB state just after doping.).

Revelation Four

k is not a good quantum number, which not only makes these discussions with k as a basis open to being challenged, it also makes the Boltzman model open to being challenged. The quasi-particle picture is not a good approximation:

$$l < \lambda_\text{F}.$$

2.1.5. *Transition Five*

A small number of carriers transit to a large number of carriers.

There exists a special concentration cut-off at 0.19 in the transition. Note that special attention must be paid to the Uchida's experiments (Fig. 2.9) [7].

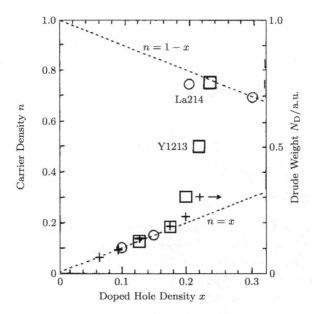

Fig. 2.9. Variation of the carrier density.

In Fig. 2.9, we see what appears to be a transition from a small number to large number, which is the result announced by Uchida in 1997. The experiment was based on Zn-doping, i.e. using Zn as impurities; then, they used the scattering formula to quantitate the carriers, finally giving a very simple expression. Figure 2.9 indicates the agreement with the experimental results.

After doping beyond the 0.19 mark, transiting to the large doping, there is a parallel emergence of a large Fermi surface, and a large number of carriers, or you can say the copper ions have already become itinerant. Is that right? We should do more experiments to confirm that how the holes in Cu localize or delocalize, or how the carriers in Cu transit from localization to delocalization in the over-doped region.

This measurement fully corresponds to the transition from the completely collapsed point of anti-ferromagnetic order to $n \sim 1-x$ (the concentration), i.e. the large Fermi surface. The measured value of the large Fermi surface electron structure in the high-doped region corresponds to the value calculated by the band theory, i.e. the FL behavior recovery.

In the under-doped region, there is a relationship worth emphasizing:

$$x \sim n \sim n_s \sim T_C \sim V_{FS}.$$

where from left to right, they are doping amounts, carrier concentration, superfluid density, the superconducting transition temperature and the Fermi surface volume, all of which have the same proportion and magnitude. This relationship might have condensed many important aspects.

This area is the so-called pseudogap region, which has attracted the most attention, but there is a divergence of views about it. The pseudogap region occupies a very important position, and its positive relationship is worthy of our attention. There has been and still us much debate about how to draw the T^* line.

Revelation Five

Electrons in the Cu site transition from localized anti-ferromagnetic short-range order to the delocalized itinerant electron, which will happen with the disappearance of superconductivity in parallel.

So, when we observe the phase diagram and find the five above transitions in high-temperature superconductors, we need to stop and think. The traditional concepts and theories may be problematic, which is why it has baffled so many people in last 20 years. We currently have not achieved a consensus about the mechanism. The emergence of high-temperature superconductivity in transition, included in the phase diagrams given above, should be the key point of the mechanism problem.

The special nature of the superconducting state plays an increasingly important role, i.e. the particularity of the body superconducting state in HTSC, from early neglect to current emphasis on the analysis of its particularity, is an important progress. Also, the inhomogeneity of research in the superconducting state should be given full attention.

2.2. Main Features of the Superconducting State

2.2.1. *Inhomogeneity of the Superconducting State*

Note the color distribution in Fig. 2.10. Figure 2.10(a) shows 60 K$< T_C =$ 64 K, the inhomogeneity of gap in real space. Local pairing in real space

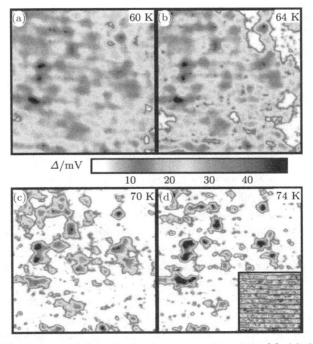

Fig. 2.10. The real spatial gap distribution measured by STM [8], (a)–(d) show the change as a function of temperature, $T_C = 65$ K. Note that color indicates the size of gap, and there is a gapless area appearing in (b). Taken from Ref. [8].

does not occur simultaneously with coherence condensation, and when coherence condensation occurs at T_C, the macroscopic quantum phenomena appears as well.

The inhomogeneity shows up when we see different colors representing different energy gap sizes in the materials below T_C. When using STM to measure local density of states with fixed point, the energy gap can be observed. When the gap is in the superconducting area, it is inhomogeneous in the full version. The material is measured for the superconducting state at 4 K, and when the temperature is increased to 60K in the inhomogeneous state, and then when it is raised to 64 K, there is actually no macroscopic phenomena: Meissenr effect, zero resistance. But some of the energy gap regions can be detected. These regions are the superconducting areas preserved during the heating process. These superconducting areas are short-range pairing, small superconducting coherent systems, and so such a continuous transition better supports the preformed pair picture in the pseudogap area.

2.2.2. Indicate that the Local Pair in Real Space and Coherence Condensation Do Not Occur Simultaneously

This experiment also tells us of such a situation where the coherence length is only in the 10 nm scale, thus giving us some evidence of the short coherent length. Figure 2.10(c) and 2.10(d) still retain energy gap areas, even coherent range on the 10 nm scale. How we should treat them is a significant issue.

Summing up the above analysis we can see that high-temperature superconductivity appears in very special conditions, and these conditions are contained in a special structure (perovskite structure), a special copper oxide. Conventional, simple models often cannot summarize its characteristics. A more sophisticated model — three-band (two-component) model — is thought to be a good starting point by most people. This was proposed by N. P. Ong in his paper, in which he and Anderson discussed the starting point, but because the mathematical treatment was much more difficult than that of the single-band model, they later did not use the three-band model. Of course, more attempts have been made (but mathematical treatment was more difficult). Some well-known, representative three-band or two-component works are given below. For Zhang–Rice state, they start from a three-band model and prove that there may be an effective single-band model. There is also Varma's cyclotron current image, but recently it is being more and more questioned [9–12].

Anderson first mentioned the RVB picture, but RVB is not anti-ferromagnetic at zero doping. Contrary to his experiment, the current experimental evidence strongly proved it to be anti-ferromagnetic, which caused Anderson to abandon the RVB picture. Later, Anderson came back and thought the doped area was still RVB. I think this idea is correct.

New concepts and the reason why superconductivity occurs in the transition area should be explored in detail.

Recently, there have been many new discoveries and controversies: is there an electron pocket in small Fermi surface measuring? Does the dual-energy gap exist? While discussing all this, the basic concept must not be forgotten: the phase diagram provides comprehensive information. We must not leave the basic properties to analyze a specific and practical problem, because doing so we will get lost.

The development and improvement of a new two-component RVB model is a step in the right direction.

In a phase diagram, there are several points worth noting: besides the anti-ferromagnetic and T_C curves, there are two lines: one is the T^* line, the other is the dividing line between the FL and non-FL regions.

The T^* line is defined as temperature line of the pseudogap opening. One view is that it is the beginning of the pseudogap and is associated with the preformed-pair, non-coherent disordered local pair. The other view is that it is the beginning of a new phase. The under-doped region below T^* has attracted the most attention, but the largest divergence of views exists in this part of the phase diagram, including the relationship between the pseudogap and superconducting energy gap — "independent" two-gap or the natural evolution — each supported by some evidence. The effects of striped superconducting phase and the other ordered phase, etc. on superconductivity also attract many researchers.

Two decades later, there are two different ideas: first, see the superconducting ground state from the unusual excited state (which was the idea a lot of people had early on); second, define the mechanism of the ground state, then observe the characteristic of the excited state. These two methods should supplement each other.

There are relatively few experiments that study the over-doped region, and different experiments give the second across line differently.

Assessing the phase diagrams from high concentration to low concentration, the strong correlation was seen in Cu sites accompanied by the localization of holes in the Cu ions, resulting in strong correlation in sites. The localization of holes in Cu is extremely important. Also, the copper ions are not replaced, which illustrates that acting as the background, it may be performing one of the most basic functions. We should strengthen the high-quality work about the over-doped region and non-superconducting region. We should conduct more serious studies about association of the emergence of the strong correlation and non-FL behavior.

There are some very interesting additional phase diagrams which also provide us with information. They are doped or distorted by the magnetic field phase diagrams.

2.3. Phase Diagram

2.3.1. Alloul Phase Diagram

Figure 2.11 shows the phase diagram given in 1991 by Alloul et al. [13].

This phase diagram uses Zn^{2+} ions, which have the same charge as Cu, i.e. replacing the Cu^{2+} ions, but the spin is zero for Zn^{2+} ($3d^{10}$), partially changing the Cu^{2+} ions' spin state. Its effect weakens the superconductivity and AFM simultaneously. Other experiments indicate that replacing 3–4% Zn ions will cause the superconductivity to disappear completely [7],

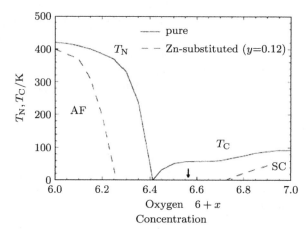

Fig. 2.11. Phase diagram of Zn-substituted (solid line) and pure sample (dotted line). Taken from Ref. [13].

which shows that the displacement effect is very strong. Although the simultaneous disappearance of AFM has not been measured (no evidence found), it has been shown to have a close relationship with the superconductivity, magnitude and spin. Replacement by many other ions, such as Fe, Ni, Ga, Sn, Li, etc., has been studied in depth, and Zn replacement of Cu is a typical representative, indicating that the Cu ions cannot be replaced.

Zn has no spin but significantly affects the AFM and destroys the super-exchange interaction. This indicates that the anti-ferromagnetic order is irreplaceable, i.e. indispensability of electrons in O and irreplaceability of spin in Cu are the necessary conditions for the anti-ferromagnetic order.

2.3.2. *Phase Diagram Under Strong Magnetic Field*

This is the so-called 0.19 relative phase diagram under strong magnetic field. It is an insulator to the left of the 0.19 mark, but a metal to the right. However, in the transition, there are very strong magnetic flux dynamics, one Hall effect temperature changes, sign changes and insulator–metal transition. There are also experiments which indicate that the superconducting density of states resumes in the spin gap, etc. It has recently become a topic of concern. Also, these problems make the issue more complicated. I think

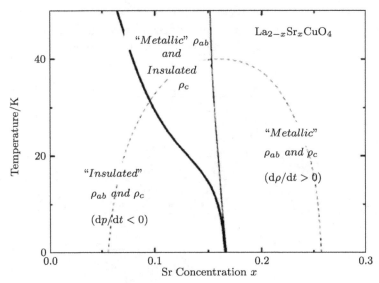

Fig. 2.12. The phase diagram after the superconductivity is inhibited by a strong magnetic field.

the questions should be fully studied in the absence of magnetic field first, and then considering the magnetic field. This clearly pointed out: 0.19 is still a special cut-off point between the two different effect of the doped regions.

2.3.3. Guo-Qing Zheng's Important Work in NMR Study

This work has not been given due attention, a proper analysis and understanding. The importance of this work is in its questioning of the single-model and 2D paring mechanism. Figure 2.13 shows the result of the Knight shift in this experiment [14, 15], and similar results were obtained in Ref. [16].

The isotope effect of ^{63}Cu and ^{17}O on the single-crystal measurements gives some very important information.

(1) Cu sites have core information and O sites have qualitative difference; the black circles represent the data of Cu (left vertical axis), while empty circles represent the O data (K_{ab} and K_c have the same trend, and so are left out altogether, see Refs. [14–16]).
(2) There are moments of the Cu nuclear signal, which has $d_{x^2-y^2}$ features: $^{63}K_c$: $^{63}K_{ab} \sim 4$.
(3) The O nuclear signal tends to zero, reflecting the superconducting state's signal: the superconducting condensation is seen in the O nucleus, the Knight shift tends to zero and the superconducting state

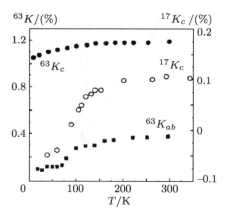

Fig. 2.13. The total Knight shift for single-crystal samples [14], including the contributions of spin and orbital.

has no moment: no spin moment and no orbital moment as well. (Currently, the most discussed $d_{x^2-y^2}$ state has moments, but it is not part of the orbital moments.) Note that in the heavy fermion superconductors, the Knight shift of materials with $d_{x^2-y^2}$ symmetry does not tend to zero.

(4) There is a sign that the Cu cores are affected by the superconductivity. (Say that there is a coupling between Cu and O, and there is a weak decline.)

I believe this will lead to a lively discussion: Such as, why is there a qualitative difference in the signal of Cu and O? And why can we only see the $d_{x^2-y^2}$ characteristic in Cu, not in O?

The conventional superconducting BCS theory was proposed 46 years after the superconductors were discovered. In fact, a critical experiment, i.e. the so-called isotope effect, has played a significant role in promoting it. In the copper oxide high-T_C superconductor experiments, there is currently a wide gap relating to which of these is representative of and closely relates to the mechanisms. There is now a general consensus that the phase diagram is one of the most important parts. The information that a careful analysis of the phase diagram can provide us is very critical.

As stated earlier, the BCS theory was proposed 46 years after the superconductors were discovered (now, it's just 21 years past, and the time is still there.) But we think that today's conditions are very different from what was seen earlier, at least in terms of the comprehensive global investment in experimental work. We believe after a comprehensive assessment of the theoretical mechanism, it is now the time to propose a new model creatively.

Questions

Question: The above mentioned transition is considered by some as a phase transformation, and what is the difference between these two?

Answer: The evidence of phase transformation was not clear. From the full phase diagram, going from left to right, it covers a large area; so, can we say it goes through a phase transition? In the transition, the insulation, strong correlation and anti-ferromagnetic order in the left part of phase diagram must all disappear. But no such evidence has been found. The phase transition can exist in the transition process.

Question: Usually, we cannot make $U \to 0$ by doping. Is this statement true?

Answer: Yes. The extreme right part of phase diagram corresponds to the energy band picture, which indicates $U \to 0$ is true. Only doping cannot do this, the collapse of the anti-ferromagnetic background is also necessary. As doping increases, a part of the copper ions, affected by the holes in O, have a certain itinerancy, forming the gap state, which has certain scalability, leading to a decrease of U. After doping higher than 0.19, the whole anti-ferromagnetic background collapses, and the electrons (holes) in the copper ions would become itinerant, contributing to the Fermi surface. The large Fermi surface and Luttinger theory tell us that the number of carriers is large, not x, but $1 - x$.

Question: How to define the Fermi surface volume in the under-doped region?

Answer: By de Hass-van Alphen oscillation experiment.

Question: The single-band model is increasingly being questioned, for the three-band model — is there clear evidence?

Answer: There is much evidence. The important evidence indicates that it is a charge-transfer insulator, not a Mott insulator. The starting point of the proof of Zhang–Rice state in the effective single-band model is the three-band model.

Question: What is the situation on hole-doping?

Answer: The doped holes are in the oxygen ions. The hybridization of the states in O and Cu form the gap state, and this state has a certain itinerancy. A bad itinerant corresponds to a bad metal. It is also the condition of RVB state turning up. The interaction of the state and anti-ferromagnetic background causes the superconductivity, and the gap state makes the anti-ferromagnetic correlation completely collapse at last. Then, the superconductivity disappears. This is the entirely of our knowledge on hole-doping. It is a double-edged sword that brings up superconductivity and eventually destroys it.

References

[1] H. A. Kramers, *Physica* **1**, 182 (1934).
[2] P. W. Anderson, *Phys. Rev.* **79**, 350 (1950).
[3] P. W. Anderso, *Phys. Rev.* **115**, 2 (1959).
[4] G. Shirane, *Phys. Rev. Lett.* **59**, 1613 (1987).
[5] K. Yamada *et al.*, *Phys. Rev. B* **57**, 6165 (1998).
[6] F. C. Zhang, T. M. Rice, *Phys. Rev. B* **37**, 3759 (1988).
[7] S. Uchida, *Physics C* **282**, 12 (1997).
[8] K. K. Gomes, *Nature* **447**, 569 (2007).
[9] M. E. Simon, C. M. Varma, *Phys. Rev. Lett.* **89**, 247003 (2002).
[10] C. M. Varma, *Phys. Rev. B* **73**, 155113 (2006).
[11] C. M. Varma, *Phys. Rev. B* **73**, 233102 (2006).
[12] M. Greiter, *Phys. Rev. Lett.* **99**, 27005 (2007).
[13] H. Alloul, *Phys. Rev. Lett.* **67**, 3140 (1991).
[14] G. Q. Zheng, 高温超导基础研究 (Shanghai Scientific & Technical Publishers, Shanghai, 1999).
[15] G. Q. Zheng, *Physica C* **260**, 197 (1996).
[16] A. P. Gerashchenko, *JETP* **88**, 545 (1999).
[17] T. Thio, *Phys. Rev. B* **38**, 905 (1988).

3
Infrared and Optical Response of High-Temperature Superconductors

Nan-Lin Wang

Institute of Physics, Chinese Academy of Sciences, Beijing 100190, China

3.1. Introduction

Optical spectroscopy is a powerful technique to probe the electronic excitations and charge dynamical properties of solids. It plays a key role in identifying an energy gap and determining the spectral weight and transport lifetime of quasi-particles. Typically, wavenumber in cm^{-1} or electron-volt (eV) is used as the unit of energy in the infrared community. The conversion between them is as follows:

1 eV = 8065 cm^{-1} = 11,400 K.
1.24 eV = 10,000 cm^{-1}.
1 wavenumber (cm^{-1} as the unit) = 10,000/λ (λ wavelength, μm as the unit).

Fourier-transform infrared (FTIR) spectrometer and grating spectrometer are among the most commonly used spectrometers in a laboratory. With the use of those spectrometers and ordinary light sources, it is easy to perform reflectance or transmission measurement in the frequency range from a dozen of wavenumbers (far infrared) to tens of thousands of wavenumbers (ultraviolet). Other light sources or experimental techniques may be used for lower or higher frequency spectral measurements. For example, in the very far infrared or terahertz frequency region, one can use the higher brilliant terahertz radiation from a synchrotron as the light source for FTIR measurement, or one can use ultrafast laser-based terahertz

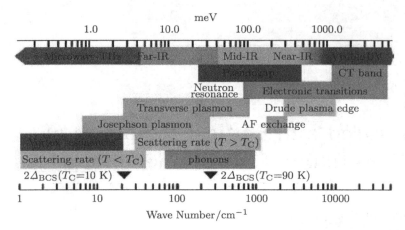

Fig. 3.1. Possible physical processes in different energy regions. Taken from Ref. [1].

time domain spectrometer to determine optical constants. Brighter light radiation in the vacuum–ultraviolet region is also available in synchrotron radiation for higher frequency optical experiment. Nevertheless, the optical measurement in an energy range accessible in an ordinary laboratory already contains a lot of interesting physical processes. Let us take high-T_C cuprate superconductors as examples (see Fig. 3.1).

In the cuprate superconductors, the charge carriers are confined in the CuO_2 planes, which are separated by insulating charge reservoir layers. The inter-layer Josephson coupling of cuprates along the c-axis in the superconducting state could lead to the formation of a Josephson plasmon from condensed superfluid carriers. The Josephson plasmon usually locates in an energy scale from a few to hundreds of wavenumbers. For the two-layer or multi-layer high-T_C superconductors, the coupling of two longitudinal Josephson plasmons could generate a transverse plasma mode, which would show a resonance peak in the conductivity spectrum. Its energy scale is also typically below a few hundred wavenumbers. From far infrared to thousands of wavenumbers, the charge carriers may couple with some collective excitations (such as the resonance mode detected in the neutron scattering experiment), which leads to a significant change in the carrier scattering rates. The scattering rate change can be probed by optical measurements. Optical spectroscopy can also probe the superconducting energy gap or the pseudogap of a superconductor in the terahertz or far infrared region. In addition, interband transitions could manifest in higher energy scales. So, there are many characteristic features that show up in the optical spectra. A comprehensive review paper by Basov and Timusk [1] can be used for further reading.

This chapter first briefly discusses the fundamental physics about optical properties of solids. Then, a short overview on the doping-induced spectral evolution in cuprates is presented. The issue of whether or not the superconducting energy gap of cuprate could be detected by the infrared spectroscopy is also addressed. This is a controversial issue in the infrared community, even after two decades of intensive studies. For conventional superconductors, the infrared probe was the first spectroscopic technique detecting the superconducting energy gap (readers are referred to Tinkham's text book on superconductivity). The high-temperature cuprate superconductors have higher T_C, then the superconducting energy gap should be observed at higher energy scale and be easier to detect. A spectral change in optical conductivity across the T_C was indeed observed. Some researchers ascribed the observed spectral change to the superconducting gap formation, while others argued that the high-T_C superconductors were in the clean limit in the superconducting state and that the superconducting energy gap is not visible in the in-plane optical spectra. The observed spectral change was explained to be due to the coupling effect of charge carriers to certain collective bosonic modes. A comment will also be provided on this issue on the basis of recent experiments.

3.2. An Introduction about Optical Properties of Solids

The following is a brief introduction to the physical quantities measured by optical experiments. These are the so-called optical constants. For electronic materials in condensed matter physics, these optical constants are closely related to the charge carrier scatterings or transport processes. Among different optical constants, the reflectivity, $R(\omega)$, is a quantity which could be directly measured experimentally. As displayed schematically in Fig. 3.2, if the reflectivity spectrum is measured over a wide energy range, we can calculate the phase shift of the reflected beam relative to the incident beam by the so-called Kramers–Kroning transformation:

$$\theta(\omega) = -\frac{\omega}{\pi} P \left[\int_0^\infty \frac{\ln R(s)}{s^2 - \omega^2} ds \right],$$

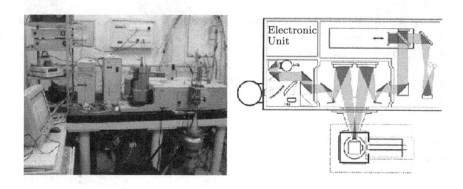

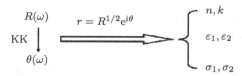

Fig. 3.2. The optical reflectance measurement system and optical path arrangement in the experimental setup in the Institute of Physics, CAS (upper panel); lower panel indicates the procedure of calculating other optical constants from the reflectance spectrum.

where the appropriate extrapolations should be made in the low- and high-frequency sides, respectively. There are simple algebraic relations among different optical response quantities, including the complex refractive index, complex dielectric function and complex conductivity. If $R(\omega)$ and $\theta(\omega)$ are known, we can easily obtain all other optical constants. For example, in the nearly normal incidence condition, the reflectance coefficient $r(\omega)$ can be expressed as:

$$r(\omega) = \frac{n(\omega) + i\kappa(\omega) - 1}{n(\omega) + i\kappa(\omega) + 1} = R(\omega)^{1/2} \exp[i\theta(\omega)],$$

where $n(\omega)$ is a refractive index and $\kappa(\omega)$ is an attenuation factor. So, we get:

$$n(\omega) = \frac{1 - R(\omega)}{1 - 2\sqrt{R(\omega)}\cos\theta(\omega) + R(\omega)},$$

$$\kappa(\omega) = \frac{2\sqrt{R(\omega)}\sin\theta(\omega)}{1 - 2\sqrt{R(\omega)}\cos\theta(\omega) + R(\omega)}.$$

The complex refractive index is $N(\omega) \equiv n(\omega) + i\kappa(\omega) = \sqrt{\varepsilon(\omega)}$, so:

$$\varepsilon_1 = n(\omega)^2 - \kappa(\omega)^2, \quad \varepsilon_2 = 2n(\omega)\kappa(\omega).$$

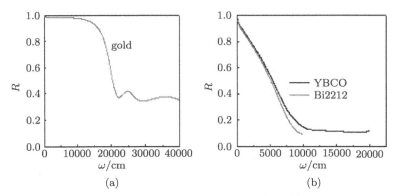

Fig. 3.3. Optical reflectivity spectrum of normal metal (a) and high-T_C superconductors (b).

On the basis of the relationship between the complex dielectric function and complex conductivity $\varepsilon(\omega) = \varepsilon_\infty + i4\pi\sigma(\omega)/\omega$, we can get $\sigma_1(\omega)$ and $\sigma_2(\omega)$ (take $\varepsilon_\infty = 1$).

Now let us compare the optical reflectance spectra between a simple metal (e.g. gold) and high-T_C cuprate superconductors. As displayed in Fig. 3.3, the reflectivity is very high in a wide energy range (0–2 eV) for gold. This is called reflection region. But at a certain energy, the reflectivity drops quickly, forming a relatively sharp edge. This reflectance edge is called screened plasma edge in optics. Beyond this edge energy, the reflectivity values become rather small, and it is named as transparent region. On the other hand, for the prototype high-T_C cuprate superconductors, such as $YBa_2Cu_3O_y$ or Bi2212, their reflectivity spectra display roughly linear frequency dependence without clearly showing a plasma edge. This structure is very different from a simple metal and is generally called overdamped behavior. In this situation, the charge carriers experience much stronger scatterings than in good metals.

What exactly does the optics detect? For the electron response in a solid, the optical spectroscopy probes the transitions of the electrons from initial states to final states with the absorption of the photon energies. The initial states must be the occupied states of the electrons (below the Fermi energy), and the final states must be the empty states (above the Fermi energy) permitted by the selection rule. The initial state and final state are allowed in either the same or different energy bands (see Fig. 3.4). If the initial state and the final state are in different bands, what the optical

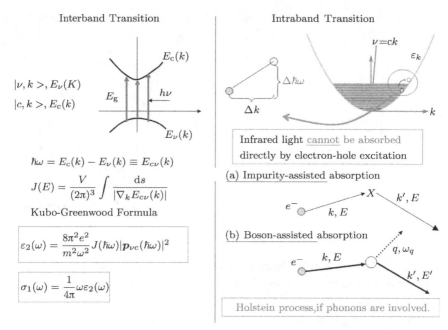

Fig. 3.4. The optical spectroscopy detects the transitions from the occupied states to the allowed unoccupied states, divided into interband transition and intraband transition.

spectroscopy detects is the interband transitions, which can provide some information about the band structure.

It is not difficult to derive the imaginary part of the dielectric function (or the real part of the conductivity) associated with the interband transitions — the so-called Kubo–Greenwood formula, which is related to the joint density of states connecting the two bands and the matrix elements of the dipole transitions. Those quantities could be calculated by the first-principle band structural calculations. Therefore, the measured optical spectrum for the interband transitions could be directly compared with first-principle band structural calculations. Taking a parent compound of the high-T_C superconductor La_2CuO_4 as an example (see Fig. 3.5), a good agreement can be found between the experimental measurements and band structural calculations for interband transitions at high energies (ignore the entirely different spectra at the low-energy region).

If the initial state and final state are in the same band, what the optics probes are the intraband excitations across the Fermi level. Because of the

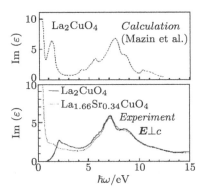

Fig. 3.5. The imaginary part of the dielectric function by experimental result and the first principle band calculation for La_2CuO_4. Taken from Ref. [2].

band dispersion, the intraband transition of electrons always has an associated change of momentum. However, the photon momentum of infrared radiation is virtually zero; a direct intraband transition by solely absorbing the phonon energy would violate the momentum conservation, which, therefore, is forbidden. There must be other scattering processes involved in the intraband transition, which makes the electron absorb photon energy simultaneously with momentum changed. It occurs through the assistance of elastic scattering of electrons from impurities or inelastic scattering process from other collective bosonic excitations (such as phonons). Thus, the intraband transitions are always involved with the elastic or inelastic scattering processes.

For free electron gas (assuming a constant relaxation time τ), a Drude model can easily be derived for its optical response. Figure 3.6 displays the characteristic features of several optical constants in the Drude model. Its reflectivity is close to 1 in the low-energy region, falling fast at screened plasma frequency and forming a plasma edge. The real part of the dielectric function crosses the horizontal axis at the screened plasma frequency. The energy loss function, defined as $Im[-1/\varepsilon(\omega)]$, shows a peak at the edge frequency, and the width of the peak is related to $1/\tau$ ($1/\tau$ is called scattering rate). The real part of the conductivity has the shape of a half Lorentz peak centered at zero frequency. The width of the Drude peak is just the scattering rate. The value at zero-frequency limit is the dc conductivity. The integral of the real part of the conductivity from zero to infinity is the square of the plasma frequency.

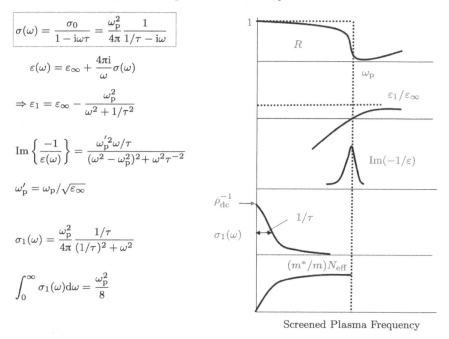

Fig. 3.6. Drude model of free electron gas.

For high-T_C superconductors and many other correlated electron systems, the optical conductivity is substantially different from the simple Drude model. Although a Drude-like peak exists in the optical conductivity at low frequency, the conductivity decays much more slowly than the Drude model. Significant spectral weight usually exists in the mid-infrared region. There are two different approaches to the optical response of high-T_C superconductors and correlated electron systems (see Fig. 3.7). The first one is the two-component approach where it is assumed that the conductivity is actually contributed by two different components: one is the low-frequency Drude component contributed by the coherent part of quasi-particles (or charge carriers), and the other is mid-infrared component contributed by the incoherent part of quasiparticles or bounded charge carriers. The second approach is the single-component analysis. The conductivity spectrum is assumed to be contributed by a single component of quasi-particles; however, the scattering rate is no longer frequency independent, but varies with temperature and frequency. Because the real part and the imaginary part of

Two possible interpretations

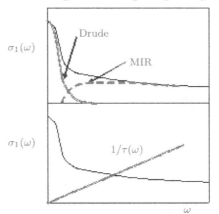

Fig. 3.7. The non-Drude-type conductivity spectrum of strongly correlated electron systems.

conductivity are linked by the Kramers–Kroning transformation, one cannot solely change the constant $1/\tau$ to a frequency-dependent one in the Drude model. An imaginary part has to be added to the damping rate, which can be interpreted as a modification of the effective mass of charge carriers.

For a superconductor at $T = 0$ K, the London electrodynamics gives:

$$\sigma = \frac{1}{8}\omega_{ps}^2 \delta(\omega) + i\omega_{ps}^2/4\pi\omega,$$

$$\Rightarrow \frac{1}{\lambda_L^2} = \frac{8}{c^2}\int_0^\infty (\sigma_1^n - \sigma_1^s)d\omega \quad \text{or} \quad \frac{1}{\lambda_L^2} = \frac{4\pi}{c^2}\omega\sigma_2(\omega),$$

i.e. in the superconducting state at zero temperature, all free electrons have condensed. The real part of conductivity becomes δ function, and its spectral weight is the condensed carrier density which is related to the London penetration depth. Because the imaginary and real part should follow the Kramers–Kronig relationship, if the real part is $\sigma_1(\omega) = A\delta(\omega)$, the corresponding imaginary part is $\sigma_2(\omega) = 2A/\pi\omega$. From the normal state to the superconducting state, we can estimate the penetration depth or superconducting carrier density from the missing spectral weight in the low-frequency real part of optical conductivity. Those quantities can also be estimated from the imaginary part of the complex conductivity in the zero-frequency limit (see Fig. 3.8).

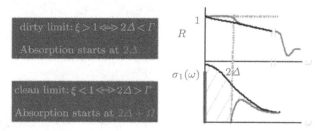

Fig. 3.8. Optical conductivity spectra of superconductors (Pippard coherence length $\xi = V_F/\pi\Delta$, $\Gamma = 1/\tau = V_F/l$).

Note that whether the superconducting gap can be clearly observed or not in optical conductivity depends on the superconductors being in the clean limit or the dirty limit. Only in the case of dirty limit, i.e. the charge carriers experience strong impurity scatterings, can the superconducting gap be observed in the conductivity spectrum. At zero temperature, the real part of conductivity starts to increase at twice the superconducting energy gap (2Δ). In the case of clean limit, $1/\tau$ arising from impurity scattering is negligible or very small (far smaller than the superconducting gap), then the conductivity has very small value at the superconducting gap energy 2Δ. It is hard to distinguish the change of spectral weight at 2Δ from the normal state to the superconducting state. In the presence of inelastic scatterings from a bosonic mode, a rapidly increasing threshold should appear in the conductivity spectrum. However, the energy scale is not at the twice superconducting energy gap, but at the energy of the sum of the boson model energy Ω and 2Δ. This is due to conservation of momentum and energy conservation requirements. For further details, the readers are suggested to refer to two good monographs [3, 4].

3.3. Optical Properties of High-T_C Superconductors

We first look at the evolution of optical spectra with doping in cuprates. The most famous work was done by the Uchida group at the university of Tokyo in the early 1990s. They measured the reflectivity $La_{2-x}Sr_xCuO_4$ system from the parent compound to the heavily doped samples (Fig. 3.9). The parent compound is an insulator. It has a rather low reflectivity. In the conductivity spectrum, there is almost no spectral weight in the low-energy region. The excitation starts to show up at 0.9 eV, and reaches a peak near 2 eV. Upon doping, the high-energy excitation is suppressed, and its

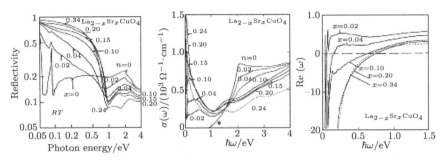

Fig. 3.9. Reflectivity, optical conductivity and the real part of dielectric function of $La_{2-x}Sr_xCuO_4$. Taken from Ref. [2].

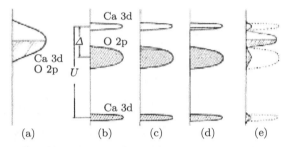

Fig. 3.10. Charge-transfer gap and band evolutions with doping concentration. Taken from Ref. [2].

spectral weight moves to the low-frequency side. In the lightly doped sample, there is no Drude response of the free electrons, but a broad peak appears in the mid-infrared region. With increased doping, the Drude component starts to emerge, and the mid-infrared component shifts to the lower energy scale. For optimally and heavily doped compounds, the mid-infrared component and the Drude component merge together.

Figure 3.10 shows the doping evolution of electronic states of the system. The density function calculation indicates that the parent compound is a metal, but in reality it is an insulator. As Prof. Ru-Shan Han addressed in his lecture, the parent compound is a charge transfer insulator. The copper 3d band is split into the upper and the lower Hubbard subbands due to the large on-site Coulomb repulsion energy U, and the oxygen 2p band is located between the two subbands. The chemical potential is located between the O2p band and the Cu3d upper band, so the parent compound is a charge transfer insulator. In terms of rigid band picture, with hole doping the chemical potential would shift down, and the holes are mainly formed in the O2p band. In contrast, with electron doping the chemical potential would shift up, and the electrons are mainly formed in the Cu3d upper Hubbard subband. However, the optical spectroscopy experiment from Uchida's group revealed that the hole doping would lead to suppression of the high-energy excitations and development of states in the mid-infrared region. This indicates the formation of the in-gap states between the O2p band and Cu3d upper Hubbard subbands. The intraband transition within the in-gap states would correspond to the new Drude component, while the mid-infrared component should come from the interband transition from the occupied in-gap states to the Cu3d upper Hubbard subband. Meanwhile, the interband transition from the O2p band to the Cu3d upper Hubbard subband (i.e. the charge transfer excitations) is reduced. With further hole doping, the in-gap states and the lower O2p band would merge together. Here, it is worth noting that the evidence for the doped holes occupying the O2p band mainly comes from the energy loss spectroscopy measurement rather than from optics probe.

By summarizing the conductivity spectral weight up to a cutoff frequency, we can get the effective carrier number below the cutoff energy (Fig. 3.11). The excited number of carriers from the optical spectral weight and number of charge carriers measured by the Hall Effect are different, because the Hall effect measures only conducting carriers, while the integration of conductivity spectrum contains the contribution from the non-conducting mid-infrared spectral weight, so it increases faster. From Fig. 3.11, we find that the spectral weight summarization up to 1.5 eV tends

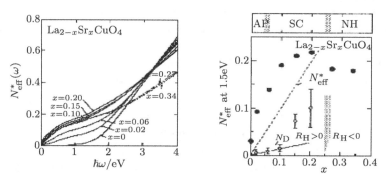

Fig. 3.11. Spectral weight sum of $La_{2-x}Sr_xCuO_4$ system and the evolution with doping. Taken from Ref. [2].

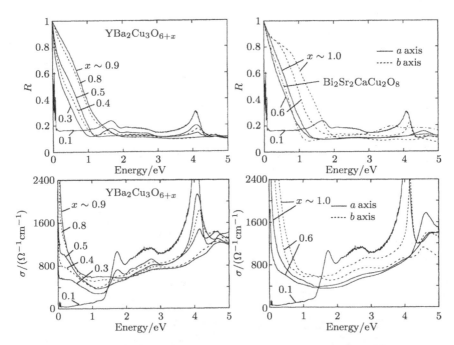

Fig. 3.12. Optical spectra of Y123 system. Taken from Ref. [5].

to saturate beyond 10% hole doping. Further doping would lead to a reduction of the spectral weight. This actually reflects the evolution of electronic structure and is also related to the choosing of cutoff frequency. Similar spectral evolutions were also observed in other hole-doped cuprate systems, e.g. YBCO (see Fig. 3.12). Its parent compound is also a charge transfer insulator but with relatively larger energy gap.

The electron-doped superconductors also show similar characteristics (see Fig. 3.13). The charge transfer energy gap of the parent compound is about 1.3 eV. With electron doping, a new peak emerges in the mid-infrared region in the conductivity. It can be ascribed to the interband transition, but actually there is a gap in the band structure located at the so-called "hot spots" in the Brillouin zone for under-doped compounds. So, it can be considered as the excitations across the energy gap near "hot spots".

Another interesting observation for the electron-doped cuprates is that the high-energy charge-transfer excitations shift to higher energy scale with electron doping. This result apparently indicates that the chemical potential moves into the Cu3d upper Hubbard subband (or formation of Cu $3d^{10}$)

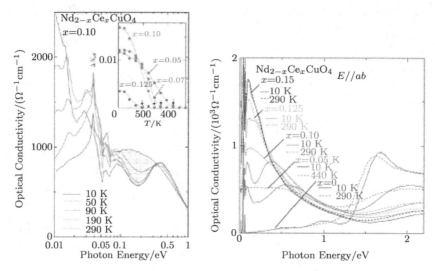

Fig. 3.13. Pseudogap and related charge dynamics in $Nd_{2-x}Ce_xCuO_4$ as a function of doping. Taken from Ref. [6].

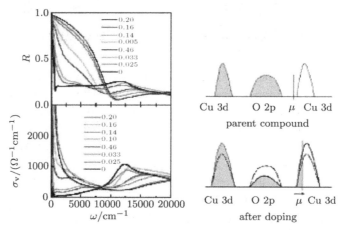

Fig. 3.14. Evolution of the chemical potential and charge dynamics with doping in $Nd_{2-x}Ce_xCuO_4$ [7]. Taken from Ref. [7].

and continuously shifts up with increasing doping, so that the interband transition from the O 2p band to the Cu 3d upper Hubbard sub-band increases in energy (Fig. 3.14).

A combination of the optical spectroscopy with the angle-resolved photoemission spectroscopy (ARPES) measurements can provide more

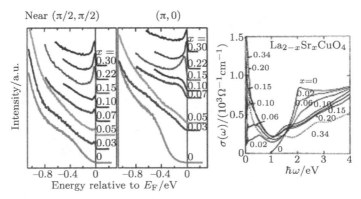

Fig. 3.15. Comparison of the ARPES and optical conductivity spectra of hole-type $La_{2-x}Sr_xCuO_4$ system. The left panel is from [8]. The right panel is from Ref. [2].

information. The optical response measures the transition from the occupied states to unoccupied states, while the ARPES measures the band structure below the Fermi level. Take the hole-doped $La_{2-x}Sr_xCuO_4$ system as an example (Fig. 3.15), the optical measurement shows that the lowest interband transition in the parent compound is 0.9 eV; however, the ARPES data indicate that the dispersive band in the parent compound is much closer to the Fermi energy. In fact, the energy difference between the two different spectroscopy probes reflects the location of chemical potential. So, for undoped La_2CuO_4 the chemical potential appears to be rather close to the top of O2p band. By contrast, for the parent compound of electron-doped cuprate $Nd_{2-x}Ce_xCuO_4$ (Fig. 3.16), the optical measurement shows that the interband transition excitation is about 1.3 eV, while the ARPES measurement indicates that the dispersive band is 1 eV away from the Fermi energy. The small difference from the two measurements indicates that the chemical potential is very close to the bottom of Cu3d upper Hubbard subband, which is significantly different from the hole-type superconductors.

3.4. Whether the Superconducting Energy Gap Can be Observed in the In-Plane Infrared Measurement

As mentioned earlier in this chapter, detecting the superconducting energy gap of cuprates from the in-plane infrared spectroscopy technique has been a controversial issue over 20 years in the infrared community. A dominant

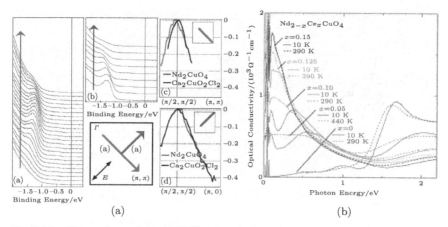

Fig. 3.16. Comparison of the ARPES and optical conductivity spectra of electron-type $Nd_{2-x}Ce_xCuO_4$ system. (a) Taken from Ref. [9] and (b) from Ref. [6].

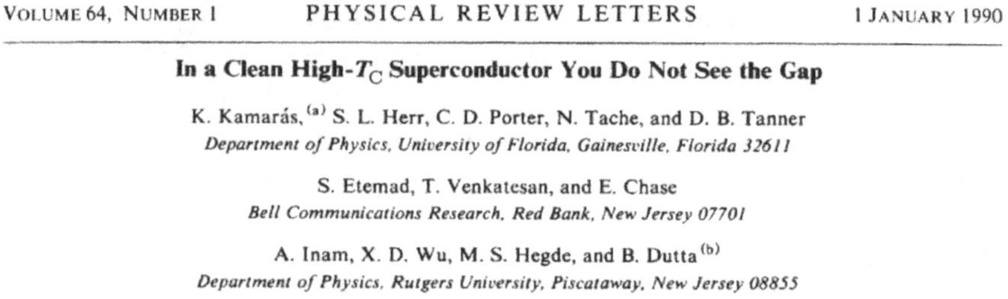

VOLUME 64, NUMBER 1   PHYSICAL REVIEW LETTERS   1 JANUARY 1990

In a Clean High-T_C Superconductor You Do Not See the Gap

K. Kamarás,[a] S. L. Herr, C. D. Porter, N. Tache, and D. B. Tanner
Department of Physics, University of Florida, Gainesville, Florida 32611

S. Etemad, T. Venkatesan, and E. Chase
Bell Communications Research, Red Bank, New Jersey 07701

A. Inam, X. D. Wu, M. S. Hegde, and B. Dutta[b]
Department of Physics, Rutgers University, Piscataway, New Jersey 08855

Fig. 3.17. The gap cannot be observed in the infrared spectra in a clean high-T_C superconductor.

view is that the CuO_2 plane of high-temperature superconductors is in the clean limit. In principle, superconducting energy gap cannot be probed by the infrared spectroscopy, and the characteristic spectral change observed from the normal to superconducting states is attributed to the mode coupling effect. However, it is worth noting that the clean limit argument for the CuO_2 plane of cuprate superconductors may not be true. Some recent experiment seems to indicate that the superconducting energy gap is directly visible in infrared probe.

Figure 3.17 displays the title of a paper published in early 1990s "In a clean high-T_C superconductor you do not see the gap". As mentioned above, in the case of clean limit with very small impurity scattering, $1/\tau$ should be much smaller than the superconducting energy gap. Even if all conductivity

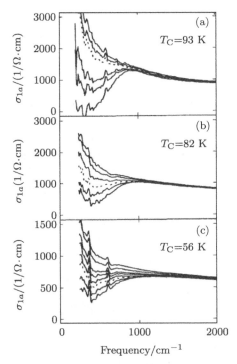

Fig. 3.18. The optical conductivity spectra of $YBa_2Cu_3O_y$ single crystal along a direction. Taken from Ref. [10].

component contributed from conducting carriers condenses into the spectral weight of $\delta(\omega)$ at zero frequency, the spectral weight change from normal to superconducting states at the energy of 2Δ is too small to detect. It is also argued theoretically that, for the Bogoliubov quasi-particle, the transition matrix element across the gap is zero. So in principle, superconducting energy gap could not be detected by infrared spectroscopy in the clean limit.

However, the absence of the superconducting energy gap in in-plane optical measurement was not accepted by all researchers. Figure 3.18 shows the conductivity spectra on detwinned single crystals of YBCO along the a-axis, which was reported in early 1990s. For an optimally doped sample, the low-frequency conductivity spectra were severely suppressed below superconducting transition temperature. The authors clearly pointed out that spectral change was caused by the opening of superconducting energy gap, and assigned the energy scale where the conductivity shows a minimum (about 500 cm^{-1}) as the 2Δ. For the under-doped samples, the suppression of low-frequency conductivity appears already at temperatures far above the T_C, which was attributed to the formation of pseudogap. Corresponding to the suppression, the reflectivity at low frequencies tends to approach unit.

Not only in $YBa_2Cu_3O_y$ but also in many other systems (see Fig. 3.19), it is found that the reflectance deviates significantly from the linear frequency dependence below a certain frequency upon cooling the samples from the normal state to the superconducting state. $Tl_2Ba_2CaCu_2O_8$ (Tl2212) has higher superconducting transition temperature than YBCO; such a spectral feature could be seen more clearly (see also Fig. 3.19). If we make the reflectance spectrum upside down, the obtained spectrum roughly has the shape of the scattering rate as extracted from the extended-Drude model. The scattering rate in the superconducting state is very low at low frequencies but increases prominently above a threshold frequency.

In fact, besides a sharp rising above a threshold frequency, there is also a peak-like structure (often called "overshoot") in the scattering rate spectrum for the optimally doped samples at the lowest temperature (see the red circle in Fig. 3.19). In the raw data of reflectance spectrum, a dip shows up the reflectivity at the corresponding frequency. Those structures are generic for optimally doped cuprate superconductors, but there are different interpretations.

One explanation is based on two-component picture: the characteristics seen are due to the condensing of free carrier spectral weight into the zero-frequency $\delta(\omega)$ function, leaving only the mid-infrared component (as schematically shown in Fig. 3.7). There are many theoretical proposals for the interpretation of the mid-infrared component in this correlated electron system. The other explanation is based on the single-component approach (see Fig. 3.20). Assuming that the conductivity spectrum over a broad frequency range is contributed by one component in which the scattering rate $1/\tau$ is not constant but changes with temperature or energy (as we mentioned in this case, an imaginary part of $1/\tau$ needs to be added so as to satisfy the Kramers–Kroning relationship), this is the so-called extended Drude model, and it can be compared with some phenomenological theories.

As we mentioned earlier, the intraband transition could occur with the assistance of impurities or bosonic collective excitations. In the single-component picture, if we assume that the superconductors were in the clean limit, i.e. the impurity scattering could be ignored, the bosonic collective excitation must be involved in the intraband transition. The experimentally observed spectral change was explained as arising from the inelastic

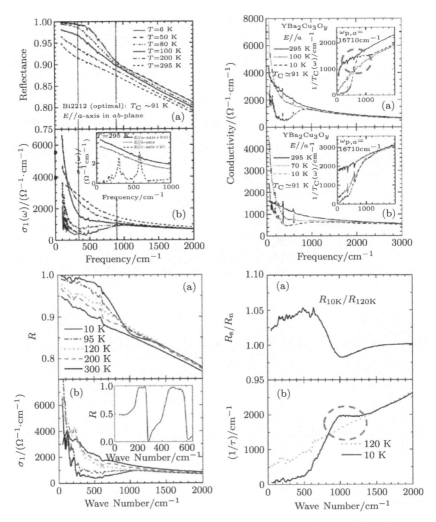

Fig. 3.19. Optical conductivity and scattering rate spectra of several high-T_C superconducting systems (Bi2212 (upper left), YBa$_2$Cu$_3$O$_y$ a direction (upper right), Tl2212 (lower) [13]). The upper left panel is taken from Ref. [11]. The upper right panel is taken from Ref. [12].

scattering process, i.e. the coupling effect between the carriers and collective modes. At very low temperature, the charge carriers experience very weak scatterings and therefore have small values of scattering rate, but at certain energy the charge carriers were suddenly scattered by some bosonic excitations, leading to a sharp increase in scattering rate.

$$\text{Drude Model } \sigma(\omega) = \frac{\omega_p^2}{4\pi} \frac{1}{1/\tau - i\omega}$$

Extended Drude Model—One component picture

Let $M(\omega, T) = 1/\tau(\omega, T) - i\omega\lambda(\omega, T)$ $\quad 1/\tau(\omega, T)$: Frequency dependent scattering rate
$\quad\quad\quad\quad\quad\quad\quad\quad\quad\quad\quad\quad\quad\quad\quad\quad$ λ: Mass enhancement $m^* = m(1+\lambda)$

$$\sigma(\omega, T) = \frac{\omega_p^2}{4\pi} \frac{1}{M(\omega,T) - i\omega}$$

$$= \frac{\omega_p^2}{4\pi} \frac{1}{1/\tau(\omega,T) - i\omega[1+\lambda(\omega,T)]}$$

$$= \frac{1}{4\pi} \frac{\omega_p^{*2}}{1/\tau^*(\omega,T) - i\omega}$$

$$1/\tau(\omega,T) = (\omega_p^2/4\pi)\text{Re}(1/\sigma(\omega))$$
$$1 + \lambda(\omega) = (\omega_p^2/4\pi\omega)\text{Im}(1/\sigma(\omega))$$

e. g. marginal Fermi liquid model
$$M(\omega,T) = 1/\tau(\omega,T) - i\omega\lambda(\omega,T)$$
$$= g^2 N^2(0)\left(\frac{\pi}{2}x + i\omega \ln\frac{x}{\omega}\right)$$

Where $x = \max(|\omega|, T)$
or $x = (\omega^2 + \alpha(\pi T)^2)^{1/2}$

And the Extended Drude Model is also commonly used optical self-energy represented

Extended Drude Model

$$\sigma(\omega, T) = \frac{\omega_p^2}{4\pi} \frac{1}{(\gamma(\omega,T) - i\omega)}$$

According to Little and Varma,

$$\gamma(\omega) = -2i\Sigma^{op}$$
$$= 2i[\Sigma_1(\omega) + i\Sigma_2(\omega)]$$

optical self-energy

$$\gamma_1(\omega) = 1/\tau(\omega) = 2\Sigma_2$$
$$\gamma_2(\omega) = \omega(1 - m^*/m) = -2\Sigma_1$$

Fig. 3.20. Extended Drude model.

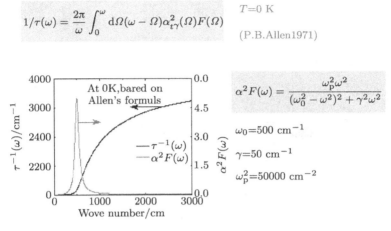

Fig. 3.21. The electron–boson (phonon) interaction at zero temperature.

For a typical electron–phonon interaction, we can use the classic Allen formula to simulate the scattering rate at zero temperature (Fig. 3.21) [14]. Given a phonon spectral function, e.g. a Lorentz shape, the scattering rate does significantly increase at the phonon energy. At the finite temperature, the Allen's formula needs to be modified. A formula given by Shulga et al. [15] could be used to simulate the scattering rate in this case (Fig. 3.21). Overall, very similar spectral shape could also be obtained in scattering rate at finite temperature, though the rising feature becomes weaker. However, we noted that the simulation could not generate an "overshoot" structure in the scattering rate spectrum, an eminent spectral feature observed experimentally.

Based on the Allen formula, we find that the scattering rate increases monotonically, and no overshoot-like characteristic can be obtained for the scattering rate. But the energy gap was not taken into consideration in the above simulations. Although a generic formula for the presence of energy gap is still lacking theoretically, the scattering rate formula for a BCS superconductor based on the electron–phonon coupling was developed by Allen. The integral includes a type-two elliptic function which contains the superconducting energy gap [14]. In this case, an overshoot-like characteristic could be obtained in $1/\tau$ spectrum, but the sudden increase of the scattering rate is not at the bosonic mode energy, but shifts to higher energy scale by a superconducting energy gap of 2Δ (Fig. 3.23).

$$\frac{1}{\tau(\omega,T)} = \frac{\pi}{\omega}\int_0^\infty d\Omega \alpha^2 F(\Omega,T)[2\omega\coth(\frac{\Omega}{2T})-$$
$$(\omega+\Omega)\coth(\frac{\omega+\Omega}{2T})+(\omega-\Omega)\coth(\frac{\omega-\Omega}{2T})]$$

Formula by Shulga

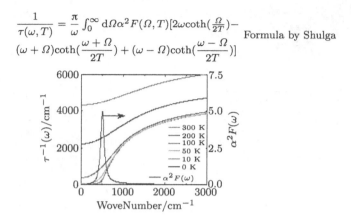

Fig. 3.22. The electron–boson (phonon) interaction at finite temperature.

$$\frac{1}{\tau(\omega)} = \frac{2\pi}{\omega}\int_0^{\omega-2\Delta} d\Omega(\omega-\Omega)\alpha^2 F(\Omega) E\left(\sqrt{1-\frac{4\Delta^2}{(\omega-\Omega)^2}}\right)$$

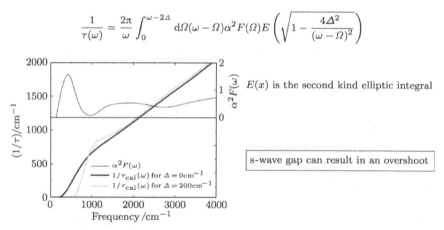

$E(x)$ is the second kind elliptic integral

s-wave gap can result in an overshoot

Fig. 3.23. S-wave superconducting energy gap could lead to an overshoot-like structure in the scattering rate. Taken from Ref. [16].

Regardless of any mode coupling, a peak structure in the scattering rate could actually be obtained from the extended Drude model as long as an energy gap opens at the Fermi level (see Fig. 3.24). There are some discussions based on the sum rule on why there must be a peak structure in the scattering rate when an energy gap exists. For electron-doped superconductors, formation of energy gap was well known at the so-called "hot spots" region (the intersection area between the anti-ferromagnetic

Infrared and Optical Response of High-Temperature Superconductors 73

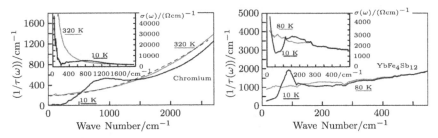

Fig. 3.24. Scattering rate spectra by extended Drude model in the partially gapped systems. Taken from Ref. [17].

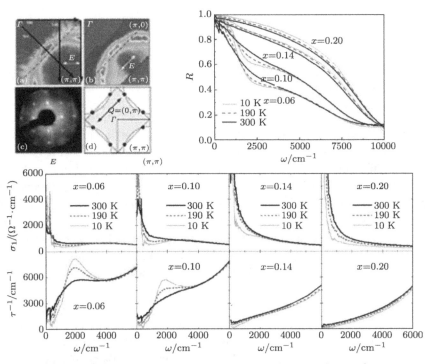

Fig. 3.25. Energy gap formation in the intersection area between the anti-ferromagnetic wave vector and Fermi surface for electron-type superconductor $Nd_{2-x}Ce_xCuO_4$ and significant peak-like structure in the scattering rate spectra. Taken from Refs. [7, 9b].

wave vector and Fermi surface). Indeed, a peak-like structure was found in the scattering rate (see Fig. 3.25) extracted from the extended Drude model. There are many other partially gapped systems showing similar spectral shape in scattering rate.

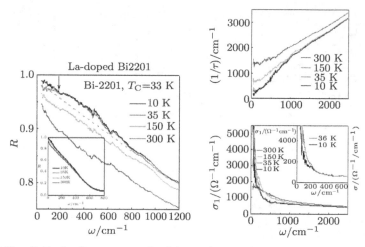

Fig. 3.26. Optical reflectance, conductivity and scattering rate spectra for La-doped Bi2201 single crystal.

Now the key question is whether the peak-like structure (or overshoot) in the scattering rate is directly caused by the energy gap formation at the Fermi surface or by the mode coupling effect in the clean limit (with a shift by the superconducting energy gap in the superconducting states).

Our recent experiments indicated that the superconducting energy gap could be indeed seen directly in ab-plane optical measurement. The data of optimally doped single crystal Bi2201 ($T_C \sim 33$ K, grown by Professor Fang Zhou from Zhong-Xian Zhao's group) is shown in Fig. 3.26. The high-temperature reflectivity is found to be linearly frequency-dependent. Cooling down the sample below 150 K, the reflectivity deviates upward from the linear dependence and develops a weak curvature at low frequency (<800 cm^{-1}). This spectral characteristic is linked to the formation of the pseudogap. Now let us compare the reflectivity at 10 K in the superconducting state with that at 35 K. The reflectivity spectra at two temperatures almost completely overlap at frequencies higher than 180 cm^{-1}, and separate only below that frequency. A new upturn in 10 K reflectivity develops. This spectral change is related to the superconducting phase transition, and therefore must be related to the pairing gap. The above results are consistent with ARPES experiments on La-doped Bi2201 samples with the same T_C (Fig. 3.27). At very high temperature (above the pseudogap formation temperature), ARPES measurement indicates a complete Fermi surface. When the temperature is between T_C and the pseudogap formation

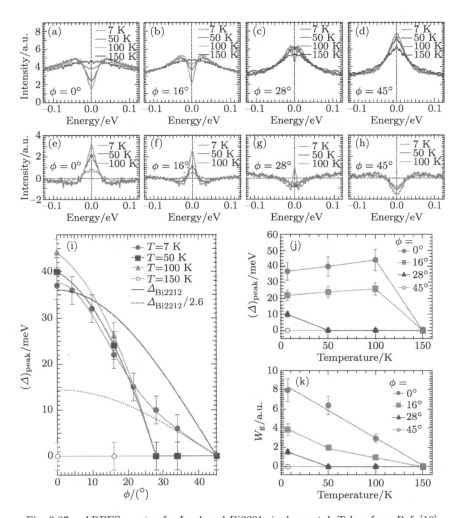

Fig. 3.27. ARPES spectra for La-doped Bi2201 single crystal. Taken from Ref. [18].

temperature, the arc-shape Fermi surface (or Fermi arc) is seen only near the nodal region, while energy gaps open at the anti-nodal region. Below T_C, a true superconducting energy gap develops in the Fermi arc.

The pseudogap is likely originated from a kind of charge density wave order, which is linked with 4 × 4 checkerboard charge modulation observed by STM. There are two energy gaps seen from ARPES: one is a pseudogap at the anti-nodal region, which opens even in the normal state; the other is the superconducting energy gap opening at the Fermi arc in the

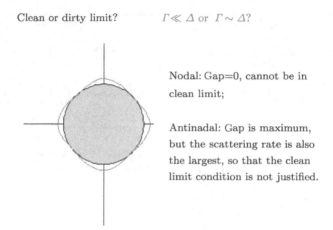

Fig. 3.28. The high-T_C superconductors have a d-wave pairing symmetry, The clean limit argument is not valid, at least in the nodal region of Fermi surface.

superconducting state. The optical experiment provides essentially the same information: the difference of reflectivity spectra at very low energy across the T_C is related to the superconducting energy gap formation, and the upward deviation from the linear frequency dependence at higher energy scale seen above T_C (but below 150 K) is related to the pseudogap formation. Of course, we need to demonstrate why the superconducting energy gap can be seen. The key issue remains whether or not the CuO_2 plane of high-T_C superconductors is in the clean limit. As the clean limit requests the charge carrier scattering rate $1/\tau$ from impurities much smaller than superconducting energy gap, the charge carriers near the nodal region clearly do not meet the clean limit condition. This is because the cuprate superconductors have a d-wave pairing symmetry and the gap size is zero at the nodal direction (Fig. 3.28). At the anti-nodal direction, the d-wave superconducting energy gap has the largest gap amplitude. However, the charge carriers experience much stronger scattering in the antinodal region, and even no well-defined quasi-particles could be found in the antinodal region, it is not clear whether or not the clean limit condition could be met in the antinodal region. So, the clean limit argument is not valid, at least in some area of the Fermi surfaces. That is why we can see the gap.

In fact, similar spectral characteristics related to the superconducting energy gap can also be detected in $La_{2-x}Ba_xCuO_4$ and $La_{2-x}Sr_xCuO_4$ systems ($T_C \approx 30$ K) in optical experiments. Figure 3.29 schematically

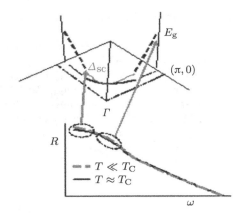

Fig. 3.29. Relationship between the optical reflection spectra and ARPES spectra.

illustrates the relationship between the superconducting gap and the pseudogap characteristics in the optical reflectance spectrum and the two energy gaps revealed by ARPES. The higher one corresponds to the pseudogap, which shows up at the anti-nodal region in the normal states, while the lower one observed below T_C corresponds to the pairing energy gap formed in the Fermi arc near the nodal direction in the superconducting state.

References

[1] D. B. Basov, T. Timusk, *Rev. Mod. Phys.* **77**, 721 (2005).
[2] S. Uchida, T. Ido, H. Takagi, *Phys. Rev. B* **43**, 7942 (1991).
[3] F. Wooten, *Optical Properties of Solids* (Academic Press, New York, 1996).
[4] M. Dressel, G. Gruner, *Electrodynamics of Solids* (Cambridge University Press, Los Angeles, 2002).
[5] S. L. Cooper, D. Reznik, A. Kotz et al., *Phys. Rev. B* **47**, 8233 (1993).
[6] (a) Y. Onose, Y. Taguchi, K. Ishizake et al., *Phys. Rev. Lett.* **87**, 217001 (2001); (b) Y. Onose, Y. Taguchi, K. Ishizake et al., *Phys. Rev. B* **69**, 024504 (2004).
[7] N. L. Wang, G. Li, D. Wu et al., *Phys. Rev. B* **73**, 184502 (2006).
[8] T. Yoshida, X. J. Zhou, T. Sasagawa et al., *Phys. Rev. Lett.* **91**, 027001 (2003).
[9] (a) N. P. Armitage, F. Ronning, D. H. Lu et al., *Phys. Rev. Lett.* **88**, 257001 (2002); (b) N. P. Armitage, D. H. Lu, C. Kim et al., *Phys. Rev. Lett.* **87**, 147003 (2001).
[10] L. D. Rotter, Z. Schlesinger, R. T. Collins et al., *Phys. Rev. Lett.* **67**, 2741 (1991).
[11] J. J. Tu, C. C. Homes, G. D. Gu et al., *Phys. Rev. B* **66**, 144514 (2002).

[12] C. C. Homes, S. V. Dordevic, D. A. Bonn *et al.*, *Phys. Rev. B* **69**, 024514 (2004).
[13] N. L. Wang, P. Zhao, J. L. Luo *et al.*, *Phys. Rev. B* **68**, 054516 (2003).
[14] P. B. Allen, *Phys. Rev. B* **3**, 305 (1971).
[15] S. V. Shulga, O. V. Dolgov, E. G. Maksimov, *Physica C* **178**, 266 (1971).
[16] S. V. Dordevic, C. C. Homes, J. J. Tu *et al.*, *Phys. Rev. B* **71**, 104529 (2005).
[17] D. N. Basov, E. J. Singley, S. V. Dordevic, *Phys. Rev. B* **65**, 054516 (2002).
[18] Takeshi Kondo, Tsunehiro Takeuchi, Adam Kaminski *et al.*, *Phys. Rev. Lett.* **98**, 267004 (2007).

4
NMR Study of High-Temperature Superconductors

Guo-Qing Zheng

Institute of Physics, Chinese Academy of Sciences, Beijing 100190, China
Department of Physics, Okayama University, Okayama 700-8530, Japan

This chapter reviews results and insights on high-T_C superconductors from the nuclear magnetic resonance (NMR) technique, with emphasis on local hole density, d-wave superconducting gap and its consequences, and the nature of the pseudogap. In addition to the total hole density in the CuO_2 plane, the partition of the holes between Cu and O orbits is an important factor in determining T_C. The temperature dependence of the spin-lattice relaxation rate and the Knight shift indicate d-wave pairing, which brings about the extended quasi-particle states outside the vortex cores. The existence of nodes in the gap function also results in strong T_C reduction by non-magnetic impurity or crystal disorder. The zero-temperature normal state has been revealed by the application of strong magnetic fields up to 45 T to be a metallic state with finite density of states (DOS) at the Fermi level, which suggests that the pseudogap and the superconducting states are coexisting states of matter. The residual DOS remains quite large even in the vicinity of the Mott insulating phase.

4.1. Introduction

A vast number of studies using various techniques have been performed to elucidate the properties and the mechanisms of the high-temperature superconductors (HTSC). These studies suggest that not only the superconducting state but also the normal state of the HTS contains rich physics. In this chapter, we review the results and insights obtained from NMR which is a microscopic experimental probe. We will focus on the local hole distribution in the CuO_2 plane, the properties of the superconducting state and the so-called pseudogap state above the transition temperature T_C. Two factors important for raising T_C will be emphasized, namely, the partition of holes between Cu and O sites, and the crystal order.

Our study has utilized the spatial resolution of NMR. As a local experimental probe, NMR can separately explore the electronic states and magnetic excitations at different atomic sites. For example, many HTS contain CuO chain and CuO_2 plane, or different CuO_2 planes. Since the Cu nucleus experiences different electric or magnetic interactions, one is able to separate the Cu signals under different environments so that the electronic state at different sites can be measured. In the superconducting state, NMR is capable of separately measuring the electronic state inside and outside the vortex cores. In addition, by analyzing the data obtained from different nuclei (such as Y, Cu, O, Tl, etc.), one can obtain momentum-resolved spin fluctuations. For more information the latter part, readers are referred to [1].

4.2. Local Hole Distribution

High-temperature superconductivity is achieved by doping carriers. Therefore, it is important to know the actual doping rate. It is also important to know where the doped carrier resides. In conventional solids, Hall coefficient can be used as a good probe for carrier density, but it does not work in high-T_C cuprates since the Hall coefficient shows a strong temperature dependence owing to electron correlations. Here, we describe a method using the nuclear quadrupole resonance (NQR) frequency to extract the local hole density.

A nuclear spin with $I > 1/2$ possesses a nuclear quadrupole moment Q, which interacts with the electric field gradient (EFG) $\partial^2 V/\partial \alpha^2 (\alpha = x, y, z)$ through:

$$H_Q = \frac{h\nu_z}{6}\left[3I_z^2 - I(I+1) + \frac{1}{2}\eta(I_+^2 + I_-^2)\right], \qquad (4.1)$$

$$\nu_z = \frac{3}{2hI(2I-1)}eQ\frac{\partial^2 V}{\partial z^2}, \qquad (4.2)$$

where $\eta \equiv |\nu_x - \nu_y|/\nu_z$ is the asymmetry parameter of the NQR frequency tensor.

The main cause for the EFG is the asymmetric electron wave functions. For example, the 3d orbital wave functions of Cu or the 2p orbital wave functions of oxygen are not symmetric about the nuclei, so they will produce a large EFG at the nucleus position. When there is a hole in a 3d orbit of Cu, it will result in a NQR frequency of $\nu_{3d}^0 = 117$ MHz. When there is a hole in a 2p orbit of oxygen, it will result in an NQR frequency of $\nu_{2p}^0 = 3.6$ MHz [2].

Figure 4.1 shows the Sr(Ce) content dependence of the NQR frequency tensor in La$_{2-x}$Sr$_x$CuO$_4$ and Nd$_{2-x}$Ce$_x$CuO$_4$. In the former compounds, ν_z increases with increasing Sr content (hole content). In the latter compounds, however, ν_z decreases with increasing Ce content (electron doping). η for ^{63}Cu is almost zero, which indicates that the holes almost reside on the 3d$_{x^2-y^2}$ orbit. On the other hand, η for ^{17}O is finite, which indicates that all the three O-2p orbits host a finite (but different) amount of holes. A quantum chemical analysis shows that [3], the hole density in Cu-3d$_{x^2-y^2}$ orbit, $n_{x^2-y^2}$, and that in the oxygen-2p$_\sigma$ orbit, n_{p_σ}, are respectively:

$$n_{x^2-y^2} = \frac{^{63}\nu_z}{\nu_{3d}^0}, \qquad (4.3)$$

$$n_{p_\sigma} = \left(1 + \frac{^{17}\eta}{3}\right)\frac{^{17}\nu_z}{\nu_{2p}^0}. \qquad (4.4)$$

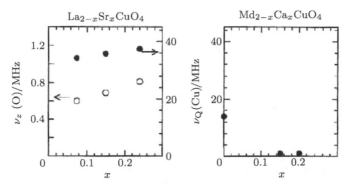

Fig. 4.1. Doping evolution of the nuclear quadrupole resonance (NQR) frequency tensor in the typical p-type superconductors, La$_{2-x}$Sr$_x$CuO$_4$, and n-type superconductors, Nd$_{2-x}$Ce$_x$CuO$_4$. For ^{63}Cu, ν_Q(Cu) $=^{63}\nu_z\sqrt{1+\eta^2/3}$.

In the above equations, the superscripts 63 and 17 stand for the isotope ^{63}Cu and ^{17}O, respectively. Calculation shows that $n_{x^2-y^2}$ is about 0.84 in La_2CuO_4. That the number is smaller than 1 is a consequence of hybridization between Cu and O; some holes are transferred to the O site. After substitution of Sr, the major part of the holes go to the oxygen-$2p_\sigma$ orbit, but a small amount of holes go to the Cu-$3d_{x^2-y^2}$ orbit.

Figure 4.2 shows the T_C dependency on the local hole distribution. The first horizontal axis is the sum of the hole density in the Cu-$3d_{x^2-y^2}$ orbit and that in the O-$2p_\sigma$ orbit in the CuO_2. By specific definition [3], the total hole density before doping is $n_{x^2-y^2} + 2n_{p_\sigma} = 1.2$. This axis represents the

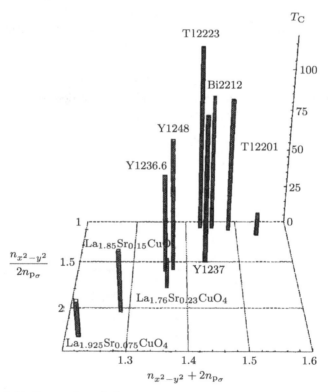

Fig. 4.2. A bird's eye view of the transition temperature of the cuprates. The first horizontal axis is the sum of the hole density in the Cu-$3d_{x^2-y^2}$ orbit and that on the O-$2p_\sigma$ orbit in the CuO_2. The other horizontal axis represents the partition of the holes between Cu and O. Keys for the materials: Y1236.6: $YBa_2Cu_3O_{6.6}$, Y1237: $YBa_2Cu_3O_7$, Y1248: $YBa_2Cu_4O_8$, Tl2223: $Tl_2Ba_2Ca_2Cu_3O_{10}$, Bi2212: $Bi_2Sr_2CaCuO_8$, Tl2201: $Tl_2Ba_2CuO_6$.

doping rate in the CuO_2 (1.2 is the origin). Another horizontal axis is the ratio of the hole density in the Cu-3$d_{x^2-y^2}$ orbit to that in the O-2p_σ orbit, which represents the partition of the holes between Cu and O. It is well known that certain amount of total hole density is required to achieve superconductivity. Figure 4.2 tells us, however, that the total hole density is not the only parameter that controls T_C. For example, for $Tl_2Ba_2Ca_2Cu_3O_{10}$ (Tl2223) and $YBa_2Cu_3O_7$ (Y123), the total hole density is about the same but T_C differs considerably. The different T_C stems from the different hole partition. For Tl2223, $n_{x^2-y^2}$ is only 0.74, the remaining holes go into the O-2p_σ orbit. The difference between Tl2223 and Y123 shows that it is necessary to move some part of the holes to the O site in order to have a higher T_C. Figure 4.2 further shows that the highest T_C is achieved for $n_{x^2-y^2}/2n_{p_\sigma} \approx 1$.

4.3. Superconducting State Properties

4.3.1. *Energy Gap and the Nuclear Spin-Lattice Relaxation Rate*

Ever since the discovery of the high-T_C superconductors, the pairing symmetry has been a central issue. NMR/NQR is one of the early experiments that provide information on the topic. In particular, NQR experiment is performed at zero magnetic field that does not reduce T_C and thus is advantageous for studying the superconducting properties. The spin-lattice relaxation rate ($1/T_1$) measured by NMR or NQR is related to the DOS and the so-called coherence factor, which can tell us about the symmetry of the orbital function of the Cooper pairs. The Knight shift measured by NMR is proportional to the spin susceptibility, which is the only probe to measure the spin symmetry in the superconducting state.

Figure 4.3 shows the temperature dependence of $1/T_1$ for various high-T_C cuprates [4]. For the sake of comparison, the horizontal axis shows the reduced temperature, and the vertical axis is the reduced $1/T_1$. The main features are(1) $1/T_1$ decreases rapidly below T_C, and its temperature dependence is identical in all materials, which is in sharp contrast to the s-wave superconductors where $1/T_1$ increases just below T_C, forming the so-called coherence peak. (2) As temperature decreases further, $1/T_1$ decreases in a power law of temperature. This is also in contrast to s-wave superconductors where $1/T_1$ decreases exponentially.

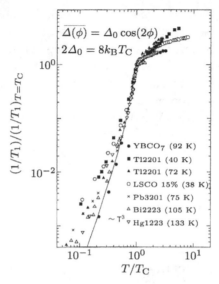

Fig. 4.3. $1/T_1$ normalized by its value at T_C as a function of reduced temperature for various high-temperature superconductors.

In the superconducting state, $1/T_{1s}$ is expressed as:

$$\frac{T_{1N}}{T_{1s}} = \frac{2}{k_B T} \iint \left(1 + \frac{\Delta^2}{EE'}\right)$$
$$\times N_s(E) N_s(E') f(E) \left[1 - f(E')\right] \delta\left(E - E'\right) \mathrm{d}E \mathrm{d}E', \quad (4.5)$$

where $1/T_{1N}$ is the nuclear spin-lattice relaxation rate in the normal state, Δ is the energy gap, $N_s(E)$ is the DOS in the superconducting state and $f(E)$ is the Fermi function. $C = 1 + \Delta^2/EE'$ is called the coherence factor [5]. For an s-wave superconductor, $N_s(E)$ diverges at $E = \Delta$, and the coherence factor is also large, which causes the coherence peak. At low temperatures, $1/T_{1s}$ is dominated by $f(E)[1-f(E)]$; therefore, $1/T_{1s} \propto \exp(-\Delta/k_B T)$.

If nodes exist in the gap function, the situation will be quite different. In such a case, Δ changes sign around the node, and its average over the entire Fermi level equals zero. Then, $C = 1$ can be obtained by writing Δ^2/EE' in the forms of $(\Delta_k \Delta_{k'}/E_k E_{k'})$, and integrating with respect to $k(k')$ over the whole Fermi surface. Furthermore, $N_s(E)$ in the case of nodal gap does not diverge so drastically as s-wave does at $E = \Delta$. For the line nodes case (e.g. four line nodes in 2D d-wave energy gap), DOS $\propto E$ at low energy, resulting in:

$$\frac{1}{T_{1s}} \propto \int E^2 f(E)[1-f(E)]dE \propto T^3$$

at low temperatures.

Figure 4.3 apparently indicates that high-temperature superconductors are not s-wave superconductors. The experimental results are consistent with d-wave gap. The theoretical curve in Fig. 4.3 is a calculation using d-wave energy gap ($\Delta(\phi) = \Delta_0 \cos(2\phi)$, $\Delta_0 = 4k_B T_C$). In clean samples (e.g. YBa$_2$Cu$_3$O$_7$), good agreement with experimental data is found. In some samples, due to impurity effect, $1/T_1$ deviates from T^3 at low temperature. We will discuss this issue again in Section 4.3.3.

4.3.2. Cooper Pair Symmetry

The Knight shift is the most accurate experimental method to measure the electronic spin susceptibility in the superconducting state. In this case, one needs to apply an external magnetic field. When the magnetic field enters a superconductor, a vortex lattice is formed. Figure 4.4 shows the spatial dependence of the magnetic field in a superconductor. M is the position of the vortex core where the field is the maximum. S is the middle of the nearest two vortices cores (saddle point) and m is the middle point of the nearest four vortices cores, where the magnetic field is the minimum.

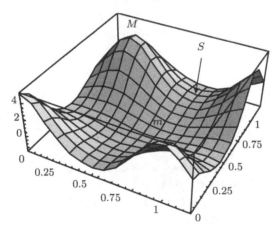

Fig. 4.4. Spatial distribution of magnetic field for a 75° vortex lattice. The unit of the horizontal axis is the interval of two vortices, and the vertical axis is the magnetic field intensity.

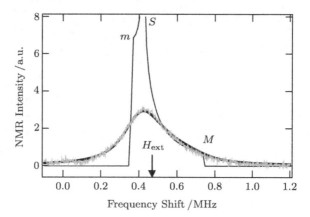

Fig. 4.5. The theoretical distribution (fine line) of the magnetic field in the vortex state. M, S and m correspond to the point in Fig. 4.4. The vertical axis corresponds to the strength of the magnetic field, and the arrow indicates the position of the external magnetic field. The thick solid curve is a convolution of the theoretical curve with a Lorentzian broadening function. The green curve is the experimental NMR spectrum.

The thin curve in Fig. 4.5 is the magnetic field distribution density, which diverges at the saddle point and forms a "knee-like" kink at M point. The NMR spectrum in the superconducting state is nothing but such a magnetic field distribution density whose peak corresponds to the saddle point and is far from the M point. In reality, however, this kind of characteristic spectrum shown in Fig. 4.5 is rarely observed because of distortion of the vortex lattice. The green curve in Fig. 4.5 is the NMR spectrum of $TlSr_2CaCu_2O_7$ [6]. The solid thick curve is a calculation by convoluting the fine curve with a Lorentzian broadening function:

$$p(H') = \frac{\sigma^2}{\sigma^2 + 4\pi^2 H'^2},$$

with a fixed value of σ/π =170 Oe that is the full-width at half maximum (FWHM) of the NMR spectrum in the normal state.

From the points mentioned above, it is clear that the NMR spectrum peak originates from the nuclear spins outside the vortex cores. Thus, the Knight shift obtained from this peak is a powerful tool to probe the spin state of the Cooper pairs. Figure 4.6 shows the typical data sets of the temperature and magnetic field dependencies of the Knight shift.

The measured K_{obs} consists of three parts:

$$K_{obs} = K_s + K_{orb} + K_{dia}, \qquad (4.6)$$

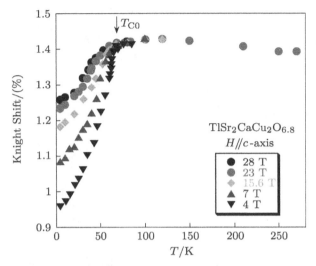

Fig. 4.6. Temperature and magnetic field dependencies of the Knight shift. The arrow indicates the superconducting transition temperature at zero magnetic field, T_{C0}.

K_s and K_{orb} are proportional to spin and orbital susceptibility, respectively. $K_{orb} \approx 1.0\%$ in $TlSr_2CaCu_2O_7$. K_{dia} is due to diamagnetic effect ($H_s < H_{ext}$, see Fig. 4.5), $K_{dia} = (H_s - H_{ext})/H_{ext}$. In $TlSr_2CaCu_2O_7$ at external field $H_{ext} = 4$ T, $K_{dia} \approx -0.3\%$ is obtained from the calculation. From the figure, it is found that K_s decreases to almost zero at $T = 4.2$ K. This result supports d-wave pairing (spin singlet).

As shown in Fig. 4.6, the Knight shift increases with increasing magnetic field at low temperatures. Figure 4.7 shows the magnetic field dependence of the Knight shift at $T = 4.2$ K. After subtracting K_{dia} for each magnetic field, K_s is found to be proportional to $\sqrt{H}$. This phenomenon is an indication of an important feature of a nodal energy gap, which is not seen in conventional s-wave superconductors.

In s-wave superconductors, the energy gap is isotropic with respect to momentum. The low-energy quasi-particles are localized in the vortex cores [7] whose radius is comparable to the coherence length ξ. In contrast, when nodes exist in the gap function, the coherence length becomes infinite at the nodal directions. So the low-energy quasi-particles can "leak" to the outside of a vortex core along the nodal directions. As a result, the low-energy quasi-particles are delocalized in d-wave superconductors. The higher the applied magnetic field, the higher the density of vortices, and as

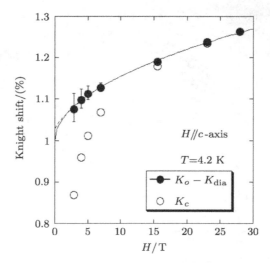

Fig. 4.7. Magnetic field dependence of the Knight shift with the field along the c-axis, K_c, at $T = 4.2$ K. The solid curve is a fit of the data to $(K_{\text{obs}} - K_{\text{orb}} - K_{\text{dia}}) \propto \sqrt{H}$.

a result, the larger the overlap of the leaked quasi-particle states outside the vortex cores. Such quasi-particles can be exactly observed by the above-mentioned NMR experiment. Volovik was the first to predict this phenomenon, which is now called the Volovik effect. Volovik found that the DOS of all low-energy quasi-particles (including inside and outside the vortices), N_V, should be:

$$\frac{N_V}{N_0} = \sqrt{\frac{H}{H_{C2}}}, \qquad (4.7)$$

where N_0 is the DOS in the normal state and H_{C2} is the upper critical field. The fitting curve in Fig. 4.7 shows that the extended quasi-particles DOS, N_{out}, outside the vortex is:

$$\frac{N_{\text{out}}}{N_0} = 0.7\sqrt{\frac{H}{H_{C2}}}, \qquad (4.8)$$

and the coefficient 0.7 was confirmed by a recent theoretical calculation [9].

Incidentally, scanning tunneling microscope (STM) is in principle the most direct method to observe the quasi-particles outside the vortex. However, this kind of quasi-particle has not been observed in STM experiment in HTS unfortunately, owing to the layered crystal structure of HTS and anisotropic tunneling matrix; the matrix element nearly vanished at the gap nodes [10].

4.3.3. *Impurity Effect*

Another consequence of d-wave pairing of HTS is that it exhibits a response to impurities or crystal disorder in a different way from s-wave superconductors. The reduction of T_C is small in s-wave superconductors with isotropic gap. On the contrary, cuprate HTS has an anisotropic gap, which is strongly influenced by isotropic impurity scattering and T_C is significantly reduced. This kind of impurity effect is very profound in heavy fermion superconductors ($T_C < 2$ K), where superconductivity can hardly survive unless the sample is very pure.

From a microscopic point of view, the different impurity effect arises from the different way to create an impurity state. Adding paramagnetic impurity into s-wave superconductors will form a localized bound state inside the energy gap, i.e. Yu–Shiba state, whereas superconductivity is basically unaffected by adding a non-magnetic impurity. As opposed to an isotropic s-wave superconductor, in HTS, scattering by non-magnetic impurity introduces a low-energy excited state, leading to the appearance of a finite DOS at the Fermi energy [13, 14].

Besides substituting Zn for Cu, crystal disorder is another source of quasi-particle scattering. It is suggested that, in order to improve T_C, not only is a flat CuO_2 plane is needed but also, a well-ordered structure outside the CuO_2 plane is important. Below is an example to illustrate the point.

Before the discovery of $HgBa_2Ca_2Cu_3O_{10}$ (T_C = 135 K), annealed $Tl_2Ba_2Ca_2Cu_3O_{10}$ (Tl2223, T_C = 125 K) was the world record holder. T_C of as-grown Tl2223 samples is only 117K. In the as-grown material, a portion of Tl substitutes for Ca, which causes a disorder of Ca layer. We found that the primary effect of annealing was to improve the lattice order of Ca layer. It is demonstrated by the change in the $^{205/203}$Tl-NMR spectrum between as-grown and annealed samples. Figure 4.8 compares the temperature dependence of $1/T_1$ of the as-grown and annealed samples [15, 16]. There are three layers in this material. The innermost CuO_2 plane is 4-fold (the open symbols in Fig. 4.8) and two outer CuO_2 planes are 5-hold pyramid-like (the closed symbols in Fig. 4.8). The Ca layer is between these two CuO_2 planes. Before annealing, $1/T_1 \propto T$ at low temperatures

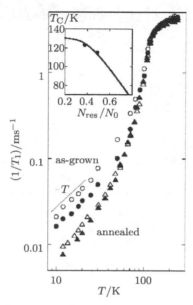

Fig. 4.8. $1/T_1$ of ^{63}Cu for as-grown (circles) and annealed (triangles) $Tl_2Ba_2Ca_2Cu_3O_{10}$. The solid line indicates the relation $1/T_1 \propto T$. The inset shows T_C as a function of residual DOS.

(or $1/T_1T =$ constant), indicating that scattering by the crystal disorder causes a residual DOS at the Fermi level (residual DOS, N_{res}). It is also noted that 4-hold CuO_2 plane is more affected than the pyramid-like CuO_2 plane, since the former is sandwiched by two Ca layers, while there is only one Ca layer adjacent to the latter.

The value of $1/T_1T$ at low temperature for the annealed sample obviously decreases, which indicates that the decrease of scattering by crystal disorder reduces the residual DOS. The relationship between T_C and residual DOS calculated by:

$$\left(\frac{N_{res}}{N_0}\right)^2 = \frac{(1/T_1T)_{low-T}}{(1/T_1T)_{T=T_C}}$$

is shown in the inset of Fig. 4.8, where the solid curve is calculated by treating scattering in the unitary limit. It is suggested that T_C of $Tl_2Ba_2Ca_2Cu_3O_{10}$ can be increased to 130 K by eliminating the crystal disorder (e.g. completely eliminating Tl/Cu inter-substitution). $HgBa_2Ca_2Cu_3O_{10}$ is the case where the value of $1/T_1T$ at low temperature is quite small [17].

4.4. Electron Correlations and the Ground State of the Pseudogap

In general, not only for a metal but also an insulator, $1/T_1$ can be expressed as [18]:

$$\frac{1}{T_1 T} = \frac{\gamma_N^2 k_B}{2\mu_B^2} \sum_q A_q^2 \frac{\mathrm{Im}\chi(q,\omega_N)}{\omega_N} \bigg|_{\omega_N \to 0}, \quad (4.9)$$

where Im $\chi(q,\omega_N)$ is the imaginary part of the dynamical susceptibility, and A_q is the hyperfine coupling between the electron and nuclear spins. The symbol $\sum$ means summation over all wave vectors. In NMR experiments, the frequency of ω_N is in the megahertz range, which is much less than the electronic energy scale. So, $1/T_1T$ probes the slope of energy dependence of Im $\chi(q,\omega_N)$ at low energy, while neutron scattering technique measures that at high energies. These two kinds of experimental methods are complementary.

The imaginary part of the dynamical susceptibility of free electrons can be written as:

$$\mathrm{Im}\chi(q,\omega) = 2\pi\mu_B^2 \sum_k [f(E_k) - f(E_{k+q})]\,\delta(E_{k+q} - E_k + \hbar\omega), \quad (4.10)$$

and the hyperfine coupling does not have wave-vector dependence. Thus, $1/T_1T$ of conventional metals follows the relation:

$$\frac{1}{T_1 T} = \pi \hbar \gamma_N^2 k_B A^2 N_0^2, \quad (4.11)$$

where N_0 is the DOS at the Fermi level.

The Knight shift K_s is proportional to the uniform susceptibility:

$$K_s = A\chi_s(q=0) = A\mu_B^2 N_0, \quad (4.12)$$

and so the following identity is obtained:

$$T_1 T K_s^2 = \frac{\mu_B^2}{\pi \hbar \gamma_N^2 k_B}, \quad (4.13)$$

which is the so-called Korringa law [19].

Figure 4.9 shows the temperature and La concentration dependence of $1/T_1T$ in $Bi_2Sr_{2-x}La_xCuO_6$(Bi2201) [20]. The crystal structure of this system is simple since it has only one CuO_2 plane. Moreover, both T_C and H_{C2} are low, so the superconducting state can be easily suppressed by applying

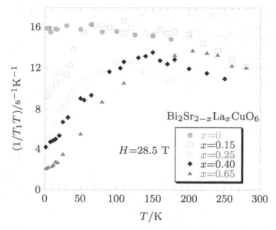

Fig. 4.9. The quantity $1/T_1T$ plotted against temperature and La concentration for $Bi_2Sr_{2-x}La_xCuO_6$.

a high magnetic field. It is therefore the most ideal system to study the normal state at the zero-temperature limit (ground state) of high-temperature superconductors. $Bi_2Sr_2CuO_6$ without La doping is an overdoped material, whose T_C is found to be 8 K. Replacing La with Sr increases the hole concentration of the system. As shown in the figure, $1/T_1T$ is essentially a constant for $Bi_2Sr_2CuO_6$. After increasing La concentration, $1/T_1T$ starts to increase as temperature decreases, which is attributed to electron correlations. Interestingly, in the samples with $x > 0$, instead of increasing to $T = T_C$, $1/T_1T$ starts to decrease at a temperature higher than T_C, leaving a broad peak in the $1/T_1T$ vs T plot. This is the so-called pseudogap phenomenon seen in the temperature dependence of $1/T_1T$. The temperature corresponding to the peak is called T^*. Below T^*, $1/T_1T$ decreases because of a loss of DOS due to the onset of the pseudogap.

Such a pseudogap phenomenon is also seen in the Knight shift, as seen in Fig. 4.10 [21]. Above T^*, the shift is a constant, but it decreases below T^*.

The doping dependence of T^* is shown in Fig. 4.11. The T^* decreases as hole concentration increases, and there is a critical point for the La concentration, $x_C \approx 0.05$, beyond which the pseudogap disappears. The critical point corresponds to the hole concentration $p_C \approx 0.21$, according to the characterization by electronic transport measurement [22]. The $x = 0$ sample is still superconducting, but its normal state is a Fermi liquid (T_1T = constant) which is also the ground state when the superconductivity

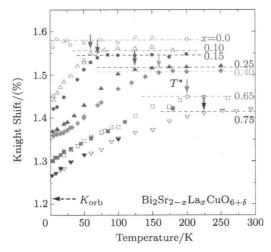

Fig. 4.10. The temperature dependence of the Knight shift in $Bi_2Sr_{2-x}La_xCuO_6$. The dotted lines are guides to the eye. The dotted arrow indicates the value of K_{orb}.

is suppressed by the application of a high magnetic field. This is similar to the case of electron-doped cuprates [23].

Then, what is the electronic state for $p < p_C$ when the superconducting state is suppressed? This question had been posted for a long time, but was not answered due to the high H_{C2} until very recently when we performed a series of experiments on Bi2201, under high magnetic fields up to 45 T [20, 21].

Figure 4.12 shows the magnetic field dependence of $1/T_1T$ for the optimally doped $Bi_2Sr_{1.6}La_{0.4}CuO_6$ [20]. Note that, at zero magnetic field, $1/T_1T$ decreases abruptly below $T_C(H = 0) = 32$ K. After applying the magnetic field of $H = 28.5$ T, the decrease is quenched, and $1/T_1T$ continues to follow the temperature dependence at high temperatures. By increasing the magnetic field to 43 T, no further field dependence is observed. This indicates that the superconductivity is already suppressed by 28.5 T. Therefore, our results for $H \geq 28.5$ T characterize the normal state when the superconductivity is removed. The same feature is seen in the Knight shift [21], as shown in Fig. 4.13. There remains a large DOS in the ground state of the pseudogap.

Since the electrical conductivity is metallic, it is concluded that the pseudogap ground state is a metallic state with finite Fermi surface. The persistence of the pseudogap when superconductivity is removed and the

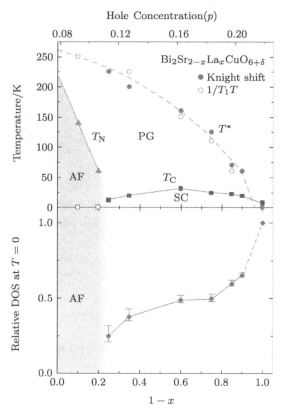

Fig. 4.11. Phase diagram of $Bi_2Sr_{2-x}La_xCuO_6$. AF, PG and SC denote the antiferromagnetically ordered, pseudogap state and superconducting state, respectively. The upper panel shows the characteristic temperatures vs. $1-x$ (hole concentration). The lower panel shows the hole-concentration dependence of relative DOS at $T = 0$.

criticality of pseudogap imply that the pseudogap and superconductivity are coexisting states of matter. The lower panel of Fig. 4.11 shows that the sizable DOS remained at the ground state of the pseudogap. Superconductivity develops out of such residual DOS.

It is noted that there is no magnetic order or charge order when superconductivity is removed. Figure 4.14 shows the FWHM of the ^{63}Cu-NMR spectrum [24]. The width, which is due to the distribution of the Knight shift owing to the electronic imhomogeneity, increases linearly with increasing field. This result indicates that there is no additional broadening due to a magnetic or charge order. Reasonably assuming that one has a resolution

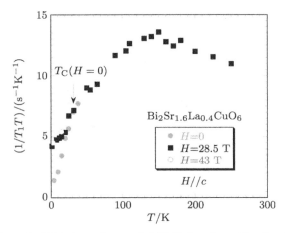

Fig. 4.12. Magnetic field dependence of $1/T_1T$ for the optimally doped $Bi_2Sr_{1.6}La_{0.4}CuO_6$.

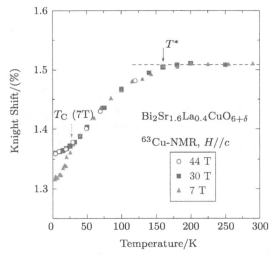

Fig. 4.13. T dependence of the ^{63}Cu Knight shift ($H//c$) for the optimally doped $Bi_2Sr_{1.6}La_{0.4}CuO_6$ measured under different magnetic fields. The solid and dotted arrows indicate the pseudogap temperature T^* and superconducting transition temperature T_C ($H = 7\,T$), respectively.

of detecting the additional broadening that is one tenth the width, one can put an upper bound for the magnetic moment, if any. The upper bound is 0.8 m μ_B, given the hyperfine coupling constant of 19.5 T/μ_B. Namely, there is no magnetic moment larger than 0.8 m μ_B. Therefore, the ground state of

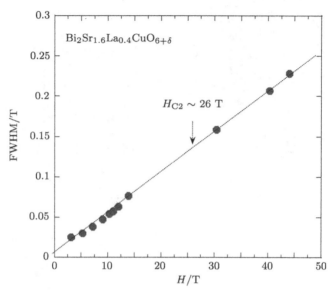

Fig. 4.14. Magnetic field dependence of the full width at half maximum (FWHM) of the ^{63}Cu-NMR spectrum.

the pseudogap is a metallic state without translational symmetry breaking, yet has only partial volume of the Fermi surface. This is a new state of matter out of which high-temperature superconductivity evolves. Hence, any theory for high-temperature superconductivity should be capable of accounting for such a new state.

4.5. Concluding Remarks

In this chapter, several aspects seen by NMR experiments of the high-temperature superconductors were reviewed. The high-temperature superconductors are hole-doped or electron-doped Mott insulators. The distribution of the hole between Cu and O is the key factor for T_C. The high-temperature superconductors are d-wave superconductors, which provided an excellent opportunity to study novel superconducting states other than s-wave BCS state. The existence of nodes in the gap function gives rise to novel low-energy excitations. The so-called Volovik effect is the new knowledge acquired in the research of high-temperature superconductors, which also greatly contributes to the studies of other unconventional superconductors, e.g. heavy-fermion superconductors. The pseudogap state setting in above T_C is not a Fermi liquid state familiar to us. The pseudogap ground state is a metallic state with reduced Fermi surface but without translational symmetry breaking. Although the mechanism of the pseudogap is still unknown, there is no doubt that the strong electron correlation plays an important role.

It is worth pointing out that, during the course of studying high-temperature superconductors, all kinds of experimental methods have enjoyed breakthrough development, which in turn promoted the exploration of new materials and the development of related research fields. For example, MgB_2 and $Na_xCoO_2 \cdot 1.3H_2O$ superconductors were discovered during this course. The recent discovery of iron-based high temperature superconductor is another typical example. The research on new phenomena such as colossal magneto resistance (CMR) and multiferroics phenomenon in transition metal oxides is also an emerging field stimulated by the study of high-T_C superconductivity in cuprates. There is no doubt that high-temperature superconductivity and the novel electromagnetic functions of transition metal oxides will lay the foundation for strongly correlated electron engineering.

The author thanks Rui Zhou of Institute of Physics for help in preparing this chapter.

References

[1] W. Z. Zhou, W. Y. Liang, *Fundamental Research in High T_C Superconductivity* (Shanghai Scientific and Technical Publishers, Shanghai, 1999).
[2] K. Hanzawa, F. Komatsu, K. Yosida, *J. Phys. Soc. Jpn.* **59**, 3345 (1990).
[3] G. Q. Zheng, Y. Kitaoka, K. Ishida et al., *J. Phys. Soc. Jpn.* **64**, 2524 (1995).
[4] K. Asayama, G. Q. Zheng, Y. Kitaoka et al., *Physica C* **178**, 281 (1991).
[5] J. R. Schriefer. *Theory of Superconductivity* (Addison-Wesley Longman Inc., Redwood, 1988).
[6] G. Q. Zheng, H. Ozaki, Y. Kitaoka et al., *Phys. Rev. Lett.* **88**, 077003 (2002).
[7] C. Caroli, P. G. de Gennes, J. Matricon, *Phys. Lett.* **75**, 2754 (1964).
[8] G. E. Volovik, *JETP Lett.* **58**, 469 (1993).
[9] N. Nakai, P. Miranović M. Ichioka et al., *Phys. Rev. B* **70**, 100503 (2004).
[10] C. Wu, T. Xiang, Z. B. Su, *Phys. Rev. B* **62**, 14427 (2000).
[11] L. Yu, *Acta Phys. Sin.* **21**, 75 (1965).
[12] H. Shiba, *Prog. Theor. Phys.* **40**, 435 (1968).
[13] S. S. Rink, K. Miyake, C. M. Varma, *Phys. Rev. Lett.* **57**, 2575 (1986).
[14] R. Fehrenbacher, M. R. Norman, *Phys. Rev. B* **50**, 3495 (1994).
[15] G. Q. Zheng, Y. Kitaoka, K. Asayama et al., *Physica C* **260**, 197 (1996).

[16] G. Q. Zheng, Y. Kitaoka, K. Asayama et al., J. Phys. Soc. Jpn. **66**, 1880 (1997).
[17] K. Magishi, Y. Kitaoka, G. Q. Zheng et al., Phys. Rev. B **53**, R8906 (1996).
[18] T. Moriya, J. Phys. Soc. Jpn. **18**, 516 (1963).
[19] J. Korriga, Physica **16**, 601 (1950).
[20] G. Q. Zheng, P. L. Kuhns, A. P. Reyes et al., Phys. Rev. Lett. **94**, 047006 (2005).
[21] S. Kawasaki, C. T. Lin, A. P. Reyes et al., Phys. Rev. Lett. **105**, 137002 (2010).
[22] S. Ono et al., Phys. Rev. Lett. **85**, 638 (2000).
[23] G. Q. Zheng, T. Sato, Y. Kitaoka et al., Phys. Rev. Lett. **90**, 197005 (2003).
[24] J. W. Mei, S. Kawasaki, G. Q. Zheng, Phys. Rev. B **85**, 134519 (2011).

5
Angle-Resolved Photoemission Spectroscopy Studies of Many-Body Effects in High-T_C Superconductors

Xing-Jiang Zhou

National Laboratory for Superconductivity, Institute of Physics, CAS, Beijing 100190, China

5.1. Introduction to Angle-Resolved Photoemission Spectroscopy

Angle-resolved photoemission spectroscopy (ARPES) is a direct and powerful tool in probing the electronic structure of materials (see Fig. 5.1). Its advantage is the ability to directly measure two microscopic quantities for describing the electronic state of materials, i.e. energy (E) and momentum (K), then to determine the basic physical quantities such as electron velocity, electron effective mass, electron scattering rate, energy gap and Fermi surface etc., which help determine the macroscopic physical properties. ARPES has emerged as a conventional band mapping tool over the last two decades, and has played a very important role in the research of high-T_C superconductors. Meanwhile, the ARPES technique itself has also been undergoing rapid development and the instrument resolution has been significantly improved, which has been gradually developed into an important means to study the many-body interaction between materials.

Under the sudden approximation, ARPES measures the single particle spectral function:

$$A(k,\omega) = \frac{1}{\pi} \frac{\mathrm{Im}\Sigma(k,\omega)}{[\omega - \varepsilon_k - \mathrm{Re}\Sigma(k,\omega)]^2 + [\mathrm{Im}\Sigma(k,\omega)]^2}, \quad (5.1)$$

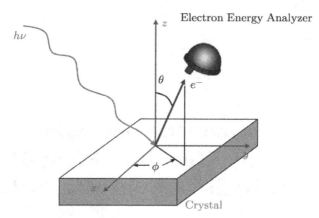

Fig. 5.1. A schematic of angle-resolved photoemission spectroscopy. With a beam of monochromatic light (which can be VUV laser, gas-discharge lamps or synchrotron radiation) hitting on the sample surface, due to the photoelectric effect, the photoelectrons escape from the sample surface, and the escaped photoelectrons, along the different angles, are measured by the energy analyzer. According to the escaped photoelectrons' energy and momentum (determined by the escaped angle), and the energy and momentum conservation law, the electronic states within the material can be determined.

where $\Sigma(k,\omega)$ describes the quasi-particle self-energy, the imaginary part $\text{Im}\Sigma(k,\omega)$ describes the quasi-particle's lifetime and the real part $\text{Re}\Sigma(k,\omega)$ describes the quasi-particle's energy. When the electrons of materials interact with other electrons, phonons and magnetons, the interaction will lead to the change of self-energy, so we can get the many-body interaction inside the materials by measuring the self-energy using ARPES.

Figure 5.2 schematically shows how to obtain the electronic self-energy from the ARPES data. While using the modern advanced ARPES technology, the measured raw data constitute a two-dimensional image: it represents the relationship between the photoelectron number (in colors) and energy, as well as momentum. At a certain energy, we can obtain the relationship between the photoelectron number and momentum from the image, which is called momentum distribution curve (MDC). The curve can be generally fitted with a Lorentzian function to get the MDC peak position and width. So, by fitting MDCS corresponding to the different energy, we can obtain the energy–momentum dispersion relationship and MDC width (FWHM) in Fig. 5.2(b). If we know the bare band (there is no many-body interactions), the self-energy caused by the many-body interaction can be

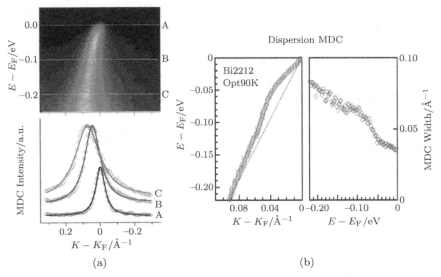

Fig. 5.2. (a) The original two-dimensional ARPES spectra, A, B, C corresponding to three energy points in the MDC; (b) the dispersion curve and the corresponding MDC width at half maximum of MDC peak by fitting the (a) MDCS. The real and imaginary parts of self-energy can be obtained from Equations (5.2) and (5.3).

obtained using the following formula:

$$\text{Re}\Sigma(k,\omega) = \omega - x_0 v_0, \tag{5.2}$$

$$\text{Im}\Sigma(k,\omega) = (\Gamma/2)v_0, \tag{5.3}$$

where ω represents energy, x_0 represents momentum k position of MDC fitting peak, Γ represents FWHM and v_0 represents the Fermi velocity corresponding to the bare band.

In 1911, Onnes first discovered the superconductivity phenomenon in mercury. Then in 1957, Bardeen, Cooper and Schrieffer established BCS superconductivity theory and successfully uncovered the origin of superconductivity. Therefore, Bardeen et al. received the Nobel Prize for Physics in 1972. According to the BCS theory, superconductivity is based on the electron–electron pairing and condensation, which is achieved through the exchange of phonons. It is worth mentioning that the BCS theory, from a proposal to general acceptance, also went through a long process. Later on, the tunneling experiments were used to analyze the boson spectral function engaged in coupling, which played a key role in the establishment of the origin of superconductivity. According to the BCS theory, the energy gap is isotropic, and there is a theoretical limit in superconducting transition temperature (the specific values depends, generally believed to be ∼40 K).

In 1986, the Swiss scientists Bednorz and Muller discovered high-T_C superconducting phenomenon in copper oxide for the first time. During the next few years, more high-T_C superconducting materials were discovered quickly, and the highest transition temperature recorded reached 160 K (Hg1223, under pressure), which is far higher than the temperature limit predicted by the original BCS theory. Meanwhile, the existing experiments have confirmed that in the new cuprate superconductors, the superconducting order parameter is not isotropic s-wave-like, but mainly has a d-wave symmetry. It indicates that the conventional BCS theory obviously cannot explain the new high-T_C superconducting phenomenon, and so exploring for a new superconducting mechanism has once again become the primary and urgent issue. Because ARPES has a unique and important role in the measurements of materials' electron structure and many-body effects as well as in other aspects, it has played and will continue to play a very important role in the mechanism study of high-T_C superconductivity.

5.2. Angle-Resolved Photoemission Spectroscopy Studies of Many-Body Effects in High-T_C Superconductors

5.2.1. *The First Observation of Many-Body Effects in High-T_C Superconductors — Nodal Direction*

Figure 5.3 shows the corresponding schematic diagram of Brillouin zone and Fermi surface in the high-T_C cuprate superconductor. As the d-wave superconducting gap along the $(0,0) \sim (\pi,\pi)$ direction is zero, this direction is generally referred to as the nodal direction, and the $(\pi,0)$ and $(0,\pi)$ regions are known as the anti-nodal regions. According to the ARPES measurements of Bi2212 along the nodal direction, there is a "kink" structure in the energy–momentum dispersion curve (see Fig. 5.4).

Undertaking more detailed and comprehensive measurements, we find that the "kink" exists in the hole-type cuprate superconductors [2]. By analysis of the different hole-type cuprate superconductors (LSCO, Bi2212 and Bi2201), with different doping concentrations (under-doped, optimally doped and over-doped) and different temperatures (above T_C and

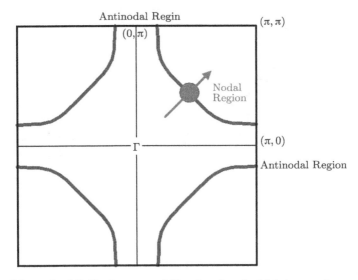

Fig. 5.3. Schematic of Brillouin zone and Fermi surface for high-temperature cuprate superconductors $(0,0) \sim (\pi,\pi)$ — nodal direction, $(0,\pi)$ and $(\pi,0)$ — anti-nodal region.

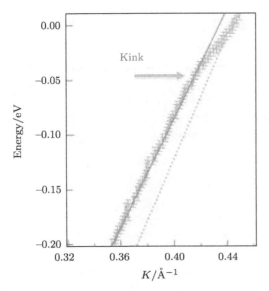

Fig. 5.4. The energy–momentum dispersion relation along nodal direction for high-T_C superconductor Bi2212. In the vicinity of ~ 70 meV, the slope of the dispersion curve changes significantly, with the emergence of a kink at the position, the so-called "kink" structure. Taken from Ref. [1].

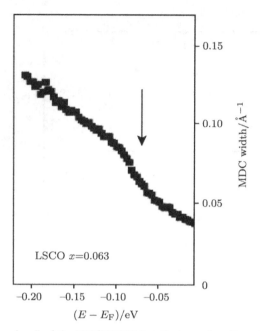

Fig. 5.5. A "steep drop" of the MDC FWHM plotted as a function of energy along the nodal direction for superconductor LSCO [3].

below T_C), we reveal that the "kink" is widespread and is about at 70 meV on the energy scale.

Further measurements indicate that in the MDC FWHM curve, reflecting the electron scattering rate, at the corresponding energy scale in the energy dispersion, when the "kink" turns up, there exists a sudden "steep drop" structure, as shown in Fig. 5.5.

5.2.2. *The Origin of the 70 meV "Kink" Structure*

As previously mentioned, there generally exists a "kink" structure in the energy–momentum dispersion of cuprate superconductors as well as a sudden "steep drop" change at the corresponding energy in the scattering rate. These characteristics are similar to traditional electron–boson coupling. Currently, the general consensus is that they are caused by the coupling between electrons and some kind collective excitations. In the high-T_C superconducting materials, it is known that there are three types of collective excitations: magnetic resonance mode, other magnetic excitations and

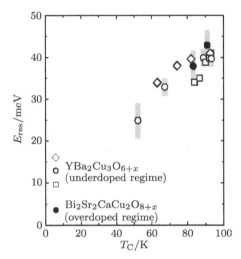

Fig. 5.6. A synopsis of the resonance peak energy E_{res} plotted as a function of the superconducting transition temperature T_C. Taken from Ref. [4].

phonon. First, the boson cannot be the magnetic resonance mode, because the general existing "kink" is inconsistent with the following characteristics of the magnetic resonance mode:

(1) The magnetic resonance mode is not observed in all superconducting systems (but the "kink" generally exists in hole-type superconductors).
(2) In the optimally doped and over-doped samples, the magnetic resonance mode is only seen in the superconducting state and disappears in the normal state (but the "kink" exists above and below T_C).
(3) There is a relationship between the energy E_{res} of magnetic resonance mode and superconducting transition temperature: $E_{\text{res}}/kT_C = 5.4$ (see Fig. 5.6). Taking Bi2201 as an example, if the mode exists, then the corresponding energy scale is about 15 meV (the energy scale of the "kink" is about 70 meV along the node direction.).
(4) The spectral weight of the magnetic resonance mode is very small, only about 1% of the overall spectral weight.

It is noteworthy that in the recent neutron scattering experiments of $(\text{La}_{2-x}\text{Sr}_x)\text{CuO}_x$ ($x = 0.16$, optimally doped), two peak-shaped magnetic structures are observed near 0–70 meV [5], which makes the magnetic excitation become an option with the magnetism forming the nodal "kink".

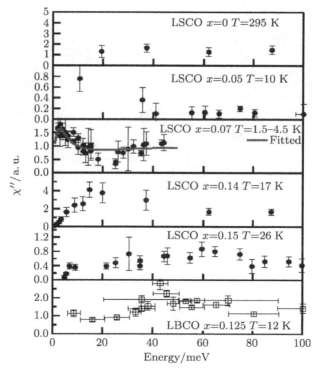

Fig. 5.7. The magnetic response curve for differently doped LSCO samples [6–10]. As can be seen, the magnetic excitation changes a lot with doping variation, and there is no obvious double-component structure. The upper panel is from Ref. [6].

However, taking into account other relevant neutron scattering results, it was shown that copper oxide in the magnetic excitation has a strong doping dependence, and this double-peak structure does not generally exist, as shown in Fig. 5.7. This is inconsistent with the general existing "kink" in the differently doped samples, so this particular magnetic structure cannot be the reason for the 70 meV "kink".

In all known collective excitations (boson), the most provable cause of formation of the 70 meV "kink" is the electron–phonon coupling. It is also consistent with the existing experimental facts. The energy scale of half-breathing mode movement between copper ions and oxygen ions in the CuO_2 plane is very close to the observed "kink" energy scale, so it is thought that this is the reason for the formation of the 70 meV "kink".

5.2.3. Non-Traditional Electron–Phonon Coupling

Figure 5.8 shows in the LSCO system the measured dispersion curves of nodal direction involve a wide range of doped samples. One can see the following points: (1) the "kink" structure widely exists in all doped samples, with energy scale about 70 meV; (2) the "kink" structure reduces as the doping concentration increases; (3) the change of the dispersion at Fermi energy reaching 70 meV is weak as the doping concentration changes; (4) the dispersion relationship of high energy (70–200 meV) shows abnormal changes with doping: as doping concentration reduces, the velocity of high energy increases. Generally speaking, the electron–electron interaction tends to narrow the original band, and reduce the electron velocity. In high-T_C superconductors, as generally considered, the low-doped samples have strong electron correlation, which is expected to be because the bands are narrower than in the high-doped ones and have a lower velocity. However, the experimental results are just the opposite of the expected conclusion.

The above results further indicate, in high-T_C superconductors, the generally existing electron–phonon (e–p) interaction is different from the one seen in conventional metals. The effect of the conventional e–p interaction

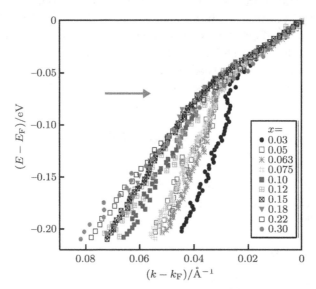

Fig. 5.8. The "kink" structure along the nodal direction for different doping concentration of LSCO samples [3].

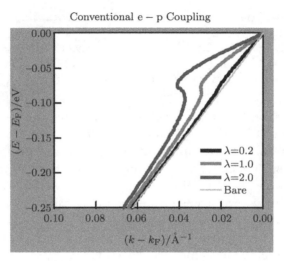

Fig. 5.9. A simulation of the "kink" structure on conventional e-p interaction: the energy–momentum dispersion relationship as a function of the strength of e-p interaction.

on the energy–momentum dispersion is related to the original dispersion relation, phonon spectral function and the intensity of e-p interaction. Considering the effects from the intensity λ of e-p interaction, we can obtain the dispersion relation by simple simulation, as shown in Fig. 5.9. We find that the dispersion close to the Fermi energy changes significantly with λ, but in the high-energy region the dispersion becomes weaker and tends to the bare band. These trends are oppositely related with the change in the LSCO experimental results (see Fig. 5.8), which reflects the peculiar nature of the e-p interaction in the high-T_C superconductors.

5.2.4. *Extracting the Boson Spectral Function from the ARPES Experimental Results*

Theoretically, in a many-body interaction system, the self-energy of the quasi-particle and the spectral function of the bosons have the following mathematical relationship:

$$\mathrm{Re}\Sigma(k,\varepsilon,T) = \int_0^\infty \mathrm{d}\omega \alpha^2 F(\omega,\varepsilon,k) K\left[\frac{\varepsilon}{kT}, \frac{\omega}{kT}\right], \quad (5.4)$$

where $\mathrm{Re}\Sigma(k,\varepsilon,T)$ is the self-energy of the quasi-particle and $\alpha^2 F(\omega,\varepsilon,k)$ is the spectral function of bosons, where $f(x+y)$ is the Fermi–Dirac distribution function.

$$K(y,y') = \int_{-\infty}^{+\infty} dx \frac{2y'}{x^2 - y'^2} f(x+y) \qquad (5.5)$$

Because the ARPES could directly measure the self-energy of quasiparticle, in principle, we can obtain the boson spectral function from the Equation (5.4), in order to determine the nature of the boson involved in coupling. However, in practice, inevitably, noise exists in the experimental data, making the direct inverse in mathematics often unstable and difficult to get reasonable results. Recently, due to both theoretical and experimental progress, the direct analysis of the boson spectral function from the self-energy has become possible.

In theory, the Maximum Entropy Method is successfully applied in the ARPES data analysis, which overcomes the effects of noise very well and finds an effective way to obtain the spectral function of boson and has been successfully in being verified in a typical e-p interaction system Be(0101) surface state (see Fig. 5.10).

In the ARPES experiments, the improvement of experimental accuracy makes extracting the spectral function of the boson from the high-T_C superconductors' ARPES data possible. As Fig. 5.11 shows, there are some

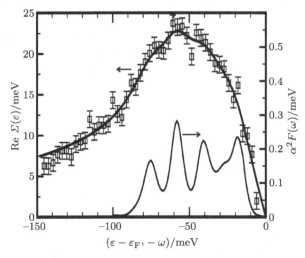

Fig. 5.10. The self-energy curve and the boson spectral function by maximum entropy method (MEM) fittings in the self-energy curve for the Be(0101) surface state. The panel is from [11].

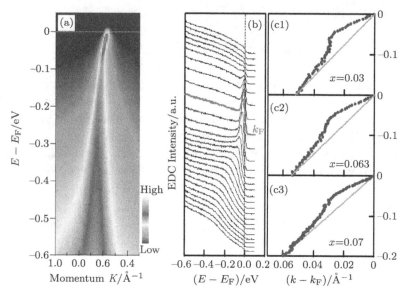

Fig. 5.11. The ARPES results of LSCO along the $(0,0)$-(π,π) nodal direction [12] In improved statistical data, in the vicinity of "kink", the finer structure can be seen.

more fine structures, which could be directly seen from the dispersion along the nodal direction in the LSCO system. By analyzing the self-energy of differently doped samples using the Maximum Entropy Method, we can see these fine structures corresponding to four characteristic peaks. These characteristic peaks from ARPES measurements correspond very well with the phonon structures of phonon DOS, which is directly measured by neutron scattering. The experimental result provides further evidence that the "kink" in the high-T_C superconductors' dispersion is related to the e-p interaction; meanwhile, there are also indications that the involved phonons have several modes rather than originally thought, i.e. on one and a half breath mode (see Fig. 5.12).

5.2.5. *Many-Body Effects in Anti-Nodal Direction*

Similar to the previous nodal direction, there is also a "kink" structure found in vicinity of the anti-node from the experiments. As shown in Fig. 5.13, for the over-doped Bi2212 samples ($T_C = 58$ K), above the T_C, the observed bonding band (Fig. 5.13(a), as shown by letter "B") is basically linear. But at the superconducting state (Fig. 5.13(b)), an obvious "kink"

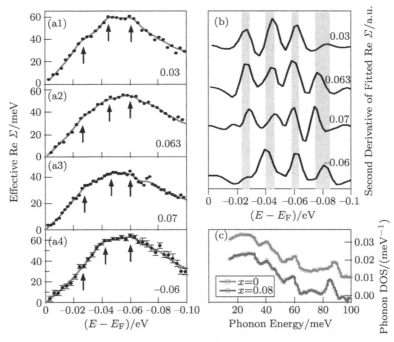

Fig. 5.12. Multiple bosonic mode coupling in the electron self-energy of LSCO [12] (a1)∼(a4) represents the self-energy of different doping samples, and the arrows in the figure mark possible fine structures in the self-energy. (b) The second-order derivative of (a), shows four main structural characteristics. (c) The phonon density of states for LSCO $x=0$ and $x=0.08$ measured from neutron scattering.

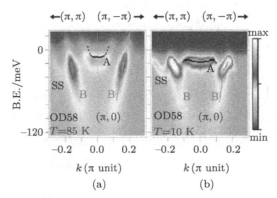

Fig. 5.13. The anti-nodal region "kink" structure for superconductor Bi2212. (a) Energy spectrum of normal state in the Bi2212 anti-nodal region. (b) Energy spectra of Bi2212 superconducting state in this region; we can clearly see an apparent "kink" structure at the energy scale of 40 meV. The panel is from [13].

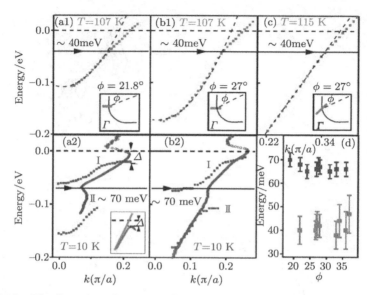

Fig. 5.14. The dispersion relationship of normal state and superconducting state in the Bi2212 anti-nodal region. (a1), (b1) and (c) are the fitting results. (a2) and (b2) adopt two fitting methods, MDC and EDC. (d) "Kink" positions as a function of ϕ in the anti-nodal region. The panel is from [14].

appears in the original band, accompanied by a strong spectral weight reconstruction. Because the "kink" turned up in the superconducting state and the energy scale was similar to the magnetic resonance mode found in the Bi2212, so Gromko et al. thought the "kink" is caused by the interaction between the electron and the magnetic resonance mode.

T. Cuk et al. have also measured the electronic structure of Bi2212 in the vicinity of the anti-node, and Fig. 5.14 shows the energy–momentum dispersion of the normal state and the superconducting state. Different from the previous results, there is a 40 meV "kink" structure observed in the normal state. As the magnetic resonance mode appears only in the superconducting state, and disappears in the normal state, the presence of the anti-node 40 meV "kink" structure in the normal state cannot be ascribed to the magnetic resonance mode. In the high-T_C superconductors, there vibration (B_{1g} phonon mode) of the oxygen in the CuO_2 plane occurs perpendicular to CuO_2 plane with the frequency ∼40 meV. T. Cuk et al. thought that the "kink" in the anti-node region of Bi2212 is caused by the electron and B_{1g} phonon mode coupling in the copper oxide plane.

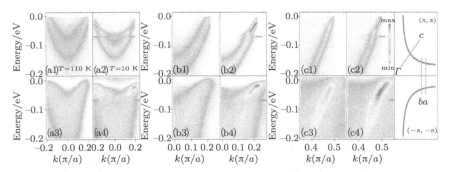

Fig. 5.15. Comparison of the theoretical calculations and experimental measurements of e-p interaction in Bi2212 (upper) the theoretical calculation of the spectrum, (lower) the experimental results, (subscript) 1,3 for the normal state; 2,4 for the superconducting state.

T. P. Devereaux et al. considered the strong anisotropy of electron structure, anisotropy of electron–phonon coupling and d-wave symmetry of superconducting gap in high-T_C superconductors, and carefully calculated the single electron spectral function of electron, 70 meV half-breath phonon mode and 40 meV B_{1g} phonon in the nodal and anti-nodal region of the normal state and superconducting state(see Fig. 5.15). The calculation results show a very good agreement corresponding to the results from the direct measurements.

5.2.6. The Possible and New Electronic Coupling Modes in High-T_C Superconductors

Recently, we have adopted a newly developed vacuum ultraviolet laser ARPES and observed some new fine structures in the high-energy region of Bi2212, and this discovery benefits from the unique advantages of accuracy and data quality in laser-ARPES. Table 5.1 shows the comparison of the main performances of VUV laser ARPES and a synchrotron.

Figure 5.16 shows the original spectrum (Fig. 5.16(a)) of photoelectron intensity versus energy and momentum measured by our newly developed laser-ARPES along the nodal direction of Bi2212, and the energy versus momentum dispersion relation is obtained by MDC fitting with the FWHM of the corresponding MDC plotted together (Fig. 5.16(b)). In the high-energy resolution measurements, the 70 meV "kink" structure in the dispersion and the 'steep drop' in the MDC width are very clear.

Table 5.1. Comparison of VUV laser ARPES and synchrotron [15].

Light source	VUV laser	Synchrotron
Energy Resolution/meV	0.26	5~15
Momentum Resolution/Å^{-1}	0.0036 (6.994 eV)	0.0091 (21.1 eV)
Photon Flux/(Photons · s^{-1})	10^{14}~10^{15}	10^{12}~10^{13}
Electron Escape Depth/Å	30~100	5~10
Photon Energy Tunability	Limited	Tunable
k-Space Coverage	Small	Large

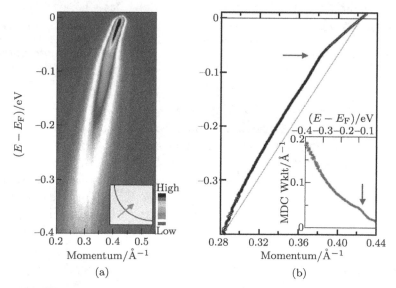

Fig. 5.16. The energy–momentum spectra, dispersion relations and MDC width along the nodal direction of optimally doped Bi2212 (17 K) by the ultra-high resolution ARPES [16]. The panel is from [16].

As the experimental precision has increased significantly, in terms of the electron self-energy along the nodal direction of Bi2212, besides 70 meV, we observed other two new structures at 115 meV and 150 meV (as shown in Fig. 5.17). These two new structures are all apparent in the vicinity of the node cross a large momentum area.

Further detailed temperature measurements show that the two new 115 meV and 150 meV structures are seen when the sample enters the superconducting state (see Fig. 5.18), but there is no apparent structure in the normal state, indicating that they are closely related with superconductivity.

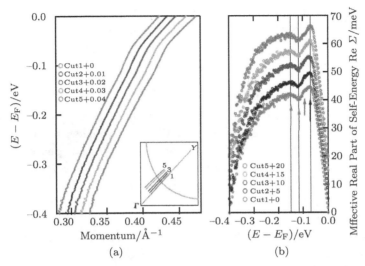

Fig. 5.17. (a) Momentum–energy curve of different cuts near the nodal region and the inset shows the corresponding cuts. (b) The self-energy curve corresponding to (a), the arrows mark structure of the corresponding energy position [16]. The panel is from [16].

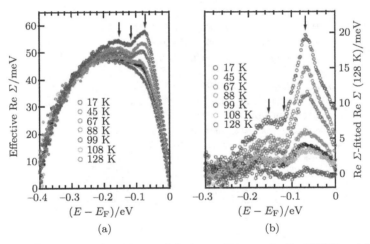

Fig. 5.18. Temperature dependence of electron self-energy in the Bi2212 nodal direction [16]. (b) Temperature dependence of the difference between the measured self-energy in (a) and the fitted one for 128 K. Taken from Ref. [16].

Unlike the previously mentioned 70 meV and 40 meV "kink" structures, the two new structures at 115 meV and 150 meV obviously cannot be simply attributed to the phonon mode or the coupling of electronic and magnetic resonance, because in the optimally doped Bi2212 system, the maximum of phonon mode is about 80 meV and the energy of magnetic resonance mode is about 42 meV, which are lower than the energy of the two new structures. This indicates that there might be a new electronic coupling mechanism. One possibility is that in the normal state, the electrons have coupled with some existing high-energy excitation phase, with the temperature changing or progressing to superconducting transition, and the high-energy excitation spectral function will redistribute and lead to two new structures. This energy may correspond to the spin fluctuations in the high-temperature superconductivity, because the neutron scattering experiments indicate that its energy can be extended to 200 meV or higher, and lead to stronger temperature dependence.

References

[1] P. V. Bogdanov et al., Phy. Rev. Lett. **85**, 2581 (2000).
[2] A. Lanzara et al., Nature **412**, 510 (2001).
[3] X. J. Zhou et al., Nature **423**, 398 (2003).
[4] H. He et al., Phys. Rev. Lett. **86**, 1610 (2001).
[5] B. Vignolle et al., Nat. Phys. **3**, 163 (2007).
[6] S. M. Hayden, Phy. Rev. Lett. **76**, 1344 (1996).
[7] H. Goka, Physica C **388–389**, 239 (2003).
[8] H. Hiraka, J. Phys. Soc. Jpn. **70**, 853 (2001).
[9] K. Yamada, J. Phys. Soc. Jpn. **64**, 2742 (1995).
[10] J. Tranquada, cond-mat/0401621.
[11] J. R. Shi et al., Phys. Rev. Lett. **92**, 186401 (2004).
[12] X. J. Zhou et al., Phys. Rev. Lett. **95**, 117001 (2005).
[13] A. D. Gromko et al., Phys. Rev. B **68**, 174520 (2003).
[14] T. Cuk et al., Phys. Rev. Lett. **93**, 117003 (2004).
[15] G. D. Liu et al., Rev. Sci. Instrum. **79**, 023105 (2008).
[16] W. T. Zhang et al., Phys. Rev. Lett. **100**, 107002 (2008).

6
ARPES Study on the High-Temperature Superconductors — Energy Gap, Pseudogap and Time-Reversal Symmetry Breaking

Shan-Cai Wang

Department of Physics, Renmin University of China, Beijing 100872, China

6.1. The Application of ARPES in the High-Temperature Superconductivity

Angle-resolved photoemission spectroscopy (ARPES) is a technique based on the photoelectric effect, and it is the only experimental method that can simultaneously measure the electron momentum and energy dispersion thus far. Its basic principle is based on the photoelectric effect, shown in Fig. 6.1. When a beam of monochromatic light ($E = h\nu$) irradiates on the sample surface, based on the photoelectric effect, the electrons in the sample absorb the photon and jump to the excited states. Parts of electrons will escape from the sample surface and become the photoelectrons. In the transition process, the electron energy and momentum, parallel to the sample surface, are conserved, so the kinetic energy and the momentum of runaway electrons are in correspondence with the electron binding energy

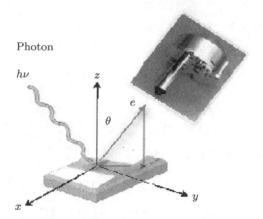

Fig. 6.1. The schematic diagram of ARPES.

and momentum of the measured sample as:

$$E_{\text{kin}} = h\nu - \Phi - |E_{\text{B}}|,$$

$$k_{\parallel} = \sqrt{\frac{2m_e E_{\text{kin}}}{\hbar^2}} \sin\theta,$$

where the Φ is the electron work function, E_{B} is the electron binding energy of sample, $k_{\parallel}$ is the electron momentum parallel to the sample surface, and θ is the included angle between the runaway electrons and the sample surface normal. By measuring the density of outgoing electrons as functions of the kinetic energy E_{kin} and angle (momentum), we can obtain the electron energy and momentum distribution in the sample. So, ARPES can directly measure the electron energy and momentum distribution of the sample. ARPES, as the electron energy and momentum function, simply consists of three parts:

$$I(k,\omega) = I_0(k,\omega) f(k,\omega) A(k,\omega),$$

where $I_0(k,\omega)$ is the single-particle matrix element, which is related to the incident photon polarization direction, the direction (momentum) of outgoing electrons and the relationship between the initial and final states of the system. $f(k,\omega) = 1/(e^{\omega/k_{\text{B}}T}+1)$ is the Fermi distribution function. $A(k,\omega)$ represents the electronic interaction term, called spectral function, which is proportional to the imaginary part of the Green's function:

$$A(k,\omega) = -\frac{1}{\pi}\mathrm{Im}G(k,\omega)$$

$$= -\frac{1}{\pi}\frac{\Sigma''(k,\omega)}{[\omega - \varepsilon_k - \Sigma'(k,\omega)]^2 + [\Sigma''(k,\omega)]^2}.$$

Under the electron–electron interaction, the ARPES spectrum can directly measure the electron dispersion ε_k, electron self-energy $\Sigma(k,\omega) = \Sigma'(k,\omega) + i\Sigma''(k,\omega)$ and energy gap function.

Matrix element effects are because of the electron initial–final state transition. According to the Fermi golden rule, the transition probability is:

$$I(k, E_{\mathrm{kin}}) = \sum_{f,i} \frac{2\pi}{\hbar} |\langle \Psi_f^N | H_{\mathrm{int}} | \Psi_i^N \rangle|^2 \delta(E_f^N - E_i^N - h\nu),$$

$$H_{\mathrm{int}} = -\frac{e}{2m}(A \cdot P + P \cdot A) = -\frac{e}{2m} A \cdot P.$$

Its intensity is determined by the initial and final states of the system and the interaction. Under the photon–electron interaction, the intensity simplifies to $H_{\mathrm{int}} = -(e/2m)AP$.

Under the sudden approximation condition, it is acknowledged that electronic escape time is much smaller than the system relaxation time. In most conditions, the photoelectron emission meets sudden approximation condition. It is thought that the total system consists of the single-electron state and the rest of the system. In the given conditions, the system's wave function is:

$$\Psi_{f,i}^N = A\phi_{f,i}^k \Psi_{f,i}^{N-1}.$$

Electron density is:

$$I(k, E_{\mathrm{kin}}) = \sum_{f,i} \langle \phi_f^k | H_{\mathrm{int}} | \phi_i^k \rangle^2 \langle \Psi_m^{N-1} | \Psi_i^{N-1} \rangle^2$$

$$= \sum_{f,i} |M_{f,i}|^2 \sum_m |c_{m,i}|^2 \delta(E_{\mathrm{kin}} + E_f^{N-1} - E_i^{N-1} - h\nu),$$

where $M_{f,i} = \langle \phi_i^k | H_{\mathrm{int}} | \phi_f^k \rangle$ represents the single-electron matrix element. The size of the matrix element depends on the density of initial and final states, symmetry and the energy and momentum of the incident light and outgoing electrons. For a high-symmetry plane, as shown in Fig. 6.2, if the electronic structure has parity relative to the plane, the density of

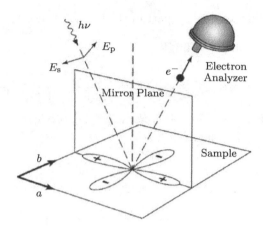

Fig. 6.2. The mirror plane in the ARPES.

outgoing electrons can reach the maximum or minimum (0), by choosing the angle of the incident electron to keep it with the outgoing electrons in the same plane and adjusting the polarization direction of the incident photons. This high-symmetry plane is called mirror plane.

ARPES plays a key role in the high-temperature superconductivity research. Because the ARPES method is a direct way to measure and research the electron dispersion, it plays a very important role in measuring the band structure of the high-T_C superconducting materials and superconducting gap, especially in the research of the symmetry of the superconducting gap and the discovery and study of the pseudogap. Kaminski *et al.* successfully showed the superconductors have a time-reversal symmetry in the pseudogap state using the high symmetry of mirror plane.

6.2. Superconducting Gap, Pseudogap and Two-Gap Problems

In 1986, when the Swiss scientists Bednorz and Muller discovered high-T_C superconducting phenomenon in the ceramic material (($LaBa)_2CuO_4$), the field high-T_C superconductivity became the main research subject in the strongly correlated system. Besides the difference of critical temperature, the high-T_C superconductors are an electron–electron strongly correlated system. In conventional BCS superconductors, the superconducting energy gap as the superconducting order parameter is isotropic, and the electron pairing is electronic singlet state of s-wave superconductors. In the high-T_C superconductors, the superconducting order parameter is not completely solved. The symmetry of the superconducting gap is different from that of the conventional superconductors, and the size of the gap has a distribution as a function of the momentum. In the under-doped region, there exists a gap above the critical temperature — this is called the pseudogap.

Debating on the superconducting pairing mechanism is the central question in the superconducting phenomenon. In the superconductors, the size of superconducting gap is about 30 meV. The measurement distribution of the energy gap in the vicinity of the Fermi surface becomes a research focus. In the early ARPES research, sample $Bi_2Sr_2CaCu_2O_{8+\delta}$'s gap is anisotropic. In the over-doped samples, as shown in Fig. 6.3, Shen et al. [1] found that above the critical temperature, in two directions (in the nodal and anti-nodal directions), there was a spectrum crossing the Fermi level, which indicates that the gap is very small or that no gap exists above the temperature. However, below this temperature, the gap is anisotropic. Along the nodal direction (near B point), the spectrum has no significant change above and below the critical temperature, indicating

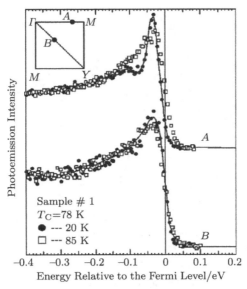

Fig. 6.3. The ARPES spectrum of the over-doped Bi2212 ($T_C \sim 73$ K) in the vicinity of Fermi surface. As shown in inset, A point is near the $(\pi, 0)$ point, and B point is in the direction of node near the Fermi surface. The panel is from Ref. [1].

that there is no clear gap opening. But along the anti-nodal direction (near A point), there is a significant gap opening below the T_C, and the quasi-particle peak. The whole spectrum becomes a peak–dip–hump structure.

The anisotropy of the Fermi surface can be expressed by measuring the gap along the Fermi surface at low temperature. Ding et al. [2] discovered in the sample Bi2212 ($T_C \sim 73$ K), that within the 1/4 Brillouin zone along the ΓY, the gap decreased as the angle between ΓY and ΓM increased, and decreased to zero at 45°. There exists a node in the whole energy gap at 45° direction. The changing behavior of the gap via the angle can be fitted by d-wave form ($\Delta = \Delta_0(\cos k_x - \cos k_y)$). The experimental data is shown in Fig. 6.4.

But the high-T_C superconducting gaps are not the pure d-wave form. By measuring the gap symmetry along the Fermi surface in the superconducting state for different doped samples of Bi2212, [3] it was shown that away from the optimally doped region, the gap varies with the angle, and there is a node at 45° (ΓY). In the under-doped samples, the band gap can be fitted by the higher order terms of trigonometric functions, such as

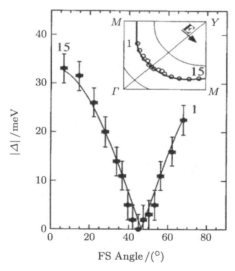

Fig. 6.4. The d-wave fitting of the superconducting gap. Within the 1/4 Brillouin zone, the size of the gap is plotted as a function of angle, the location in the Fermi surface of the data point is shown in the inset, the sample is Bi2212, $T_C \sim 73$ K. The panel is from Ref. [2].

$\Delta_k = \Delta_{\max}[B\cos(2\phi) + 9(1-B)\cos(6\phi)]$, where the angle ϕ is the intersection angle between the momentum and the ΓM axis.

6.3. Pseudogap

Another difference between the high-T_C superconducting gap and the conventional superconducting gap is the temperature behavior. In the conventional superconductors, the gap is the superconducting order parameter, below the T_C the gap opens, and above the T_C the gap disappears. However in the under-doped region, as the temperature is increased, the high-T_C superconducting gap is not completely closed. There exists a gap in the non-superconducting state $T > T_C$. The difference between the energy gap measured in the non-superconducting state along the Fermi surface and the superconducting energy gap is that the pseudogap is measured above the superconducting transition temperature. Ding et al. [4] discovered (as shown in Fig. 6.5(a)), in the under-doped region (Bi2212, $T_C = 83$ K, and 10 K), along the anti-nodal direction $(0,0) \sim (\pi,\pi)$, the quasi-particle spectrum in the Fermi surface varies with the temperature. At low temperature, there existed an energy gap of finite size. As temperature increases, and

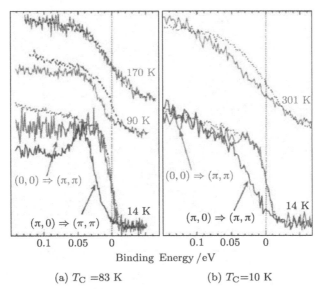

(a) $T_C = 83$ K (b) $T_C = 10$ K

Fig. 6.5. The ARPES spectrum of under-doped sample Bi2212 along $(0,0) \sim (\pi,0)$ and $(0,0) \sim (\pi,\pi)$ direction in the Fermi surface. The panel is from Ref. [4].

becomes higher than the superconducting transition temperature (90K), although the quasi-particle peak disappeared, the energy gap still existed, according to the midpoint of the leading edge definition. When the temperature increased even higher, up to the about 170 K mark, the forefront of spectrum overlapped with the reference material Pt, marking that the energy gap closes. In the case of lower doping ($T_C \sim 10$ K), the energy gap defined by forefront of spectrum would close at higher temperature ($\sim$300 K). However in the Fermi surface along the $(0,0) \sim (\pi,\pi)$ direction, there exists no gap above the superconducting transition temperature. This energy gap is called pseudogap, and the temperature at which the pseudogap disappears is called T^*.

Pseudogap variations of several under-doped Bi2212 samples with temperature near $(\pi, 0)$ are shown in Fig. 6.6, indicating that as temperature increases, the energy gap gradually decreases, but there is no significant change near T_C, even up to higher temperature ($T^* \sim 200$ K) becoming zero [5].

Debate on the symmetry of the energy gap, temperature variation and pseudogap in the under-doped region are always the topics high-T_C superconductivity research focuses on. One of the important reasons is in the under-doped region, there is no clear, consensus approach to extract the superconducting gap from the ARPES spectrum. The typical

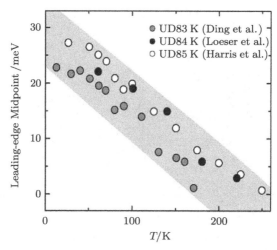

Fig. 6.6. The temperature variation behavior of pseudogap in the under-doped sample, near $(\pi, 0)$ point. Samples are UD83K, UD84K and UD85K. The panel is from Ref. [5].

high-T_C superconductors' ARPES spectrum near $(\pi, 0)$ has been shown in Figs. 6.7(a)–6.7(c), which indicates that at low temperature the system has clear quasi-particle peak, and the difference from the quasi-particle to Fermi energy, i.e. energy gap, can be determined by the location of the quasi-particle. As temperature increases, the quasi-particle peak becomes smaller, and disappears at about T_C. But at this moment, the ARPES spectrum does not coincide with the reference spectrum of polycrystalline metals, and the leading edge of the spectrum has a significant difference compared to the Fermi level (as shown in Fig. 6.7(a), the third ARPES spectrum). The system still had an energy gap, but there was no longer a recognized method to determine its size. And the size of the energy gap, closing temperature and symmetry defined by different methods are completely different. M. Norman et al. [6] assumed that there exists particle–hole symmetry in the Fermi level near the tens of milli-electron volts range, and ARPES showed the spectrum:

$$I(\omega) = \Sigma_k I_0 f(\omega) A(k, \omega),$$

according to the particle–hole symmetry:

$$A(k, \omega) = A(k, -\omega),$$

and the Fermi function effect could be removed by symmetrizing the spectrum:

$$I(\omega) + I(-\omega) = \Sigma_k I_0 A(k, \omega).$$

Even carrying out the specific method shown in Fig. 6.7, symmetrically treating the EDC, at low temperatures the effect of symmetry was not obvious, but above the superconducting transition temperature, after the symmetrical treatment of the ARPES spectrum, near the Fermi level, there was weight loss. It can be concluded that in the system there existed a pseudogap at this temperature, and until a higher temperature, when the energy gap completely closed.

Based on the above observations, in the under-doped region there is a pseudogap and the pseudogap temperature T^* is much higher than the superconducting transition temperature. According to the observed pseudogap closure temperature with the hole doping rate changes, summarized in the phase diagram, we can obtain the T^*-line, besides the T_C curve. Ding et al. [4] summarized the under-doped Bi2212 samples ($T_C = 83$ K, 10 K), as seen in the phase diagram shown in the Fig. 6.8. In the under-doped region,

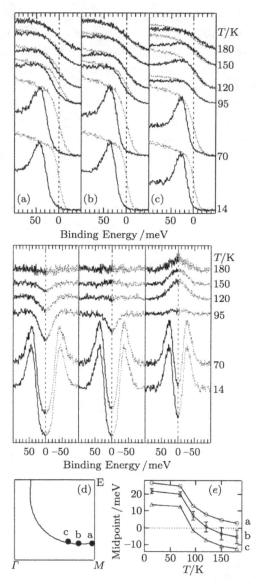

Fig. 6.7. (a)–(c) the EDC of three different k points in the Brillouin zone, their locations marked in (d), respectively, below corresponding to the result of after symmetrization. The panels are from Ref. [6].

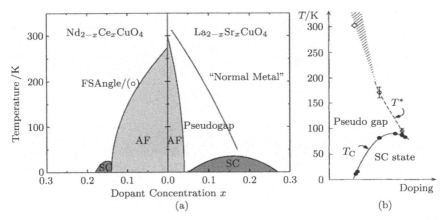

Fig. 6.8. High-temperature superconducting phase diagram. The panel is from Ref. [4]

besides the T_C curve, there existed T^* of energy scale much higher than T_C. With the hole doping rate increasing, the pseudogap temperature T^* gradually decreased slightly until it was touching the T_C curve at the optimally doped point. Above this doping rate, the behavior of the pseudogap temperature was similar to that of the conventional superconducting gap. The specific relationship of the pseudogap and the superconducting gap is still an important research focus. To sum up the electrical and magnetic properties of the electron-doped and hole-doped high-temperature superconducting materials, a system phase diagram is shown in Fig. 6.8(b).

The symmetry of the pseudogap is another question. Based on works by Harris, Ding, Loeser et al., getting the pseudogap variation with the angle is shown in Fig. 6.9. As can be seen, the angle change of the pseudogap coincided with the form of the superconducting gap. But because the quasiparticle peak of the under-doped region was not obvious, and there was a problem of the energy gap definition, the obtained gap value covered a certain range, as shown in the gray area of Fig. 6.9.

For the corresponding physical state of the pseudogap, there is no clear conclusion yet, which is also one of the difficulties in high-T_C superconductivity research. Several specific types of conjectures have been presented including Varma's theory (who thought it was because of the formation of the ring current in the pseudogap, but which maintained the systematic translational symmetry), d-density wave (DDW) theory, CDW theory, stagger flux phase and unklapp scattering. But the question needs further

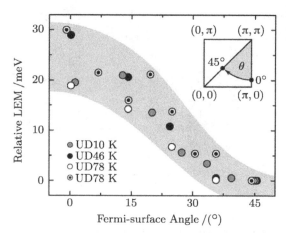

Fig. 6.9. The size of the Bi2212 materials' pseudogap as a function of angle. The results are selected from the UD10K, UD46K, UD78K, and the size of energy gap above the UD78K is obtained by the cutting-edge method. The panel is from Ref. [5].

experimental studies to confirm if the pseudogap and the superconducting gap are different gaps or if the pseudogap is just the non-coherent state above the superconducting transition temperature.

The development of the energy gap, or pseudogap, is mainly due to the improvement of the experimental accuracy, including the application of the high-resolution Scienta detector. Lee et al. [7] from the Shen group, discovered that in the slightly under-doped sample Bi2212 ($T_C = 92$ K), the temperature change curve of the ARPES spectrum from the different angles along the Fermi surface indicated that there existed two gaps: one was on the pseudogap scale, the other was at the superconducting scale. They respectively collected C1~C8 ARPES data in the Fermi surface along the direction shown in Fig. 6.10(d). Figure 6.10(a) showed the results of the ARPES data divided by the Fermi function under the corresponding temperature. It could be seen that except for the nodal direction, there was a superconducting gap opening along the Fermi surface, when the temperature was slightly lower than the transition temperature (82 K), and the energy still existed around the Fermi surface. When the temperature was higher than the T_C (102 K), the energy gap of the Fermi surface near the node closed (C1~C4), but the pseudogap still existed near the $(\pi, 0)$ point. So the energy gap of the node and the anti-node have different behaviors, and they may indeed be different gaps.

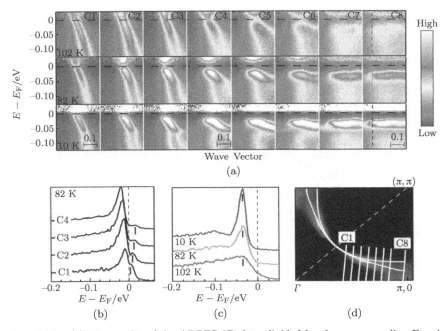

Fig. 6.10. (a) The results of the ARPES 2D data divided by the corresponding Fermi function along the cuts C1–C8 in the Fermi surface; (b) The EDC curve of the C1–C4; (c) EDC as function of temperature by the dotted line. The panel is from Ref. [7].

As for the nodal direction, Figs. 6.11(a) and 6.11(c) show the ARPES spectrum of A point and C point and the spectrum, after removing the Fermi function effect by the symmetrization method, with temperature variation. It can be clearly seen that at the nodal direction the ARPES spectrum decreased as the temperature increased, and the energy gap closed at the T_C. The size of the energy gap variation with the temperature is shown in Fig. 6.11(d), and the dotted line represents the BCS theory calculation results as a reference. It can be clearly seen that it has a good match with the energy gap of BCS form. So at the nodal direction, the energy gap has a BCS form, but it has a pseudogap form at the anti-nodal direction as mentioned earlier.

At different angles, the energy gap as a function of temperature is shown in Fig. 6.12(a), indicating that at C1, C2, C3, it cuts close to the nodal direction, the energy gap turned to zero at near T_C, but at C5, C6, C7, it cuts close to anti-nodal direction, and the energy gap obviously had a limited value above the T_C.

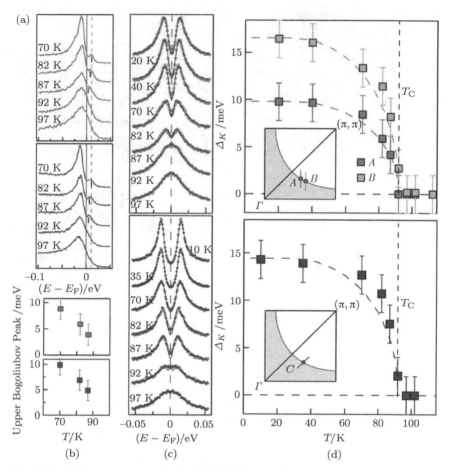

Fig. 6.11. Existing two gaps in the Bi2212 (a) and (c) correspond to the EDC and results of A and C point symmetrization. The panel is from Ref. [7].

Figures 6.12(b), 6.12(c), 6.12(d), respectively show the energy gap symmetry of different doped samples above and below the T_C. Below the T_C, the energy gap symmetry had a good match with the d-wave, so it was thought the superconducting gap was d-wave form. Above the transition temperature, the gap of Fermi surface partially closed to form the so-called Fermi arc. At that moment, the symmetry of the energy gap was far deviated from d-wave form.

Simultaneously, Kondo et al. [8] discovered two energy scale gaps in the optimal state in the optimally doped Bi2201 ($T_C = 35$ K): one

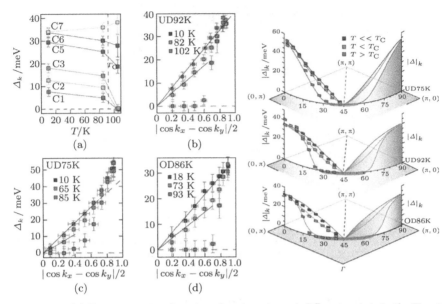

Fig. 6.12. (a) Energy gap as a function of temperature at different points in the Fermi surface, (b)–(d) energy gap as a function of angle of UD92K, UD75K, OD86K and Bi2212. The panels are from Ref. [7].

corresponding to the superconducting gap of the BCS form, another corresponding to the pseudogap. Figure 6.13(a) shows the ARPES spectrum was the EDC curve in the superconducting state and also shows the symmetry results. The peak position or the size of the gap was shown in Fig. 6.13 in the arrow. In this state, such as Fig. 6.13(b), the gap of the system can no longer be fitted by a simple d-wave. Integrating the angle change of the gap at different temperatures, they found two different energy scales in the superconducting state. As shown in Fig. 6.13(c), the symmetry of the gap cannot be fitted by a simple d-wave, but can be fitted by the combination of a superconducting gap and pseudogap. In the superconducting state, the energy scale is the result of joint action of superconducting gap and pseudogap.

The two energy scales observed by the ARPES experiments indicate that there exist superconducting gap and pseudogap in the under-doped region, but there is no conclusion about the origin of pseudogap, or the relationship of the pseudogap and superconducting gap specifically.

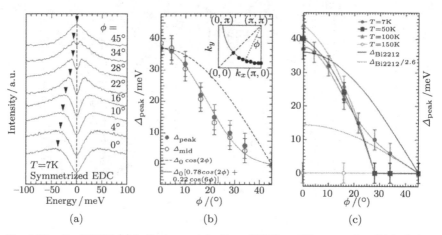

Fig. 6.13. For Bi2212 (a) is the symmetrization of EDC at different angles; (b) is fitting for the energy gap symmetry; (c) is energy gap as a function of angle. The panels are from Ref. [8].

6.4. Time-Reversal Symmetry Breaking

The origin of pseudogap and nature of the ground state are always important issues in the high-T_C superconductivity physics. Microscopic theory has no unified understanding yet. One of the theories is the marginal Fermi liquid (MFL) of Varma et al. [9]. According to the MFL theory, the high-T_C superconducting phase diagram in the hole doping region is shown in Fig. 6.14. Close to the optimally-doped point there exists the quantum critical point, $x = x_C$. The non-superconducting region can be divided into several parts. Region I: existing Fermi surface with no quasi-particle, called MFL region. Region II: the pseudogap region. Region III: with Fermi liquid behavior. There exists systematic symmetry transition between I and II. In Varma's theory, in Region II, the current has four-degree symmetry within each unit cell, called circulating circuit. And through calculations it is found that the circulating circuit caused the energy gap anisotropy. Current distribution is shown in Fig. 6.15. At this state, the system still has translational symmetry, no time-reversal symmetry and fourth-degree rotational symmetry. But the product of time-reversal symmetry and fourth-degree rotational symmetry remains unchanged.

Kaminski et al. [10] make use of ARPES to verify time-reversal symmetry of the system in the pseudogap state. The specific experiments are based on the matrix element effect described above. In the specific high

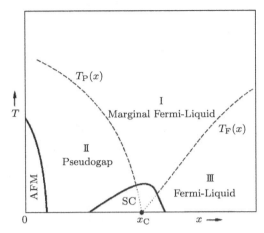

Fig. 6.14. The high-T_C superconducting phase diagram under the Marginal Fermi liquid [9].

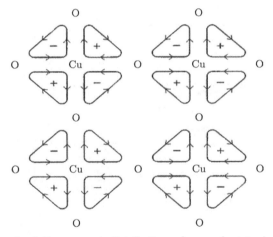

Fig. 6.15. The circulating current distribution of ground state in the pseudogap region. Arrow represents the current direction, "+/−" represents the magnetic field direction [9].

symmetry condition, for the incident whose frequency is ω, the electronic state of the first Brillouin zone is $|k\rangle$, and the photoelectron momentum and energy are p and E_{kin}, respectively. The matrix element can be written as follows:

$$\langle p|M|k\rangle = \frac{ie}{2mc}\int \mathrm{d}r \Phi_p A \cdot \nabla \Psi_k(r).$$

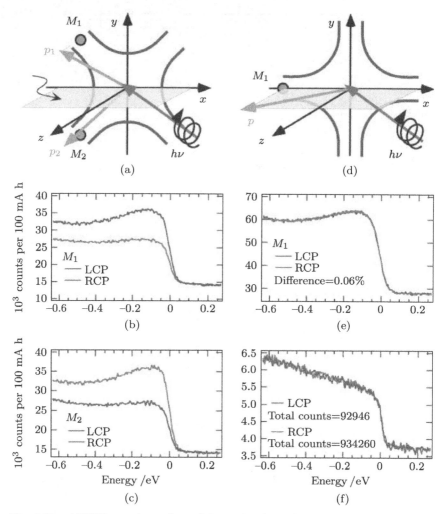

Fig. 6.16. ARPES experimental conditions, the ARPES spectrum of the optically-doped Bi2212 samples at $T=300$ K. (a) The incident light along the ΓX direction, but the direction of runaway electrons along M_1 and M_2; the ARPES spectrum of the runaway electrons along these two directions shown in (b) and (c); blue and red lines, respectively, indicate the EDC at the M_1 and M_2 point, when the incident light are the left circularly polarized light and right circularly polarized light; (d) incident light along the ΓM direction, but the direction of runaway electrons along M_1. The panel is from Ref. [10].

For the left circular polarized light (LCP) and right circular polarized light (RCP), their vectors are:

$$A_{(l,r)} = A_0(-x' \pm iy').$$

When the electron state is in the mirror plane, according to the momentum conservation, the momentum of the outgoing electrons are also in the mirror plane. Relative to the mirror plane, the electron states in the system have symmetric and anti-symmetric form:

$$|p\rangle = \alpha_m|pe\rangle + \beta_m|p,o\rangle, \quad |k\rangle = \mu_m|ke\rangle + \nu_m|k,o\rangle.$$

Under the mirror symmetry operation:

$$R_m|p\rangle = \alpha_m|pe\rangle - \beta_m|p,o\rangle, \quad R_m|k\rangle = \mu_m|ke\rangle - \nu_m|k,o\rangle.$$

In the case of time-reversal symmetry, $\nu_m = 0$; in the case of time-reversal symmetry breaking, $\nu_m \neq 0$. Ultimately, the measured ARPES intensity is determined by the matrix element, so in the case of LCP and RCP, the measured ARPES density is $M_l = \langle p|R^{-1}A_r \cdot \nabla R|k\rangle$ and the ARPES intensity difference between LCP and RCP is:

$$\begin{aligned}D_m &= |M_l|^2 - |M_r|^2 \\ &= 4\Re(\alpha_m^*\beta_m|\mu_m|^2\langle p,e|M_r^*|k,e\rangle\langle k,e|M_r|p,o\rangle \\ &\quad + \alpha_m\beta_m^*|\nu_m|^2\langle p,o|M_r^*|k,o\rangle\langle k,o|M_r|p,e\rangle \\ &\quad + \mu_m\nu_m^*|\alpha_m|^2\langle p,e|M_r^*|k,e\rangle\langle k,o|M_r|p,e\rangle \\ &\quad + \nu_m\mu_m^*|\beta_m|^2\langle p,e|M_r^*|k,e\rangle\langle k,e|M_r|p,o\rangle).\end{aligned}$$

Before the experiment, at high temperature, in the non-time-reversal symmetry breaking state, their inspection of the experimental system, the results are shown in Fig. 6.17. Choosing the ΓX direction as a mirror plane, the incident light is in the mirror plane. The experimental EDC of the $M_1(\pi,0)$ and $M_2(\pi,0)$ points are shown in Figs. 6.17(b) and 6.17(c). From Fig. 6.17(b) at the M_1 point, the EDC intensity of LCP is stronger than the RCP's. But at M_2 point, the reverse happens, because M_1 and M_2 are mirror symmetry. Choosing the ΓM direction (Cu-O band direction) as a mirror plane, the incident light is in the mirror plane. Because of no dichotomy, the EDC intensity of LCP and RCP are the same, indicating in this state no time-reversal symmetry breaking.

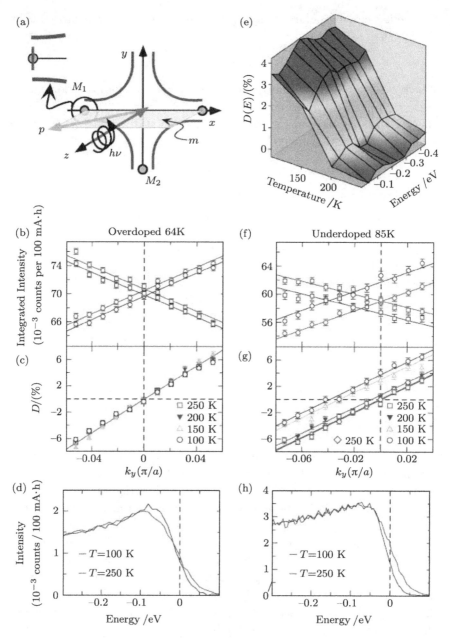

Fig. 6.17.

Fig. 6.17. (*figure on facing page*) Dichotomy experiments. (a) Schematic diagram of experimental arrangement, incident light and runaway electrons are all in the same plane; (b) EDC intensity integration as a function of momentum normal to the mirror plane. OD64K sample, relative intensity as a function of momentum from the result of (c) and (d); (e) relative intensity as functions of energy and temperature; (f) and (g) the result of dichotomy experiments show the significant difference at M_1 for the UD85K sample; which indicates the system has time-reversal symmetry breaking at this point. The panel is from Ref. [10].

However, in the case of time-reversal symmetry breaking, i.e. the initial state $|\Psi(k)\rangle$ obeying the time-reversal symmetry, $|\Psi(k)\rangle$ is not the eigenfunction of mirror operation. Even if in some high-symmetry mirror plane, the intensity of photoelectron spectroscopy of LCP and RCP are still different. As shown in Fig. 6.17(a), when incident light is parallel to ΓM direction and the direction of outgoing electrons is near M_1, in this mirror plane, there is no intensity difference from the geometric factors. EDC curve is shown in Fig. 6.17, in order to facilitate comparison, they do integration in the vicinity of the Fermi level within the range of $-600 \sim 100$ meV, get the current strength I, and define $D = (I_R - I_L)/(I_R + I_L)$ to describe the system dichotomy. From the Fig. 6.17(b), it can be seen that in the case of non-time-reversal symmetry breaking state (Bi2212, OD64K), along the direction perpendicular to the mirror direction, the intensity I linearly increases or decreases, but the intensities of LCP and RCP are equal at the mirror plane point (M_1). So the D value is equal to zero at this point, shown in Fig. 6.17(c). From Fig. 6.17(f), for the under-doped Bi2212 (UD85K), above the T^*, the intensities of LCP and RCP are the same at M_1, and the D value is zero. When the temperature drops lower than T^*, in the pseudogap region, the intensities are different at M_1, and D value has a difference of about 3%.

In order to better determine the pseudogap region of the time-reversal symmetry breaking behavior, Kaminski *et al.* continued to study, in the under-doped region, the temperature behavior above and below the T^*. By measuring the symmetry of M_1 and M_2 points, they found that in the under-doped samples, the symmetry is odd relative to mirror along the ΓX direction. Figures 6.18(a) and 6.18(b) show the asymmetric D at low temperature, at $M_1(\pi, 0)$ point, RCP intensity I_R is stronger than LCP intensity I_L, but after being rotated 90°, at $M_2(0, \pi)$ point, I_L is stronger than I_R. And in the pseudogap region, the asymmetric D as a function of temperature is shown in Fig. 6.18(c). Below T^*, D value is non-zero, but above

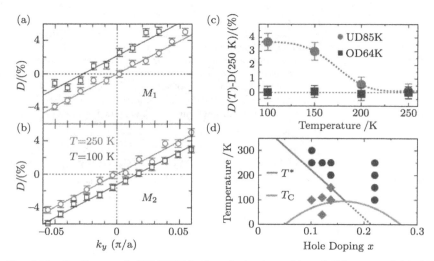

Fig. 6.18. In the sample UD Bi2212, the relative intensities of dichotomy of (a) M_1 and (b) 90°-rotation M_2; (c) The relative intensity difference of dichotomy in UD85K and OD64K samples; (d) in Bi2212 phase diagram, the dichotomy exists in red region and doesn't exist in blue region. The panel is from Ref. [10].

the T^*, it disappears, which indicates that there exists time-reversal symmetry breaking at low temperature in the pseudogap region. However, in the over-doped samples (OD64K), because there do not exist pseudogap and time-reversal symmetry breaking, so the dichotomy cannot be observed. Summary of sample temperature changes are shown in Fig. 6.18, time-reversal symmetry breaking takes place in pseudogap region below the T^* even in the superconducting states, which also have time-reversal symmetry breaking.

Because a series of Bi2212 samples contains 5-times the superstructure phenomenon along the b-axis, the electronic structure has a 1/5 cycle band fold along ΓY direction. The folded band and Fermi surface will pollute the systematic basic bands. In order to eliminate the pollution caused by the superstructure, they use the Bi2212 superconducting film as experimental samples, which are sputtered to the surface of $SrTiO_3$. The thin films have achieved atomic-level flatness, so it can ensure the difference during the experiment was not caused by the external sample difference. The superstructure makes only 3% contribution to the ARPES spectrum. Borisenko et al. [11] use the Pb-Bi2212 samples, in which the Pb atoms can eliminate the superstructure, and do not find the signal of time-reversal symmetry

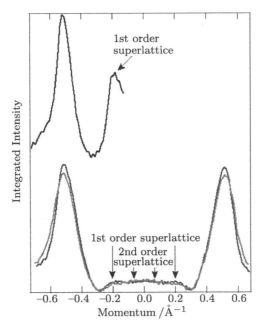

Fig. 6.19. Signal intensity of the Bi2212 sample and Bi2212 film along the ΓY. In the Bi2212 film sample, the super-lattice gives a less than 3% contribution to the basic signal intensity. The panel is from Ref. [12].

breaking within the experimental error. So we think Kaminski's experiment results come from the contribution of the super-lattice structure, which is not completely eliminated. Kaminski *et al.* [12] have done the measurement along the super-lattice direction (ΓY), shown in Fig. 6.19, and found that in the general samples Bi2212, about 50% spectrum are caused by superstructure, but in the thin film samples, only less than 3% are, so the superstructure contribution to the measurement can be ignored.

The experiments about the time-reversal symmetry in the pseudogap state give different results. Different methods also give different results. The debate about the existence of time-reversal symmetry breaking has no clear experimental results. Borisenko *et al.* did not find the signal of dichotomy by using the no-superstructure samples. However, the YBCO neutron scattering experiments tends to see the time-reversal symmetry breaking in the system, but the NMR results of $La_{2-x}Sr_xCuO_4$ indicate there is non-time-reversal symmetry breaking in the pseudogap region. Therefore, whether there is time-reversal symmetry breaking in the pseudogap region or not is still a major problem worthy of study. While in the energy gap and pseudogap state, the system's basic nature is still an important research direction of high-temperature superconductivity up to now.

References

[1] Z. X. Shen, D. S. Dessau, B. O. Wells et al., *Phys. Rev. Lett.* **70**, 1553 (1993).
[2] H. Ding, M. R. Norman, J. C. Campuzano et al., *Phys. Rev. B* **54**, R9678 (1996).
[3] J. Mesot, M. R. Norman, H. Ding et al., *Phys. Rev. Lett.* **83**, 840 (1999).
[4] H. Ding, T. Yokoya, J. Campuzano et al., *Nature* **382**, 51 (1996).
[5] A. Damascelli, Z. Hussain, Z. Shen, *Rev. Mod. Phys.* **75**, 473 (2003).
[6] M. Norman, H. Ding, M. Randeria et al., *Nature* **392**, 157 (1998).
[7] W. S. Lee, I. M. Vishik, K. Tanaka et al., *Nature* **450**, 81 (2007).
[8] T. Kondo, T. Takeuchi, A. Kaminski et al., *Phys. Rev. Lett.* **98**, 267004 (2007).
[9] C. M. Varma, *Phys. Rev. B* **55**, 14554 (1997).
[10] A. Kaminski, S. Rosenkranz, H. M. Fretwell et al., *Nature* **416**, 610 (2002).
[11] S. V. Borisenko, A. A. Kordyuk, A. Koitzsch et al., *Nature* **431**, 1 (2004).
[12] J. C. Campuzano, A. Kaminski, S. Rosenkranz et al., *Nature* **431**, 7004 (2004).

7
Progress in the Scanning Tunneling Microscopy Study of High-Temperature Superconductors

Qiang-Hua Wang

National Laboratory of Solid State Microstructure, Nanjing University, Jiangsu 210093, China

In this chapter, we review the recent progress in the study of high-temperature superconductors using scanning tunneling microscopy (STM). We begin with a brief introduction to the principle of STM, and then describe various novel phenomena observed by STM in high-temperature superconductors and discuss the physics therein. These include the impurity resonance state, the electronic structure of vortex core, the energy gap inhomogeneity, the quasi-particle scattering interference, the density waves, the energy gap closing versus temperature and the coupling to bosonic modes. Open questions and challenges are also presented in the discussions.

7.1. Introduction to Scanning Tunneling Microscopy

STM is a high-resolution tool to measure the local density of states (LDOS) of the sample. The STM tip is near but separated from the sample vacuum. Under a fixed bias voltage V, the tunneling conductivity (in the low-temperature limit) is proportional to the LDOS of the sample at the energy eV. The proportionality factor, or the tunneling matrix element, $F(z)$ depends on the distance z between the tip and the sample, since the latter determines the overlap between the electron wave functions. In practical microscopy applications, two operational modes are often used. The first is the iso-current mode. Under a fixed bias voltage V_0, the tunneling current is tuned to I_0 by tuning the height z. Thus, $z(x,y)$ provides a surface topography. The second mode is the iso-height mode. With fixed bias voltage V_0 and height z, the tunneling current $I(x,y)$ also reflects the surface topography. Apart from topography, the most important application of STM is to measure the LDOS with high resolution in (x,y). However, because of the tunneling matrix element $F(z)$, the tunneling conductivity cannot be used as the absolute value of LDOS. In practice, one first uses the iso-current mode to set the height z, and then measures the tunneling conductivity $G(\text{eV})$ with z fixed. This is related to LDOS as $G(\text{eV}) = F(z)\rho(\text{eV}) = I_0 \rho(\text{eV}) / \int_0^{V_0} \rho(\text{eV}) dV$, where in the second equality, the iso-current calibration is used. This relation tells us that upon calibration, the tunneling conductivity is given by LDOS divided by the integrated DOS within the calibration bias voltage. This is sometimes important while performing quantitative analysis of the STM data. With the above caution in mind, in the following we use LDOS and tunneling conductivity (or tunneling spectra) interchangeably.

STM with atomic resolution plays important roles in the study of high-temperature superconductors, and we will describe this in the following sections.

7.2. Impurity Resonance States

The majority of STM experiments are done on bismuth samples. The reason is that such samples have easy cleave planes. Yazdani *et al.* [1] found that in some locations the tunneling conductivity develops a low-energy peak within the pairing gap. They attribute the in-gap peak to the surface impurities and the d-wave nature of the pairing gap function.

Indeed, theoretical studies [2] reveal that in d-wave superconductors, in the unitary limit (with strong impurity), there is a quasi-bound state, or resonance state, near the impurity, with an energy close to the Fermi level. The zinc impurity in the copper oxide plane turns out to satisfy the unitary limit. The STM results of Pan *et al.* [3] near such impurities can be explained well by the theory. They found a very sharp resonance peak near the impurity. The spatial distribution of the tunneling conductivity with the bias voltage set at the resonance energy reveals a four-lobe structure, which is in good agreement with the theory of impurity states in d-wave superconductors. However, in experiments the maximal LDOS appears right on the impurity, and then on the next-nearest neighbor, while in theory the maximum should occur on the nearest-neighbor site. Such a disagreement might be related to the so-called filter effect. The d-electrons on the copper atom cannot tunnel directly in the vertical direction. They must hop to other obits on neighboring atoms before eventually entering the STM tip [4]. Therefore, the current through the tip comes not from the copper atom right below the tip, but from those in the neighborhood of it. The theory and experiment on the impurity resonance state were believed to be one of the strong evidences of d-wave pairing in cuprates. However, recent STM reveals that almost equally sharp resonance states also appear in the pseudogap normal state [5]. Theoretically, the resonance state requires good superconducting phase coherence, which is lost in the normal state. If the above experiment is robust, it is a challenge to the mechanism of the resonance state.

Later on, Hudson *et al.* [6] performed STM measurements near nickle (Ni) impurities. They observed that Ni only induces a weak peak at the gap edge, and the energy of this peak depends on the distance away from the impurity. It can be either above or below the Fermi level. They concluded that Ni does not induce low energy bound states and does not suppress the superconducting gap. In other words, the pairing is insensitive to magnetic scattering. They implied that this is consistent with the magnetic correlation-driven pairing mechanism. They also emphasized that the spatial distribution of LDOS is again a four-fold lobe, and, moreover, that the lobe rotates by 45 degrees when the energy changes from above to below the Fermi level.

7.3. Vortex Core States

There is now no doubt that the pairing symmetry is d-wave in high-temperature superconductors. However, is a d-wave BCS theory sufficient to describe the low-energy excitations? The answer seems to be NO. This can be seen from the quasi-particle excitations in the vortex core. Starting from a d-wave BCS theory, numerical calculations reveal that there should be a zero energy state in the vortex core which contributes a zero-bias peak in the tunneling conductivity [7], similar to the impurity resonance state. (The vortex behaves as an impurity with phase winding.)

However, the experimental STM measurements did not find such a zero-bias peak. The STM result on Y123 single crystal [8] shows that the LDOS at the vortex core develops a mini-gap feature around the Fermi level. It is not clear whether this mini-gap is a reflection of an induced second order [9] or if it is due to other mechanisms. Similar results have been observed by many groups in bismuth samples [10].

7.4. Gap Inhomogeneity

Theoretically, a uniform superconductor is the first research priority because of its theoretical simplicity. However, this seems to be an unrealistic starting point for high-temperature superconductors. Pan et al. observed in Bi2212 [11] that there exists considerable spatial inhomogeneity in the electronic structure. Defining the peak-to-peak energy in LDOS as twice that of the energy gap, one sees that the gap is also highly inhomogeneous. A spatial correlation analysis of the data shows that the correlation length is about 3–4 nms. A further analysis of the line shape of the spectra reveals that the coherence peak is higher (lower) where the gap is smaller (larger). In fact, for the largest gap the coherence peak is hardly visible. This implies that there may exist two types of gaps. There seems to be an anti-correlation between the size of the energy gap and the integrated DOS up to a positive bias (roughly proportional to the hole density). Later on, Lang et al. made further measurements on gap inhomogeneity using Ni as a probe [12]. What they found is that only if Ni is located in a region with a small gap size (thus higher coherence peaks) does it induce impurity resonance states. Therefore, the smaller gaps are related to superconductivity, while the larger gaps may be related to the pseudogaps.

7.5. Quasi-particle Scattering Interference

The anomalies in high temperature superconductors are so common, it is not strange if a new one is found. In this sense, it is very prestigious to have a beautiful explanation of a phenomenon in terms of a usual picture. The quasi-particle scattering interference is such phenomenon. The basic picture is as follows. The impurities cause elastic scattering between quasi-particles, and the resulting wave function is a linear combination of the parent eigen states. If in the momentum space there are regions with high joint density of states, the scattering between such regions enjoys the largest phase space, leading to standing-wave like modulation in the spatial distribution of DOS. The modulation wave vector is determined by the momentum transfer during the scattering between the above regions. Since the quasi-particle energy depends on momentum, the modulation vector depends on quasi-particle energy, and is thus dispersive. Wang and Lee [13] predicted theoretically the dispersive quasi-particle interference. They realized that the LDOS in momentum space is largest at the points where the equal energy contour of the quasi-particles in d-wave superconducting phase intersects the normal state Fermi surface. Therefore, the vectors connecting to such intersections are the modulation vectors [13]. This is verified by the experiment of Hoffman et al. [14]. To pin down the modulation vectors, they performed Fourier transform of the real space patterns of LDOS. The power spectra at a given bias voltage peaks at some momenta along directions $(1,0)$ and $(1,1)$. From the peak positions, they read out the modulation vectors, which was in good agreement with the result of calculations according to the theory [13], using inputs from the Fermi surface and energy gap obtained by angle-resolved photoemission measurements. The STM measurement together with Fourier analysis is now called FT-STM. From the modulation vectors and the symmetry of the band structure, FT-STM enables recovering of the normal-state Fermi surface as well as the gap dispersion along the Fermi surface, given the fact that the modulation vector connects the intersection of the equal energy contour and the normal-state Fermi surface. This technique is widely used in later experiments.

7.6. Density Waves

Hoffman et al. [15] first observed that in the spatial distribution of LDOS under a small bias voltage, each vortex is associated with a dark cloud. The modulation is quite obvious, with a period of roughly $4a$, where a is the lattice constant. This kind of modulation exists within a range of bias voltage and is believed to be the evidence of vortex-induced density wave order.

The above modulation was also observed in the pseudogap normal state by Vershinin et al. under zero magnetic field. Moreover, they found that the modulation vector does not change with the bias voltage [16]. Therefore, it is believed to be related to the pseudogap. Recently, Hanaguri et al. observed clearer modulations in underdoped ($x = 10\%$) NaCaCuOCl, with both $4a$ by $4a$ as well as $4a/3$ by $4a/3$ patterns [17].

The important question is whether the modulation is the evidence of charge density wave. Kohsaka et al. tried to answer such a question recently [18]. From the principle of STM, it is known that because of tunneling matrix element, the tunneling conductivity is proportional to but is different from the exact local density of states. To get around the difficulty, one realizes that for a doped Mott insulator, the spectral weight above (or below) the Fermi level is proportional to $2x$ (or $1-x$), where x is the hole density. Since the matrix element depends on energy only weakly, for a good approximation it can be taken as the same for energies near the Fermi level. Under this approximation, the ratio between the integrated DOS above and below the Fermi level is $R = 2x/(1-x)$, independent of the tunneling matrix element. This ratio can be used to determine the local hole density. In practice, the ratio R is calculated by the tunneling currents at positive and negative 150 meV. This is not exact but is perhaps the best one can do. In the spatial dependence of R in Na-CCOC and Dy-Bi2212 superconductors, there is a clear stripe structure with period $4a$. The authors also claimed that if only the copper atoms are analyzed there also exists $4a$ by $4a$ structures consistent with earlier literature. They call such a complex stripe structure the electronic cluster glass. One caution is in order, however. Since the correlation on the oxygen atoms is weak, a lateral application of the expression (for R) in the Mott limit is not robust theoretically.

7.7. Gap Closing Versus Temperature

In the previous sections, we discussed the gap inhomogeneity and the coherence peak versus the gap size. In order to know which type of gap is related to superconductivity, the best method is to monitor the closing of the gaps with increasing temperature. Technically, this is a challenge since it is difficult to fix the tip against thermal drift. Gomes et al. found a clever statistical method to get around the difficulty [19]. They first obtained the statistical distribution of the gap at various temperatures, and the corresponding real space distribution. They read out the relative area in which the gap closes. If all gaps close at a temperature proportional to the zero temperature (in reality at low temperatures) gap size, then the probability of finding a gap below a given value at zero temperature should exactly correspond to the relative area for closed gaps at a given temperature. They find that the correspondence indeed holds in overdoped Bi2212 samples. The BCS ratio $2\Delta_0/k_B T_C$ turns out to be roughly 8, irrespective of the gap inhomogeneity. They also analyzed systematically the results in overdoped, optimally doped and underdoped samples and concluded that the correspondence applies for over and optimal doping, but fails for underdoping, where the large gaps survive up to very high temperatures. This further strengthens the view that the two types of gaps are different in origin. The large gap is very likely linked to the pseudogap.

7.8. Bosonic Modes

If electrons are coupled to bosonic modes, there will be satellite peaks in the second derivative of the tunneling current, and the energy difference between such peaks and the superconducting coherence peak is just given by the energy of the boson mode. Traditionally, this method enables us to reveal the change of the phonon spectrum across the superconducting transition temperature and provides a justification of the electron–phonon pairing mechanism in conventional superconductors. It is rather natural that one is attracted to this method while tackling the pairing mechanism of high temperature superconductors. Lee *et al.* indeed observed evidence of bosonic features from the differential conductivity [20]. They looked into the statistical distribution of the energy gap in a series of samples, and the corresponding statistics of the boson-mode energy. Although the gap variance is large and the statistics changes from sample to sample, the boson energy distribution hardly changes. Therefore, these modes may be independent of the electrons. They may originate from some optical modes of lattice vibration. To make progress, they performed isotope substitution of oxygen and found isotope shifts of the boson energy. On the other hand, they claimed to be able to identify the oxygen site by the STM characteristics. This enables them to analyze the correlations among the gap size, the oxygen position and the boson energy. They concluded that the gap is larger where there is oxygen doping, that the gap is smaller where the boson energy is larger, and finally that there is no correlation between the boson energy and the oxygen doping. This could be understood as follows: The boson mode is the vibration of the oxygen octahedral, the oxygen doping causes local distortions to the band structure and leads to a larger gap, and finally the coupling to electrons causes phonon softening. The experimental results, as well as the relation to the pairing mechanism, are subject to further analysis.

References

[1] A. Yazdani, C. M. Howald, C. P. Lutz et al., Phys. Rev. Lett. **83**, 176 (1999).
[2] M. I. Salkola, A. V. Balatsky, D. J. Scalapino, Phys. Rev. Lett. **77**, 1841 (1996).
[3] S. H. Pan, E. W. Hudson, K. M. Lang et al., Nature **403**, 746 (2000).
[4] T. Xiang, J. M. Wheatley, Phys. Rev. B **51**, 11721 (1995).
[5] K. Chatterjee, M. C. Boyer, W. D. Wise et al., Nat. Phys. **4**, 108 (2008).
[6] E. W. Hudson, K. M. Lang, V. Madhavan et al., Nature **411**, 920 (2001).
[7] Y. Wang, A. H. MacDonald, Phys. Rev. B **52**, R3876 (1995).
[8] I. Maggio-Aprile, Ch. Renner, A. Erb et al., Phys. Rev. Lett. **75**, 2754 (1995).
[9] Q. H. Wang, J. H. Han, D. H. Lee, Phys. Rev. Lett. **87**, 167004 (2001).
[10] S. H. Pan, E. W. Hudson, A. K. Gupta et al., Phys. Rev. Lett. **85**,1536 (2000); G. Levy, M. Kugler, A. A. Manuel et al., Phys. Rev. Lett. **95**, 257005 (2005).
[11] S. H. Pan, J. P. O'Neal, R. L. Badzey et al., Nature **413**, 282 (2001).
[12] K. M. Lang, V. Madhavan, J. E. Hoffman et al., Nature **415**, 412 (2002).
[13] Q. H. Wang, D. H. Lee, Phys. Rev. B **67**, 020511 (2003).
[14] J. E. Hoffman, K. McElroy, D. H. Lee et al., Science **297**, 1148 (2002).
[15] J. E. Hoffman, E. W. Hudson, K. M. Lang et al., Science **295**, 466 (2002).
[16] M. Vershinin, S. Misra, S. Ono et al., Science **303**, 1995 (2004).
[17] T. Hanaguri, C. Lupien, Y. Kohsaka et al., Nature **430**, 1001 (2004).
[18] Y. Kohsaka, C. Taylor, K. Fujita et al., Science **315**, 1380 (2007).
[19] K. K. Gomes, A. N. Pasupathy, A. Pushp et al., Nature **447**, 569 (2007).
[20] J. Lee, K. Fujita, K. McElroy et al., Nature **442**, 546 (2006).

8
Intrinsic Tunneling Spectroscopy of $Bi_2Sr_2CaCu_2O_{8+\delta}$ Cuprate Superconductors

Shi-Ping Zhao

Beijing National Laboratory for Condensed Matter Physics,
Institute of Physics, Chinese Academy of Sciences,
Beijing 100190, China

Tunneling studies of $Bi_2Sr_2CaCu_2O_{8+\delta}$ (Bi2212) cuprate superconductors using submicron intrinsic Josephson junctions (IJJs) are reviewed. We begin with a brief overview of the early tunneling studies of BCS conventional superconductors and the recent experiments on the Bi2212 superconductors using the scanning tunneling microscope (STM), break junctions (BJs) and the IJJs techniques. This is followed by an introduction to the IJJs fabrication process and the elimination of IJJs self-heating through surface–layer contact improvement and IJJs size reduction to the submicron level. The resulting temperature- and magnetic-field-dependent tunneling spectra from the submicron IJJs are then described and discussed, in particular with respect to the possible superconducting pairing glue mechanism, the formation of Fermi arcs in the nodal region and the relationship between the cuprate pseudogap and the superconducting gap. Comparisons of these results with the STM, BJs, micron-sized IJJs, angle-resolved photoemission spectroscopy and optical experiments are also presented.

8.1. Tunneling Studies of BCS Superconductors

Single-electron tunneling spectroscopy played a central role in our understanding of conventional superconductors and in verifying the BCS theory and the superconducting pairing mechanism [1–6]. In the early 1960s, Giaever first discovered the tunneling phenomenon between Al and Pb in the planar-type tunnel junctions (see Fig. 8.1), which were fabricated by evaporating an Al film, forming a natural insulating tunnel barrier on the film surface via oxidation and finally evaporating a Pb superconducting film as the counter electrode [1, 2]. As is shown in Fig. 8.1, the tunnel junctions may have a superconductor/insulator/superconductor (S-I-S) configuration or a superconductor/insulator/normal-metal (S-I-N) configuration. In the S-I-S case, the junctions will exhibit rich Josephson effects resulting from the Cooper-pair tunneling, so they are also called Josephson junctions. Compared to the high-T_C cuprate superconductors like $Bi_2Sr_2CaCu_2O_{8+\delta}$ (Bi2212) to be discussed below, the BCS superconductors are relatively simple and easy to handle: they are chemically stable, their normal state is a metal which has non-localized electronic states well described by the energy band theory and their superconducting pairing gap is isotropic, with pair coherence length on the order of 10–1000 nm. The tunnel junctions of BCS superconductors are usually made with film thickness of tens to hundreds of nanometer and barrier thickness of about 1–2 nm.

8.1.1. *Characteristics of Electron Tunneling Spectroscopy*

Electron tunneling process depends on the properties of the materials on both sides of the tunnel barrier. The I–V characteristics of S-I-S and

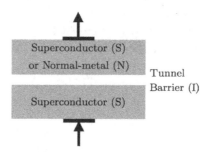

Fig. 8.1. Giaever's planar-type tunnel junctions made up of two superconducting (or one normal metal) films separated by a thin insulating tunnel barrier.

S-I-N junctions can be calculated from the following formula considering the densities of states $N(\omega)$ and the Fermi distribution $f(\omega)$ at finite temperatures [3]:

$$I(V) = \frac{1}{eR_\mathrm{N}} \int_{-\infty}^{\infty} N(\omega)N(\omega+eV)[f(\omega)-f(\omega+eV)]d\omega,$$

where R_N is the normal-state resistance of the junction. Figure 8.2 shows the typical results calculated using the BCS density of states for the S-I-S (upper panel) and S-I-N (lower panel) junctions at various temperatures below the superconducting transition temperature T_C. The I–V curves of the S-I-S junction have a quite steep current rise at twice the gap voltage $2\Delta/e$, which can be used to determine the superconducting energy gap Δ. As temperature increases, the height of the current rise becomes smaller, but a steep rise can still be seen near and below T_C.

For the S-I-N junction, the density of state of one electrode becomes unity. In this case, if we take the derivative dI/dV we will find from the above equation that at low temperatures it is directly proportional to the density of state since the derivative of the Fermi function part with respect to V tends to be a δ-function. This means that by measuring the low-temperature dI/dV tunneling spectra of S-I-N junctions, one directly obtains the electronic density of states of the superconductors [3], as can be seen in the inset of Fig. 8.2.

8.1.2. *Strong-coupling Theory and Eliashberg Equations*

For the BCS conventional superconductors, electron pairing arises from the exchange of phonons between the electrons [3, 4]. For materials such as Pb, the electron–phonon interaction is in the strong coupling regime, so the retardation effect and the lifetime effect of quasiparticles have to be considered [3–6]. In the description of the strong-coupling Eliashberg theory, there are two experimentally determined parameters: the Coulomb pseudopotential $N(0)U_\mathrm{c}$ and the effective phonon spectrum $\alpha_\lambda^2(\nu)F_\lambda(\nu)$. Using these parameters, one is able to obtain the energy gap function $\Delta(\omega)$ and the renormalization function $Z(\omega)$ from the following Eliashberg equations:

$$\Delta(\omega) = \frac{1}{Z(\omega)} \int_0^{\omega_c} d\omega' \mathrm{Re} \left\{ \frac{\Delta(\omega')}{(\omega'^2 - \Delta^2(\omega'))^{1/2}} \right\} [K_+(\omega',\omega) - N(0)U_\mathrm{c}],$$

$$[1-Z(\omega)]\omega = \int_0^{\omega_c} d\omega' \mathrm{Re} \left\{ \frac{\Delta(\omega')}{(\omega'^2 - \Delta^2(\omega'))^{1/2}} \right\} K_-(\omega',\omega),$$

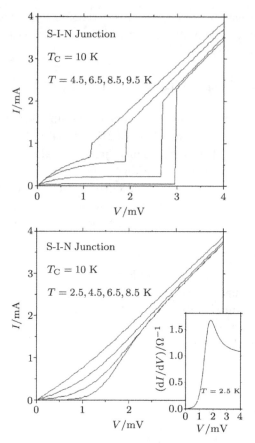

Fig. 8.2. Calculated I–V characteristics of the S-I-S (upper panel) and S-I-N (lower panel) type tunnel junctions at various temperatures below T_C. The junctions' R_N is taken to be 1 Ω for simplicity. The superconducting films have a T_C of 10 K for both cases, and the energy gap Δ that enters the BCS density of states $N(\omega) = \omega/(\omega^2 - \Delta^2)^{1/2}$ is obtained from the BCS gap equation. Inset shows the derivative dI/dV of the 2.5 K curve of the lower panel.

where ω_c is a cutoff frequency and the kernels are given by:

$$K_\pm(\omega',\omega) = \sum_\lambda \int_0^\infty d\nu \alpha_\lambda^2(\nu) F_\lambda(\nu) \left[\frac{1}{\omega' + \omega + \nu + i\delta} \pm \frac{1}{\omega' - \omega + \nu - i\delta} \right].$$

The density of states can be computed from:

$$N(\omega) = \text{Re}\left\{ \frac{\Delta(\omega)}{(\omega^2 - \Delta^2(\omega))^{1/2}} \right\},$$

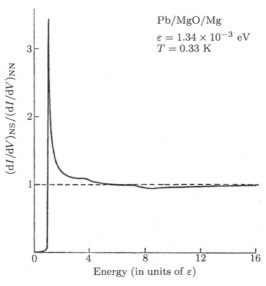

Fig. 8.3. Normalized dI/dV tunneling spectrum of Pb/MgO/Mg junction at 0.33 K. The spectrum shows an extremely sharp gap edge. The small but clear structures above the gap edge result from the strong coupling of the electrons to the phonons. The panel is from Ref. [2].

which can be used to calculate the $I-V$ characteristics as discussed above.

Figure 8.3 shows the normalized tunneling spectrum of a Pb/MgO/Mg junction (S-I-N type) measured at 0.33 K [2]. The overall shape resembles the BCS density of states, as is shown in the inset of Fig. 8.2, and has a very sharp gap structure. The curve also shows additional small concave–convex structures above the gap edge. These structures were found to be excellently reproduced by the Eliashberg strong-coupling theory [5]. Using the theory, the effective phonon spectrum was obtained, which was consistent with the measured results from the inelastic neutron scattering experiments (see Fig. 8.4). These results provided strong support for phonon-mediated pairing in the conventional superconductors.

Similar experiments and theoretical analysis were performed using the S-I-S type Pb junctions and the same results including the effective phonon spectrum were found [6].

Eliashberg analysis of the tunneling data of Pb was also carried out for temperatures near T_C. Although the phonon-induced structure appeared much smeared in dI/dV near T_C, excellent agreement was again found between experiment and theory [7].

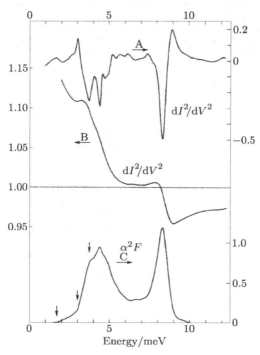

Fig. 8.4. Analysis of the tunneling data using the strong-coupling Eliashberg theory. Curves A, B and C correspond to the second derivative, the first derivative and the effective phonon spectrum for Pb, respectively. The panel is from Ref. [5].

8.1.3. *Other Effects Beyond BCS*

In Fig. 8.2, we see that the current rise at the gap voltage is very sharp for the S-I-S type junctions even for temperatures quite close to T_C. Experimentally, however, the gap feature is often smeared as temperature increases. An example is shown in Fig. 8.5 from the measurement of an S-I-S type PbBi junction [8]. The authors of this work were able to explain their measured results by introducing a finite quasiparticle lifetime parameter $\tau \sim 1/T$ and adding an imaginary part to the energy variable in the BCS density of states [8]. The fitted values of τ due to quasi-particle recombination were consistent with the theoretical calculations, as is shown in Fig. 8.5.

Superconducting fluctuations have been found to persist well above T_C in the high-T_C cuprate superconductors, as will be discussed below. We point out that they have also been observed in the BCS conventional superconductors. In Fig. 8.6, tunneling results measured in Al clearly show the

Intrinsic Tunneling Spectroscopy of $Bi_2Sr_2CaCu_2O_{8+\delta}$ Cuprate Superconductors 155

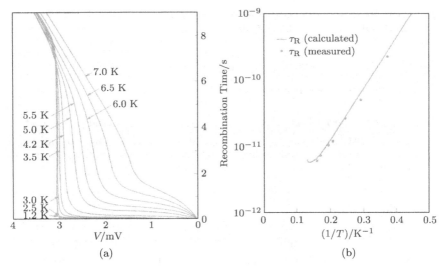

Fig. 8.5. (a) $I-V$ characteristics of PbBi tunnel junctions at various temperatures. The increase of gap smearing with increasing temperature can be seen. (b) Fitted quasiparticle recombination lifetime compared with the calculated values. The panels are from Ref. [8].

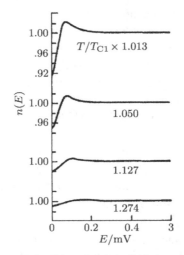

Fig. 8.6. Measured electronic densities of states of Al at various temperatures normalized to T_C. It can be seen that superconducting pairing gap structure persists well above T_C due to pair fluctuation. The panel is from Ref. [9].

gap structure for temperatures as high as $T/T_C \sim 1.3$, which is attributed to superconducting fluctuations above T_C [9].

8.2. Tunneling Studies of Bi2212 Cuprate Superconductors

In contrast to the BCS superconductors, high-T_C cuprate superconductors are doped Mott insulators [10]. Electrons in the materials are strongly correlated and highly localized. The materials are also quasi-2D, and in some cases can be highly anisotropic, making it possible to study the tunneling process between the neighboring superconducting CuO_2 planes within a crystal, which will be the subject of the present study. Cuprate superconductors are also identified as having a d-wave pairing gap, with pairing coherence lengths of 1–4 nm in the ab planes and 0.1–1 nm in the c direction.

So far, some key issues remain to be settled before a complete understanding of cuprate superconductivity can be reached. Among them we mention: (1) the so-called bosonic glue, like phonons in the BCS case, via which the Cooper pairs form [11]; (2) the formation of a Fermi arc in the nodal region in the pseudogap phase and (3) the nature of the pseudogap and its relation to the superconducting gap [12]. Tunneling could provide an effective tool to study these issues as described above for the BCS superconductors. However, the cuprate materials are chemically very unstable with very short coherence length, so Giaever's planar-type tunnel junctions made up of thin films are so far not available. Various techniques like the scanning tunneling microscope (STM) [13, 14], break junctions (BJs) [15–17] and intrinsic Josephson junctions (IJJs) [18–27] have been frequently used.

In the present work, we focus on the most commonly studied Bi2212 cuprate superconductors. The crystal structure of this material is shown in Fig. 8.7(a), in which the formation of IJJs is illustrated. In Fig. 8.7(b), we show schematically the STM and BJ measurements performed on the Bi2212 crystals. An STM measurement is simply performed between a metal tip and the single crystal. BJ measurement can be performed by pressing the tip slightly into the crystal, such that a tunnel junction forms along the crack. In Fig. 8.7(c), we show part of the mesa-type IJJs near the surface, which are fabricated from bulk crystals, as will be described below. As can be seen in Figs. 8.7(a) and (c), tunneling occurs between two adjacent CuO_2 double layers with four BiO and SrO layers in between as

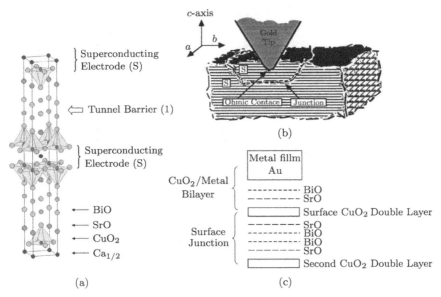

Fig. 8.7. (a) Crystal structure of Bi2212 cuprate superconductors with S-I-S type intrinsic Josephson junctions illustrated. (b) Schematics of the STM and BJs experiments. The panel is from Ref. [16]. (c) Structure of the mesa-type IJJs near the cleaved crystal surface. Inner IJJs below the surface junction are not shown for simplicity.

the tunnel barrier, which can be viewed as naturally formed planar-type tunnel junctions in series. Metal films are coated on the cleaved crystal surface for measurements.

8.2.1. STM Experiments: Precursor Pairing Scenario of the Pseudogap

There are numerous reports on STM experiments for the Bi2212 cuprate superconductors, including various studies of the sample's spatial inhomogeneity [13]. Figure 8.8 shows the temperature dependence of the tunneling spectra measured by the Geneva STM group, which were taken on an underdoped Bi2212 single crystal yielding nearly space-independent spectroscopic results [14]. In the figure, the energy gap structure is seen to evolve continuously, starting from low temperatures through T_C, and can still be seen up to a T^* near room temperature. These results suggest a pseudogap between T_C and T^* which originates from precursor pairing in the Bi2212 material. In fact, one of their important findings is that using the measured

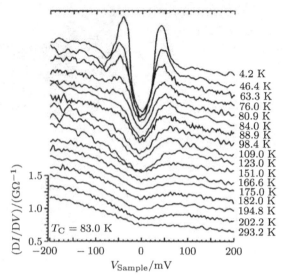

Fig. 8.8. Tunneling spectra measured as a function of temperature on underdoped Bi2212. The conductance scale corresponds to the 293 K spectrum, the other spectra are offset vertically for clarity. The panel is from Ref. [14].

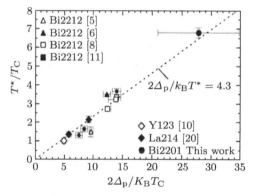

Fig. 8.9. Normalized pseudogap opening temperature T^*/T_C vs. $2\Delta_p/k_B T_C$ for various cuprates compared to the mean-field relation $2\Delta_p/k_B T^* = 4.3$, where T^* replaces T_C. The panel is from Ref. [14].

gap size Δ_p and replacing T_C with the pseudogap opening temperature T^*, one obtains a good scaling for the BCS-like ratio: $2\Delta_p/k_B T^* = 4.3$ (Fig. 8.9). This leads the authors to believe that above T_C, the pseudogap is formed from pre-pairing, and T^* is the temperature where pairing starts.

In order to explain the tunneling results from STM experiments, ideas such as van Hove singularity, bilayer splitting, collective mode, anisotropic tunneling matrix element, marginal Fermi liquid and quasiparticle scattering rate Γ have been invoked [28]. Xiaogang Wen et al. [29] and Anderson et al. [30] have also considered a SU(2)-boson mode and used the Gutzwiller-RVB theory for interpreting the experimental data.

8.2.2. BJ Experiments: Superconducting Pairing Glue Mechanism

Figures 8.10–8.12 are the experimental tunneling spectra of Bi2212 BJs reported by Zasadzinski and coworkers. Figure 8.10(a) shows the results of an optimally doped sample at different temperatures [15] and of different samples as a function of doping in Fig. 8.10(b) [17]. A clear difference as compared to the STM results from samples with similar doping strength is

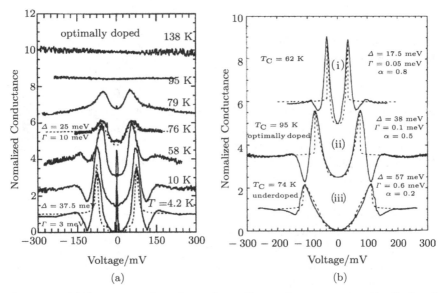

Fig. 8.10. (a) Temperature dependence of tunneling conductance on an optimally doped Bi2212 break junction. The dashed lines show the BCS d-wave fits considering a quasi-partcle scattering rate Γ fit with $\Delta = 37.5$ meV, $\Gamma = 3$ meV and $\Delta = 25$ meV, $\Gamma = 10$ meV for the tunneling conductance at 4.2 and 76 K, respectively. (b) Spectra for (i) overdoped, (ii) optimally doped and (iii) underdoped Bi2212. Dashed lines are BCS d-wave fits considering a quasi-particle scattering rate Γ as indicated. The panels are from Refs. [15, 17].

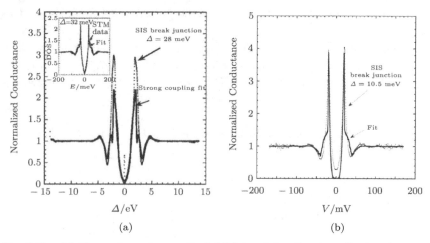

Fig. 8.11. (a) Comparison of normalized S-I-S break-junction tunneling conductance (dots) and d-wave Eliashberg fit (solid line) for a BJ with $\Delta = 28$ meV. Inset: the same $\alpha^2 F(\omega)$ is used to fit an S-I-N conductance obtained by STM. (b) Similar data for a BJ of overdoped Bi2212 with $\Delta = 10.5$ meV. The panel is from Ref. [17].

that the energy gap will disappear near T_C. Namely, the energy gap exists for temperatures below T_C and the superconductivity disappears above T_C, which is very different from the STM results.

The tunneling spectra of BJs shown in Fig. 8.10 displaying peak–dip–hump features have been fitted by the Eliashberg strong-coupling theory [17]. In Fig. 8.11, the dashed lines represent the measured data, and the solid lines represent the fitting results. The slight differences between them are the peak height of the gap structure and the fine features outside the gap edge. Figure 8.12 shows the effective bosonic spectra used (a) and the real parts of the self-energy calculated (b), which are similar to the angle-resolved photoemission spectroscopy (ARPES) results along the nodal direction. The satisfactory fits using Eliashberg theory are in favor of a pairing glue mechanism for the materials. However, the effective bosonic spectra should not result from the phonons since they extend to a high energy range up to 160 meV.

8.2.3. IJJ Experiments: Superconducting Gap and Pseudogap from Different Origins

In some IJJ experiments, it was found that the spectral peak continues to move to the low voltage range as temperature increases and it disappears

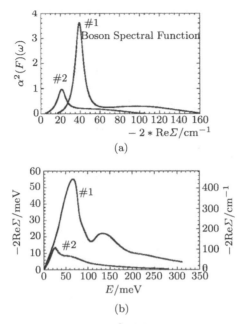

Fig. 8.12. (a) Electron–boson functions $\alpha^2 F(\omega)$, which result from the strong coupling fits to the near optimal S-I-N data in the inset of Fig. 8.11 and the overdoped S-I-S data of Fig. 8.11, labeled as #1 and #2, respectively. (b) Corresponding real part of the diagonal self-energies, $-2\mathrm{Re}\Sigma(\omega)$, obtained from the $\alpha^2 F(\omega)$ shown in (a). The panel is from Ref. [17].

at T_C [20, 21]. Figure 8.13(a) shows the tunneling spectra from which the superconducting gap and pseudogap as a function of temperature seem to show a coexistence region in Fig. 8.13(b). Based on these results, the authors believed that the superconducting gap and the pseudogap have different origins, with the latter caused probably by CDW or SDW.

Tunneling spectra from IJJs are known to be often affected by serious self-heating due to the poor thermal conductivity of the cuprate materials and relatively large current densities required to reach the gap voltage. The spectra shown in Fig. 8.13 appear different from those in Figs. 8.8 and 8.10 from the STM and BJs experiments, respectively, in that they display much sharper spectral peaks when similar doping samples are compared and as they do not show a clear dip structure. These are believed to be caused by self-heating since for samples using intercalated crystals [19] or measurements taken with short-pulse technique [21], the dip structure would

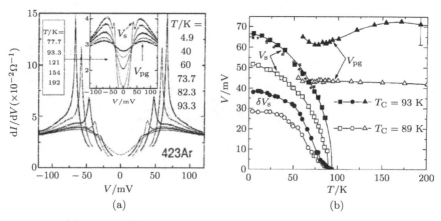

Fig. 8.13. (a) Tunneling conductance at different temperatures for an IJJ mesa. Inset shows detailed curves for high temperatures. Coexistence of the superconducting peak and the pseudogap hump is clearly visible at $T = 77.7$ K. (b) Temperature dependence of parameters of the optimally doped (solid symbols) and overdoped (open symbols) samples: the superconducting peak voltage, $V_s = 2\Delta_s/e$, the spacing between QP branches, δV_s, and the pseudogap hump voltage, V_{pg}. It is seen that the superconducting gap vanishes at T_C, while the pseudogap exists both above and below T_C. The panels are from Ref. [20].

appear and peaks become less sharp. Another effective way to reduce heating is to reduce the IJJ size so that during measurement the ratio of the heat generated to the heat escaped from the IJJ mesa decreases [20]. In the following section, we will present our results of tunneling studies using submicron IJJs and compare them with those from the other experiments.

8.3. Intrinsic Tunneling Studies of Submicron Bi2212 IJJs

8.3.1. Sample Fabrication

Our IJJs were fabricated on near optimally doped Bi2212 single crystals grown by the traveling solvent floating zone method [22, 23]. Thin pieces of the crystals with typical size of 0.5–1.0 mm were glued onto Si substrates. In our experiment, special care was taken to avoid the deterioration of the superconducting properties of the topmost surface CuO_2 double layers in the mesa during preparation. This was realized through the *in situ* low-temperature cleavage of the crystals, followed immediately

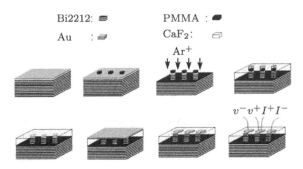

Fig. 8.14. Schematic IJJ fabrication process.

by the evaporation of Au films at high speed. For the IJJ patterning, polymethylmethacrylate (PMMA) resist was used and mesa structures with lateral dimensions from 3.5 down to 0.3 μm were fabricated by electron beam lithography and Ar-ion milling. In the latter process, samples were kept at liquid-nitrogen temperature to avoid possible oxygen loss in the mesa. Deposition and subsequent lift-off of the CaF_2 insulating layer were important in our experiment and could be successfully performed by careful resist treatment and thickness control of the CaF_2 insulating layer. A schematic sample fabrication process is shown in Fig. 8.14.

8.3.2. *Surface Layer Characterization and Control*

Using the process described above, samples with Bi2212/Au contact resistance as low as $\rho = 1.5 \times 10^{-8}$ Ω · cm^2 were fabricated. In addition, T_C of the top CuO_2 double layers could be well controlled (Fig. 8.15). This is important since a large contact resistance will result in extra heating during measurement and extra voltage drop that will be superimposed on the $I-V$ curves [22, 24].

8.3.3. *Junction Size Versus Self-heating*

To further reduce self-heating, IJJs with sizes as small as 0.3 μm were fabric- ated [23–27]. We found that when the IJJ sizes decreased from a few microns down to the submicron level, the tunneling peak position would change and gradually became saturated as shown in Fig. 8.16. Detailed analysis indicated that for the IJJ mesas with sizes of 0.4–0.6 μm, heating was already negligible below the gap voltage. Noticeable temperature rise on the order of a few K could still exist above the gap voltage, but it should

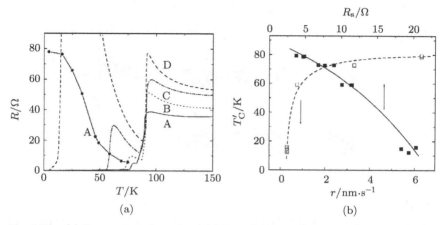

Fig. 8.15. (a) Temperature dependence of the resistance across mesas from four sample crystals (A to D). The line with symbols is the zero-bias resistance taken from the first branch I–V curves of sample A. (b) Superconducting transition temperature of the surface CuO_2 double layer versus the evaporation rate r of Au films and surface IJJ resistance. Lines are guides to the eye [22].

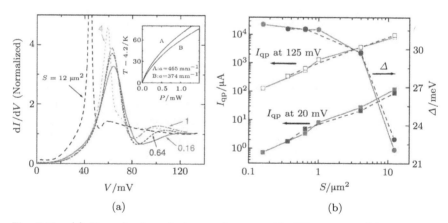

Fig. 8.16. (a) Normalized conductance of mesas with different area S in sample B measured at $T = 4.2$ K. Inset: calculated temperature rise against power for two samples A and B. (b) Junction-area dependence of the current at $V = 20$ and 125 mV (left scale), and the energy gap Δ as determined from the conductance peak (right scale). The data are taken at $T = 4.2$ K. The data with dashed and solid lines correspond to samples A and B, respectively. The panel is from Ref. [23].

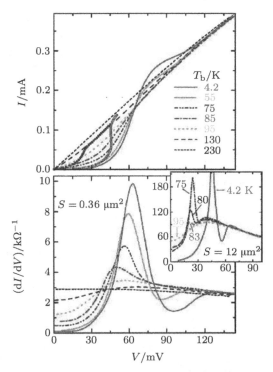

Fig. 8.17. I–V (a) and dI/dV (b) curves at various temperatures for the mesa of $S = 0.36$ μm^2 in sample A of Fig. 8.16. The inset shows the dI/dV data of the mesa with $S = 12$ μm^2 in the same sample. The I–V curves of this larger mesa at $T = 4.2$ K (open squares) and 75 K (solid squares) are plotted in the upper panel with their current divided by the area ratio of the two mesas. The upper inset shows the temperature rise of the larger mesa at $T = 75$ K against power as obtained directly from the comparison of the experimental I–V curves. The panel is from Ref. [23].

have marginal impact on the tunneling spectra. These results demonstrated that reducing the IJJ size is an effective way to reduce heating, which leads to the genuine tunneling spectra with negligible distortion in the IJJs experiments [23].

Figure 8.17 shows the I–V and dI/dV characteristics of an IJJ mesa with area of 0.36 μm^2, and the inset shows the dI/dV characteristics of a larger mesa with area of 12 μm^2 on the same crystal for comparison. It can be seen that the tunneling spectral peaks become less sharp as IJJs area decreases, indicating again the reduction of heating in the mesa.

It is interesting to compare the data in Fig. 8.17 with those in Fig. 8.13 obtained from IJJ samples of a few microns in size. We see that for the larger IJJ mesa, as temperature increases, the tunneling peak would shift toward low voltages and then disappear at T_C. In contrast, for the smaller IJJ mesa, the spectral peak would move in the same way toward T_C. However, it would not disappear. Instead, it exhibits a continuous evolution and moves in an opposite direction above T_C. This kind of spectral evolution is very similar to the results from the STM experiments that are shown in Fig. 8.8.

8.3.4. Temperature- and Magnetic-field-dependent Tunneling Spectra

Figure 8.18 shows the temperature dependences of (a) $I–V$, (b) dI/dV, and (c) d^2I/dV^2 curves of an IJJ mesa with area of 0.09 μm^2, which is the smallest sample in our experiment [26, 27]. The tunneling spectra of this sample

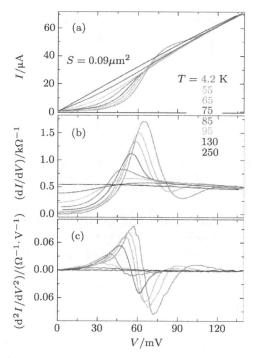

Fig. 8.18. Temperature dependences of (a) $I–V$, (b) dI/dV and (c) d^2I/dV^2 characteristics of a Bi2212 mesa containing 10 IJJs, with area $S = 0.09$ μm^2 and $T_C = 89$ K [26, 27].

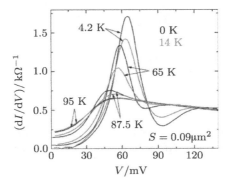

Fig. 8.19. Magnetic field dependence of the tunneling spectra at several temperatures for the IJJ mesa in Fig. 18. The panel is from Ref. [23].

have the following typical features: (1) There exists a well-defined normal-state resistance and a clear "dip" structure at low temperatures; (2) The superconducting peak changes continuously to the pseudogap structure in the vicinity of T_C, which persists up to a $T^* = 230$ K.

The magnetic-field dependence of the dI/dV spectra of this sample is shown in Fig. 8.19. Below T_C, both the superconducting peak and dip structures appear suppressed when the magnetic field is applied, which can be seen from the difference of red and black lines in the figure. Above T_C, however, the tunneling spectra become relatively insensitive to the applied magnetic field [25].

8.3.5. *Tunneling and Transport Properties Along the C Direction*

Unlike Giaever's tunnel junctions made up of BCS superconductors (normally polycrystalline thin films), Bi2212 IJJs involve electron tunneling between the adjacent CuO_2 double layers along the c direction within a single crystal (see Fig. 8.7(a)). So some particular considerations such as the coherent transport, d-wave symmetric energy gap, the angular dependence of tunneling matrix element and the quasi-particle scattering rate Γ need to be taken into account. Tunneling current from coherent process can be calculated from:

$$I(V) = \frac{C}{eR_N} \int_0^{2\pi} t^2(\theta) d\theta \int_{-\infty}^{\infty} n_1(\theta,\omega) n_2(\theta,\omega+eV)$$
$$\times [f(\omega) - f(\omega+eV)] d\omega,$$

where the density of states considering a d-wave gap and Γ in the Dynes form [8] reads:

$$n(\theta, \omega) = \text{Re}\left[\frac{\omega - i\Gamma}{\sqrt{(\omega - i\Gamma)^2 - \Delta^2(T, \theta)}}\right], \quad \Delta(T, \theta) = \Delta(T)\cos(2\theta).$$

The angular dependence of tunneling matrix element may have three cases:

$$\begin{aligned}\text{case 1}: t(\theta) &= 1, & C_1 &= 1/2\pi, \\ \text{case 2}: t(\theta) &= \cos^2(2\theta), & C_2 &= 4/3\pi, \\ \text{case 3}: t(\theta) &= \sin^2(4\theta), & C_3 &= 4/3\pi.\end{aligned}$$

It can be seen that case 1 corresponds to the simple isotropic samples. Case 2 describes the materials such as $HgBa_2CaCu_2O_6$, in which Cu atoms in the adjacent CuO_2 double layers are collinear. For the present Bi2212 crystals in which the Cu atom alignment is not collinear, case 3 should be applicable [31]. The result shown in Fig. 8.20 is the zero-bias conductivity data of both the surface and inner IJJs with different doping strengths [22], which follows well the third power of temperature as predicted by the theory considering coherent process [31].

The conductivity data in Fig. 8.20 are the results at zero-bias voltage. For the tunneling spectra of Bi2212 IJJs, the bias voltage extends to tens of mV, far above the gap edge. In this case, we found that the fits considering coherent tunneling are not satisfactory. The best fits came

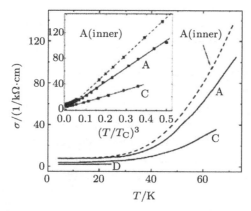

Fig. 8.20. Zero-voltage conductivity vs. temperature for three surface junctions labeled A, C and D (solid lines). The data for the inner junction are shown as a dashed line. In the inset, the results are plotted for samples A and C against $(T/T_C)^3$, where $T_C = 91.5\,\text{K}$ [22].

from considering an effective density of states $N_{\text{eff}}(\omega)$, which is built from the angle-dependent density of states $n(\theta,\omega)$ by averaging over θ with a weight of $\cos^2(2\theta)$. In the high-temperature range, $n(\theta,\omega)$ described above can be used. In the low-temperature range, those obtained from the strong coupling theory can be used. Namely, we have:

$$N_{\text{eff}}(\omega) = \frac{1}{\pi}\int_0^{2\pi} n(\theta,\omega)\cos^2(2\theta)\mathrm{d}\theta$$

with:

$$n(\theta,\omega) = \text{Re}[\omega/\sqrt{\omega^2 - \Delta(\omega)^2\cos^2(2\theta)}],$$

in which the energy gap function $\Delta(\omega)$ is obtained from solving Eliashberg equations. Note that $N_{\text{eff}}(\omega)$ is different from the density of states averaged simply with respect to θ:

$$N_{\text{d}}(\omega) = \frac{1}{2\pi}\int_0^{2\pi} n(\theta,\omega)\mathrm{d}\theta.$$

The effective bosonic spectrum used in the Eliashberg analysis can be written as:

$$\alpha^2_{k,k'}F(\Omega) = \alpha^2[c_s + c_d\cos(2\theta)\cos(2\theta')]\mathcal{F}(\Omega)$$

with $F(\Omega)$ expressed as a cut-off Lorentzian function:

$$\mathcal{F}(\Omega) = \begin{cases} C\left[\dfrac{1}{(\Omega-\Omega_0)^2 + \eta^2} - \dfrac{1}{\Omega_1^2 + \eta^2}\right], & |\Omega - \Omega_0| < \Omega_1, \\ 0, & |\Omega - \Omega_0| > \Omega_1. \end{cases}$$

With these, the $I-V$ curves can be computed from:

$$I(V) = \frac{1}{eR_{\text{N}}}\int_{-\infty}^{\infty} N_{\text{eff}}(\omega)N_{\text{eff}}(\omega + eV)[f(\omega) - f(\omega + eV)]\mathrm{d}\omega.$$

8.3.6. Tunneling Spectra at Low Temperatures

Detailed Eliashberg analysis for the low-temperature tunneling spectra in Fig. 8.18 has been performed [26]. Figures 8.21(a) and 8.21(b) show the fitted results of the $I-V$ and $\mathrm{d}I/\mathrm{d}V$ curves at 4.2 K, in which the red symbols represent the experimental data and the dotted and dashed lines are the fitting results using $N_{\text{d}}(\omega)$ and $N_{\text{eff}}(\omega)$. Both fits show the right spectral peak and dip positions consistent with the experimental data. However,

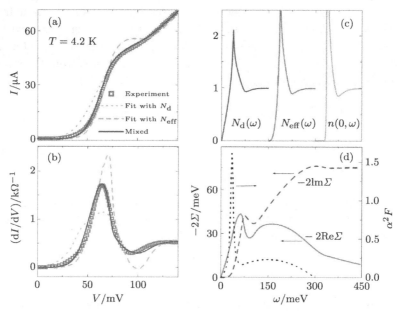

Fig. 8.21. Fits of the experimental (a) $I-V$ and (b) dI/dV curves at 4.2 K using the d-wave Eliashberg theory. The calculated densities of states and self-energy Σ are shown in (c) and (d). Note the horizontal shifts of 150 meV and 300 meV for N_{eff} and $n(0,\omega)$ in (c) for clarity. $\alpha^2 F(\omega)$ used in the calculation is shown in (d) [26].

a better fit (solid lines) was obtained considering the two densities of states simultaneously [26]. In the calculation, we chose the effective bosonic spectrum that included a resonance part and a widespread part (dotted line in Fig. 8.21(d)). We extended the wide spectrum up to 300 meV so that the fit appeared satisfactory. Wide bosonic spectra have also been used in the optical experiments for Bi2212 [32]. The calculated density of states $n(0,\omega)$ along the anti-nodal direction, the angle-averaged density of states $N_d(\omega)$, the effective density of states $N_{\text{eff}}(\omega)$ and the real and imaginary parts of the self-energy are shown in Fig. 8.21(c) and 8.21(d) [26].

When we reduced the energy range of the wide spectrum down to 160 meV, which still led to a rather satisfactory fit, the self-energy would change to the shape as shown in Fig. 8.22. The results look similar to those of BJs in Fig. 8.12, and also to the optical [32] and ARPES experiments [33]. Nevertheless, we found that this energy range cannot be further reduced in order to have a satisfactory fit. The fact that in addition to the resonance peak the bosonic spectrum should contain a wide spectrum extending up to 160 meV and above indicates its origin to be from a

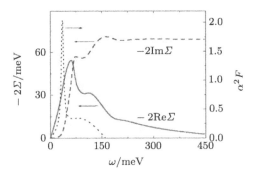

Fig. 8.22. Data similar to those in Fig. 21(d), but with a reduced energy range up to 160 meV for $\alpha^2 F(\omega)$.

source other than phonons, except from bosonic excitations such as the spin fluctuations.

8.3.7. Tunneling Spectra at High Temperatures

Understanding of the tunneling spectra in the temperature range near and above T_C can provide some important information about the Fermi arc and the pseudogap [12] in the Bi2212 superconductors. It can be seen in Fig. 8.18 that the spectral dip feature in the dI/dV curves gradually disappears as temperature approaches T_C. Thus, for the spectra near and above T_C, the density of states considering a d-wave gap and Γ in the Dynes form [8] should be sufficient to fit the experimental data. To do so, we extended the theoretical formulation based on a self-energy model developed to account for precursor pairing in the cuprate materials [34]:

$$\Sigma(k,\omega) = -i\Gamma + [\Delta_k^2/(\omega + \varepsilon_k + i\Gamma_\Delta)],$$

where ε_k is the energy of bare electrons relative to the value at the Fermi surface and Γ_Δ comes from the pair fluctuations. It can be shown that the corresponding spectral function is:

$$A(\theta,\omega) = \frac{1}{\pi} \frac{\Gamma_\Delta[\Delta^2 \cos^2(2\theta) + \Gamma\Gamma_\Delta] + \Gamma\omega^2}{[\omega^2 - \Delta^2 \cos^2(2\theta) - \Gamma\Gamma_\Delta]^2 + \omega^2(\Gamma + \Gamma_\Delta)^2}$$

and the density of states has the Dynes form [8] described above:

$$n(\theta,\omega) = \mathrm{Re}\left[\frac{\omega + i\gamma_+}{\sqrt{(\omega + i\gamma_+)^2 - \Delta^2 \cos^2(2\theta)}}\right]$$

with $\gamma_+ = (\Gamma + \Gamma_\Delta)/2$ replacing Γ [27].

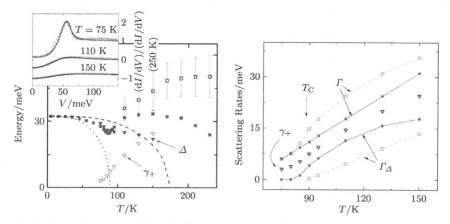

Fig. 8.23. Left panel: half the dI/dV peak energy Δ_p (solid squares). Open squares are those from the spectra normalized to the one at 250 K. Solid and open triangles are Δ and γ_+ from fitting the spectra using $N_{\text{eff}}(\omega)$. The dashed and dotted lines are the BCS d-wave $\Delta_p(T)$ with $2\Delta_p(0)/k_B T_C = 4.28$ and that with temperature normalized to T_C, respectively. The inset shows the normalized spectra at three temperatures (symbols), and the fitted results using N_{eff} (solid lines) and $N_d(\omega)$ (dashed line at 75 K). Right panel: two sets of the scattering rates Γ and Γ_Δ obtained from experimentally determined γ_+ parameter assuming a linear temperature-dependent Γ_Δ (dashed lines with open symbols), and a linear temperature-dependent Γ (solid lines with full symbols) [27].

The symbols in the inset of Fig. 8.23 are the experimental spectra, and the solid lines represent the fits using $N_{\text{eff}}(\omega)$ with the above $n(\theta, \omega)$. The dashed line at 75 K is a fit using $N_d(\omega)$. We can see that the results are fair by using $N_{\text{eff}}(\omega)$. In the left panel of Fig. 8.23, the solid squares are half the peak positions of the experimental dI/dV tunneling spectra. Open squares are those from the dI/dV curves normalized to that at 250 K. The triangles represent the fitted Δ and γ_+ parameters using $N_{\text{eff}}(\omega)$ [27].

Once γ_+ is determined experimentally, Γ and Γ_Δ can be estimated in one of the following two ways. The first is to assume Γ_Δ having a linear temperature dependence above T_C:

$$\Gamma_\Delta = \begin{cases} 0, & \text{for } T < T_C, \\ \dfrac{8}{\pi} k_B (T - T_C), & \text{for } T > T_C, \end{cases}$$

and Γ can be estimated from $\gamma_+ = (\Gamma + \Gamma_\Delta)/2$. The other is to assume a linear temperature dependence of Γ, and Γ_Δ can be estimated accordingly. In the right panel of Fig. 8.23, the Γ and Γ_Δ parameters thus estimated are shown as open and solid squares, respectively.

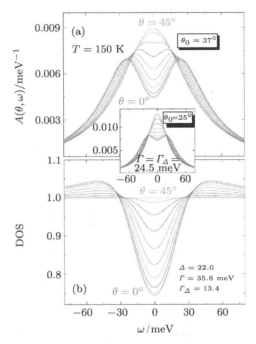

Fig. 8.24. (a) Spectral function $A(\theta,\omega)$ on the Fermi surface at 150 K calculated using Δ, Γ and Γ_Δ parameters as indicated. Curves are plotted with a θ step of 5°. (b) Corresponding results of densities of states. The inset shows $A(\theta,\omega)$ with equal Γ and Γ_Δ at the same temperature [27].

Figure 8.24(a) shows the calculated $A(\theta,\omega)$ at 150 K for 10 different θ values ranging from 0° (antinodal direction) to 45° (nodal direction) with the Δ, Γ and Γ_Δ parameters indicated. For the small θ value, $A(\theta,\omega)$ shows two peaks at $\pm\omega_p$ symmetric to $\omega = 0$. As θ increases, the two peaks shift to smaller $|\omega|$ and between $\theta = 35°$ and 40°, they merge into a single peak at $\omega = 0$, thus forming the Fermi arc as discussed in the ARPES experiment. Figure 8.24(b) shows the corresponding densities of states, in which the gapped structure exists for all θ except the nodal point of $\theta = 45°$. In Fig. 8.25, the ARPES gap ω_p vs. θ and the Fermi arc length vs. temperature are plotted.

We note that Norman et al. [35] and Chubukov et al. [36] have also considered the Fermi arc problem based on the above phenomenological self-energy model for the simpler case of $\Gamma = \Gamma_\Delta$. On the experimental side, ARPES measurements have indicated the existence of a Fermi arc

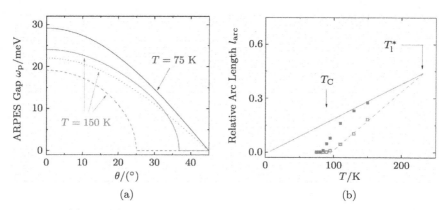

Fig. 8.25. (a) ARPES gap at 75 and 150 K calculated using Γ and Γ_Δ from the dashed lines with open symbols in Fig. 8.23 (solid lines). The dashed line is the result at $T = 150$ K with $\Gamma = \Gamma_\Delta = 24.5$ meV. The dotted line is the d-wave gap at 150 K for comparison. (b) Relative arc length l_{arc} calculated using Δ, Γ and Γ_Δ in Fig. 8.23 [27].

above T_C with its length increasing linearly with increasing temperature [37], and also a new gap opening on the Fermi arc below T_C [33, 38]. The gap opening may occur suddenly, jumping to a full d-wave gap [33], or can be gradual and follow the BCS-like temperature dependence [38]. The present results in Fig. 8.25 show that the tunneling and ARPES results are surprisingly similar concerning the Fermi arc in the pseudogap state. The results of the arc length in Fig. 8.25 exhibit two possible behaviors when a linear temperature dependence of Γ_Δ or that of Γ is assumed. In the latter case, the result bears a close resemblance to the ARPES observations in Ref. [33].

In the above discussion, the pseudogap is naturally understood as precursor pairing above T_C. Then what is the temperature at which the pairing starts? We see in the left panel of Fig. 8.23 that the data of half the peak position (solid squares) show a slope change near 150 K, which is reminiscent of the experiments suggesting two pseudogap temperatures T_1^* and T_2^* (230 K and 150 K, respectively, in the present case), such as Nernst experiment [39]. On one hand, we see the fitted Δ values in the figure (solid triangles) tends to be finite above $T_2^* = 150$ K, which would suggest T_1^* to be the pairing start temperature as discussed in the STM experiments. In this case, in light of the above discussion, a pseudogap resulting from pre-pairing will first form in the antinodal region as temperature decreases below T_1^*, with a significant length of the Fermi arc present near the node. The arc length will decrease as temperature further reduces, and the decreasing rate accelerates as T_C is approached. However, uncertainties exist in the view of this scenario. First, what is the physics behind T_2^*? Second, the temperature dependence of the fitted Δ shows a strange behavior near T_C, which deviates significantly from the expected BCS temperature dependence of the energy gap. Further studies are required to clarify the situation.

8.4. Conclusion

We successfully fabricated submicron-sized IJJs from near optimally doped Bi2212 superconductors, in which self-heating, a notorious obstacle in the IJJs studies, was proved to be negligible. The following valuable results from systematic tunneling studies of these IJJs were obtained:

(1) At low temperatures, the tunneling spectra exhibited the familiar peak–dip–hump structure. The spectral peak evolved continuously in the entire temperature range, while the dip feature disappeared gradually as temperature approached T_C from below. In the d-wave Eliashberg analysis, tunneling spectra at 4.2 K were found to be well fitted considering the electron coupling to a bosonic spectrum consisting of a resonance mode and a broad high-energy continuum, which supported the pairing glue mechanism for the cuprates.

(2) The spectra in the temperature range near and above T_C were analyzed using a self-energy model in which both quasi-particle scattering rate Γ and pair decay rate Γ_Δ were considered. The density of states derived from the model had the familiar Dynes' form with a simple replacement of Γ by $\gamma_+ = (\Gamma + \Gamma_\Delta)/2$. The γ_+ parameter obtained from fitting the experimental spectra exerted a strong constraint on the relation between Γ and Γ_Δ. We discussed and compared the Fermi arc behavior in the pseudogap region from the tunneling and ARPES experiments. Our results demonstrated an excellent agreement between the two experiments, which were in favor of the precursor pairing view of the pseudogap.

References

[1] I. Giaever, *Phys. Rev. Lett.* **5**, 147 (1960).
[2] I. Giaever, H. R. Hart, and K. Megerle, *Phys. Rev.* **126**, 941 (1962).

[3] L. Wolf, *Principles of Electron Tunneling Spectroscopy* (Oxford University Press, New York, 1985).
[4] J. R. Schrieffer, *Theory of Superconductivity* (Benjamin, New York, 1964).
[5] W. L. McMillan and J. M. Rowell, *Phys. Rev. Lett.* **14**, 108 (1965).
[6] W. L. McMillan and J. M. Rowell, *Superconductivity* (Marcel Dekker, New York, 1969).
[7] P. Vashishta and J. P. Carbotte, *Solid State Commun.* **8**, 161 (1970).
[8] R. C. Dynes, V. Narayanamurti, and J. P. Garno, *Phys. Rev. Lett.* **41**, 1509 (1978).
[9] R. W. Cohen, B. Abeles, and C. R. Fuselier, *Phys. Rev. Lett.* **23**, 377 (1969).
[10] P. A. Lee, N. Nagaosa, and X.-G. Wen, *Rev. Mod. Phys.* **78**, 17 (2006).
[11] P. W. Anderson, *Science* **316**, 1705 (2007).
[12] M. R. Norman, D. Pines, and C. Kallin, *Adv. Phys.* **54**, 715 (2005).
[13] ϕ. Fischer *et al.*, *Rev. Mod. Phys.* **79**, 353 (2007).
[14] C. Renner *et al.*, *Phys. Rev. Lett.* **80**, 149 (1998); M. Kugler *et al.*, *ibid.* **86**, 4911 (2001).
[15] N. Miyakawa *et al.*, *Phys. Rev. Lett.* **80**, 157 (1998); **83**, 1018 (1999).
[16] L. Ozyuzer *et al.*, *Phys. Rev. B* **61**, 3629 (2000).
[17] J. F. Zasadzinski *et al.*, *Phys. Rev. Lett.* **87**, 067005 (2001); **96**, 017004 (2006).
[18] R. Kleiner, F. Steinmeyer, G. Kunkel, and P. Mueller, *Phys. Rev. Lett.* **68**, 2394 (1992).
[19] A. Yurgens *et al.*, *Int. J. Mod. Phys. B* **13**, 3758 (1999).
[20] V. M. Krasnov *et al.*, *Phys. Rev. Lett.* **84**, 5860 (2000); **94**, 077003 (2005).
[21] M. Suzuki *et al.*, *Phys. Rev. Lett.* **85**, 4787 (2000); *Appl. Phys. Lett.* **83**, 2381 (2003).
[22] S. P. Zhao *et al.*, *Phys. Rev. B* **72**, 184511 (2005).
[23] X. B. Zhu *et al.*, *Phys. Rev. B* **73**, 224501 (2006).
[24] S. X. Li *et al.*, *Phys. Rev. Lett.* **99**, 037002 (2007).
[25] X. B. Zhu *et al.*, *Physica C* **460**, 963 (2007).
[26] S. P. Zhao, X. B. Zhu, Y. F. Wei, arXiv:cond-mat/0703177.
[27] S. P. Zhao, X. B. Zhu, and H. Tang, *Eur. Phys. J. B* **71**, 195 (2009).
[28] B. W. Hoogenboom *et al.*, *Phys. Rev. B* **67**, 224502 (2003).
[29] W. Rantner and X. G. Wen, *Phys. Rev. Lett.* **85**, 3692 (2000).
[30] P. W. Anderson and N. P. Ong, *J. Phys. Chem. Solids* **67**, 1 (2006).
[31] T. Xiang and J. M. Wheatley, *Phys. Rev. Lett.* **77**, 4632 (1996); T. Xiang, C. Panagopoulos, and J. R. Cooper, *Int. J. Mod. Phys. B* **12**, 1007 (1998).
[32] J. Hwang, T. Timusk, and G. D. Gu, *Nature (London)* **427**, 714 (2004).
[33] A. Kanigel *et al.*, *Phys. Rev. Lett.* **99**, 157001 (2007).
[34] M. R. Norman *et al.*, *Phys. Rev. B* **57**, R11093 (1998).
[35] M. R. Norman *et al.*, *Phys. Rev. B* **76**, 174501 (2007).
[36] A. V. Chubukov *et al.*, *Phys. Rev. B* **76**, R180501 (2007).
[37] A. Kanigel *et al.*, *Nat. Phys.* **2**, 447 (2006).
[38] W. S. Lee *et al.*, *Nature* **450**, 81 (2007).
[39] Yayu Wang *et al.*, *Phys. Rev. Lett.* **95**, 247002 (2005).

9
The Transport Properties of High-Temperature Cuprate Superconductors

Xian-Hui Chen

School of Physical Science, University of Science and Technology of China, Heifei, Anhui 230026, China

9.1. Magnetic Structure and Phase Diagram

The high-temperature superconductors (HTSC) are commonly categorized into two groups: the p-type (hole-doped) and the n-type (electron-doped). Considering the basic structure and magnetic structure, as shown in Fig. 9.1, La_2CuO_4 is an antiferromagnetic parent compound, the in-plane spin arrangement is antiferromagnetic, which is the same as the n-type HTSC. In the n-type HTSC, the magnetic structure between adjacent layers is non-collinear, in other words, the directions of spin moments in the neighboring CuO_2 planes are perpendicular. But the p-type HTSC have a 3D collinear magnetic configuration, which is a significant difference from n-type HTSC. As for crystal structure, the copper and oxygen atoms could form fourfold, fivefold and sixfold coordination. In the hole-type HTSC, when the CuO_2 planes have sixfold coordination, there will be strong distortion. But this distortion will not emerge in electron-type HTSC, in which the CuO_2 planes are rather flat. This is extremely important for the following magnetic structure analysis.

Figure 9.2 is the electronic phase diagram for the high-temperature superconductor. For hole doping, the ground state evolves fast from a long-range antiferromagnetic (AFM) order to a superconductivity (SC)

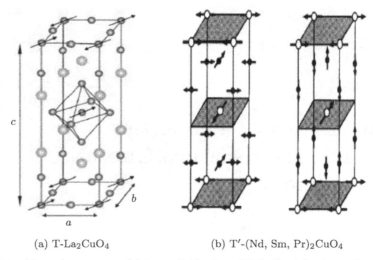

(a) T-La$_2$CuO$_4$ (b) T′-(Nd, Sm, Pr)$_2$CuO$_4$

Fig. 9.1. Magnetic structure. (a) is parallel/antiparallel (collinear) spin configuration in hole-type cuprates, (b) is perpendicular (non-collinear) spin configuration in electron-type cuprates.

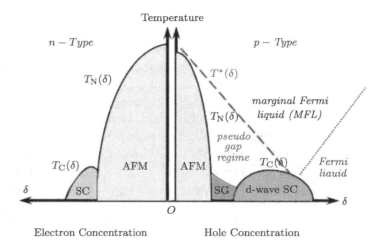

Fig. 9.2. Phase diagrams of n-type and p-type superconductors.

order. But for electron doping, the AFM order is more stable, and it needs a relatively high doping level to suppress the magnetic order and introduce superconductivity. For the hole-type doping phase diagram, there is a pseudogap region below a special temperature, and it extends from the parent

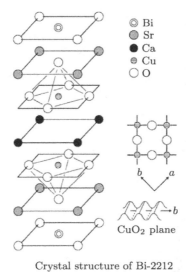

Crystal structure of Bi-2212

Fig. 9.3. Schematic layered crystal structure of HTSC.

compound till the superconducting region. As for electron-type doping, there is no report of an explicit pseudogap phenomenon to date. For both hole- and electron-type HTSC, the doping dependence of superconducting critical temperature T_C has a dome-like behavior. Another important characteristic is the layered crystal structure. As shown in Fig. 9.3, the CuO_2 planes and the reservoir layers stack alternatively along the c axis. The layered structure causes strong anisotropy in transport properties, which will be discussed later. Because of the strong anisotropy, there will be a non-homogeneous problem in electrical transport measurements, and it is a problem that we should try to avoid in experiments.

9.2. Resistivity and Hall Coefficient

In this section, we will discuss the resistivity and the Hall coefficient of HTSC first. An abnormal phenomenon in high-temperature superconductors is that the resistivity shows a linear temperature dependence, which could extend to rather low temperatures. For the material $Bi_{2+x}Sr_{2-y}CuO_{6+\delta}$, as shown in Fig. 9.4 (presents two samples with different composition), the T_C is 10 K, and the in-plane resistivity shows linear temperature dependence from 10 K to 700 K. The data of resistivity along

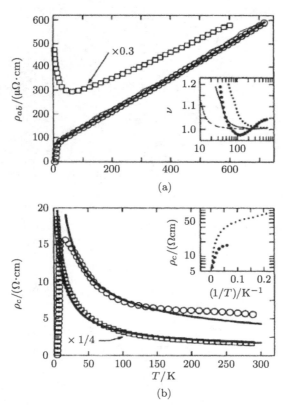

Fig. 9.4. The temperature dependence of electrical resistivity of $Bi_{2+x}Sr_{2-y}CuO_{6+\delta}$. The panel is from Ref. [1].

the c-direction has a completely different behavior compared with that of the in-plane resistivity, which will be discussed in later section.

Figure 9.5 shows the temperature dependence of the Hall coefficient, and the Hall coefficient varies strongly with temperature. This result was first reported by N. P. Ong's group. In the vicinity of optimal doping, the Hall coefficient is approximately proportional to $1/T$. This behavior is considered to be related to the separation of relaxation time. The T-linear dependent resistivity is kept in the vicinity of optimal doping, but the Hall angle follows an unusual T^2 dependence. The result of Hall angle in Fig. 9.6 was also reported by Ong's group.

P. W. Anderson has put forward a possible explanation of the T^2 dependence of the Hall angle, which was described by different relaxation times

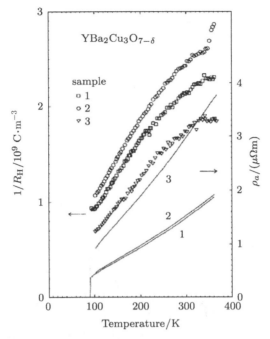

Fig. 9.5. The temperature dependence of the Hall coefficient of $YBa_2Cu_3O_{7-\delta}$. The panel is from Ref. [2].

as follows:

$$\rho_{ab} \sim \tau_{tr}^{-1} \sim T,$$

$$\cot\theta_H = \rho_{ab}/HR_H \sim \tau_H^{-1} \sim T^2,$$

where τ_{tr} and τ_H are the transport and transverse ("Hall") relaxation times. These two relaxation times have different temperature dependences, which is consistent with experimental results. Thus, there is strong evidence for the existence of two relaxation times. The situation of hole-doped YBCO is shown in Fig. 9.7. We can see that its resistivity deviates from T-linear dependence only in extremely underdoped region, and it is rather difficult to prepare overdoped YBCO sample. In the heavily overdoped LSCO sample, the T-linear relationship is also broken, and the resistivity shows T^2 dependence as a normal Fermi liquid. However, the T^2 temperature dependence of Hall angle does not change a lot with doping and is kept in a wide region. While in the Bi-2201 system, we can find a systematic change in the temperature dependence of Hall angle with carrier doping.

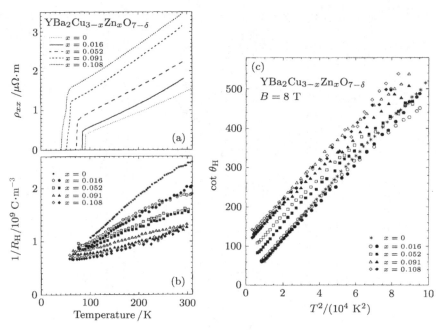

Fig. 9.6. The temperature dependence of the in-plane resistivity, Hall coefficient and Hall angle of Zn-doped YBCO. The panels are from Ref. [3].

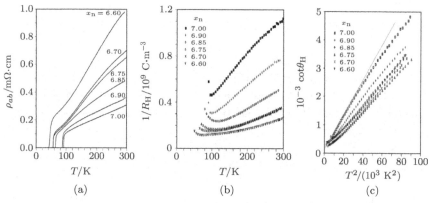

Fig. 9.7. The temperature dependence of the resistivity, Hall coefficient and Hall angle of YBCO film. The panels are from Ref. [4].

Thus far, we have discussed about the hole-type superconductors. Figure 9.8 presents the behavior of Hall coefficient and Hall angle of electron-type superconductors. This work was reported by the author's group.

We revealed that in the extremely underdoped samples, the Hall angle has a T^4 dependence rather than T^2, which is different from the hole-type superconductor. In the overdoped region, the T^2 dependence is recovered. Combined with the results of ARPES, it could be proved that the behavior of Hall angle is related to the type of carrier to some certain extent: hole and electron corresponding to T^2 and T^4 dependence, respectively.

Figure 9.9 shows the result of Hall measurement of optimally doped $Bi_2Sr_{1.51}La_{0.49}CuO_6$ by Balakirev's group, which has a transition temperature $T_C = 33$ K. In this paper, they gave the variation of the temperature dependence of Hall coefficient with different doping ratios. Here, we focus on the optimal doping, from the relationship between Hall resistivity and applied magnetic field, and it can be concluded that T_C has a linear relationship with the effective carrier concentration in the CuO_2 plane. In the electron phase diagram, there exists an anomaly of Hall coefficient at the optimal doping, where the effective carrier concentration shows a maximum.

As for the case of the overdoped samples, there is a system, LSCO, that could realize extreme overdoping as to $x \sim 0.4$. While the doping level changes from underdoped to overdoped, there is an evolution of Hall coefficient from positive to negative, and we obtain a minus sign at the heavily overdoped side, as is shown in Fig. 9.10. It is possible that the electron structure evolves from hole pocket to electron pocket in this non-superconducting overdoped region. Besides, the resistivity has a T^2 dependence in this region (even though there is a small deviation from T^2 dependence in the heavily overdoped sample). The negative Hall coefficient and the T^2 dependence of resistivity are typical behaviors of a Fermi liquid.

Figure 9.11 presents the results of the magic number doping levels obtained by Ando's group. They prepared a series of high-quality LSCO single crystals. They carefully tuned the carrier concentration by changing the Sr content and measured resistivity and Hall coefficient to obtain carrier mobility. They revealed that when the resistivity was normalized to the value at 300 K, there are several "magic dopings" in the carrier concentration. This result is similar to the idea of Prof. Zhongxian Zhao (Institute of Physics, Chinese Academy of Sciences, China). Inspired by the results of Hall measurements, some people explained the origin of magic numbers by involving the ordered occupancy of charges in charge density wave state.

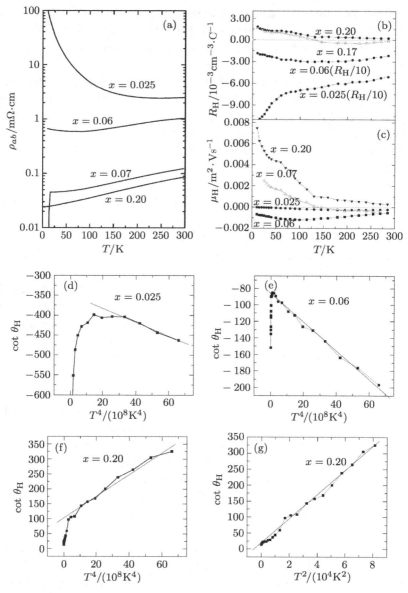

Fig. 9.8. The temperature dependence of the resistivity, Hall coefficient, mobility and Hall angle of $Nd_{2-x}Ce_xCuO_4$ system. The panels are from Ref. [5].

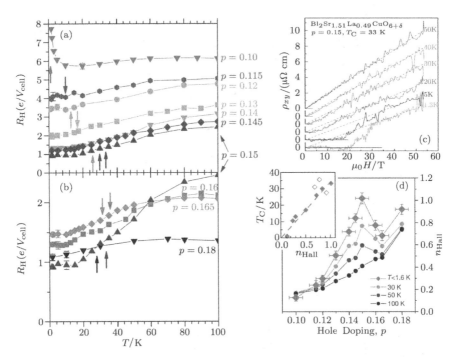

Fig. 9.9. The temperature dependence and doping dependence of the Hall coefficient under high magnetic field in hole-type Bi2201 system. The panels are from Ref. [6].

From the transport measurements mentioned above, we can make some phenomenological conclusions: the T-linear resistivity seems to come from electron–electron interaction, and it is only observed in the vicinity of optimal doping; in overdoped and underdoped samples, the resistivity does not obey the T-linear behavior, and the Hall coefficient has a strong temperature dependence. For the hole-doped superconductors, the Hall angle seems to follow the T^2 dependence in a wide doping range, which might be related to the hole pockets, while for the overdoped sample it evolves to electron pockets. Recently, we found that the temperature dependence of Hall angle for the n-type superconductor actually has a T^4 form.

Besides, the resistivity and the Hall angle have obviously different power-law temperature dependences, which indicate that there is possibly separation of relaxation times. The two relaxation times scenario was put forward by P. W. Anderson. Its main idea was that for each momentum k, there are two different relaxation times on the Fermi surface. The details can

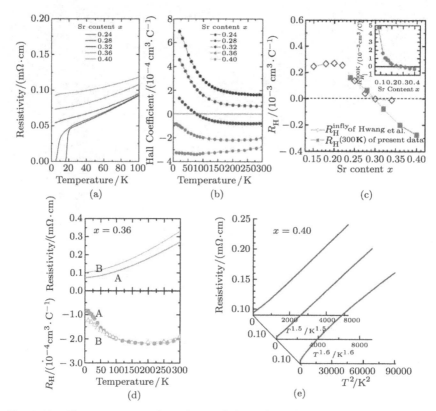

Fig. 9.10. The temperature dependence of the Hall coefficient of heavily overdoped La$_{2-x}$Sr$_x$CuO$_4$ system. The panels are from Ref. [7].

be seen in P. W. Anderson's paper. Another explanation is the anisotropic relaxation times, that is one τ corresponds to one k and will change when across the Fermi surface. There are many theoretical explanations regarding this speculation.

It is generally known that it was expected to be a Fermi arc before Fermi pocket was proposed. In extremely underdoped samples, only the Fermi arc is related to transport properties. If there is a separation of two relaxation times at the Fermi-arc, two types of power-law behavior should be observed in underdoped samples. Actually, in the experimental results of YBCO reported by Ando's group, the resistivity follows a T^2 behavior. As a result, Fermi arc has only one relaxation time in the extremely underdoped samples. At the same time, the power-law behavior can deviate from T^2

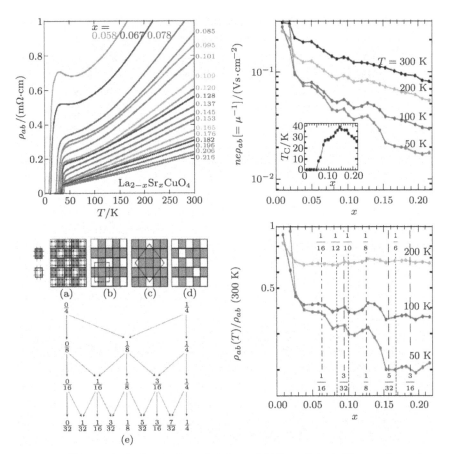

Fig. 9.11. The magic doping in the resistivity and Hall coefficient. The panels are from Ref. [8].

form at a rather high temperature. Therefore, when defining the power-law as T-square or T-linear relationship and describing the variation, it will vary with doping concentration.

Another experimental fact for the Hall coefficient and Hall angle was also completed by Ando's group. They carefully measured the LSCO samples from extremely underdoped to overdoped region. For underdoped samples, the T-square law fits well. But for overdoped samples, there is an obvious deviation from the T-square law (as the doping level is more than 0.15).

Next, we discuss the normalized Hall coefficient. In the case of extremely underdoped samples, there is only a single Fermi arc on the Fermi surface; therefore, its effective carrier density is proportional to the carrier density filling in the Fermi arc, which means the carrier density is equal to the doping level x. But this rule is not established for heavily overdoped samples. It was declared by Ando's group that the normalized behavior disappears when the doping level x is larger than 0.6 in the LSCO system. In fact, the effective carrier density is $1-x$ rather than x, which is easy to explain.

As for overdoped samples, for instance LSCO with $x = 0.25$, the Hall coefficient is independent of the temperature when it rises to above 150 K. Some related works believed that a closed electron-type Fermi surface exists in overdoped samples, which was different from the previous situation discussed. Therefore, the temperature effect was also different, which performs Fermi liquid behavior just like the normal metal.

Here, we make a conclusion about the experimental results of the Hall coefficient: in the extremely underdoped and heavily overdoped samples, Hall coefficient demonstrates a classical behavior. Besides, Hall coefficient has a strong temperature dependence, which reflects the variation of Fermi surface. These results are given in more detail in the discussions in Ando's paper. Then, we discussed the evolution of the Hall coefficient from positive to negative in LSCO samples and attempted to explain it in the two-band model. We also provided perfect explanation for the results of Hall coefficient with Fermi arc and broader band.

9.3. Charge Confinement and c-Axis Transport

We have discussed the experimental results of resistivity and Hall coefficient in the previous section. Now, we discuss the charge confinement and c-axis transport, and will focus on the latter. Considering only the results of band calculation, the transfer matrix elements for both c-axis and ab-plane have no significant difference ($t_{//} \sim 0.25$ eV for ab-plane and $t_\perp \sim 0.1$ eV for c-axis). Speculating on this point, the transport anisotropy should not be large. But the experimental results are exactly the opposite. It was suggested that the transfer matrix for c-axis could be effectively renormalized to a much smaller value. Thus, the conclusion of charge confinement was proposed here.

We give the experimental fact that the anisotropy of resistivity of c-axis and ab-plane in Bi2201 system could reach to about 10^6, which was different from the earlier results of the simplest band calculation. The resistivity of Bi2201 along the c-axis is shown in Fig. 9.12. Figure 9.13 demonstrates the c-axis resistivity and anisotropy for LSCO single crystals. One can find that the anisotropy is also strong and the ratio reached is 10^3.

There are two kinds of explanations for charge confinement in the model mentioned above. The first one is non-Fermi liquid, possibly being spin-charge separation. In this case, the transport may be incoherent as current is applied along c-axis and within ab-plane. The other explanation is based on Fermi liquid theory, in which there exists intralayer scattering and the transition along c-axis is phonon-assisted. In the latter case, both the intraplane and interplane transport are coherent.

In conclusion, the resistivity along c-axis is a result of two possible mechanisms which cause the rather small value of $t_\perp$ matrix elements in c direction. The first comes from charge confinement effect. The other possible explanation is the combined effect of coherent intraplane transport and incoherent interplane transport. As just discussed about the underdoped situation, one point is certain that charge confinement is related to AFM. Spin plays an important role in the transport properties of extremely underdoped samples, which will be discussed later.

9.4. Pseudogap and Resistivity

Now we begin the third part: pseudogap and resistivity. Many authors have mentioned pseudogap in former chapters, and it could also be observed in our transport measurements. In Fig. 9.14, it can be seen that in YBCO there is an evolution from underdoped to optimal doping, where the in-plane resistivity deviates from T-linear behavior below a certain temperature. Actually, as the in-plane resistivity decreases below this temperature, at which point pseudogap is opened, the resistivity along c-axis is enhanced below the same temperature. We can analyze the value of $\rho(T)-\rho(0)$, which shows a strong divergence in the underdoped region. Such a strong divergence can be regarded as a result of opening an energy gap. Due to the existence of a pseudogap, resistivity is enhanced in the c-direction, while it is suppressed within the ab-plane.

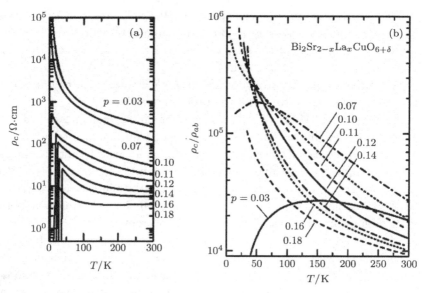

Fig. 9.12. The temperature dependence of ρ_c and ρ_c/ρ_{ab} in Bi2201. The panels are from Ref. [22].

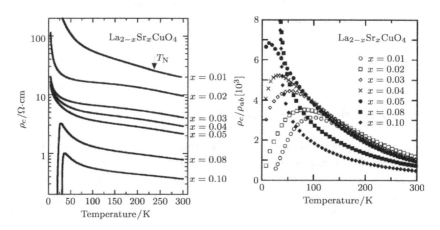

Fig. 9.13. The temperature dependence of ρ_c and ρ_c/ρ_{ab} in LSCO.

For the c-axis enhancement of resistivity with pseudogap opening, many studies have been conducted by T. Xiang and others. We will not attempt to elaborate these. (please see: T. Xiang, PRL, 1996)

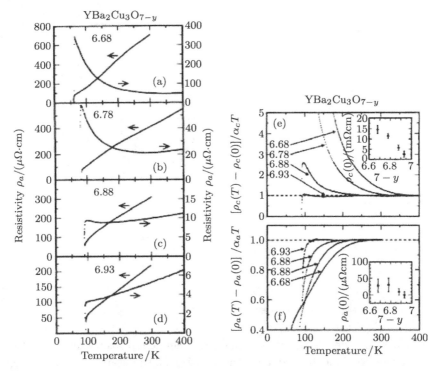

Fig. 9.14. The anomalies of transport properties due to the pseudogap in YBCO system. The panels are from Ref. [9].

Below is an explanation for the hole-type superconductor. Its conductivity along the c-axis has the following expression:

$$\sigma_C(T) = \int dk t_\perp(k)^2 N(k) Z^2(k)/2\Gamma(k),$$

while the transfer matrix element holds the relation:

$$t_\perp = t_0 [\cos(k_x) - \cos(k_y)]^2/4.$$

For the hole-type superconductor, the energy gap opens at $(\pi/2, \pi/2)$ in Fermi surface, so that such a constant along c-axis is suppressed, and then the resistivity along the c-axis is enhanced. Therefore, c-axis resistivity is a perfect probe to detect the pseudogap.

Different from the hole-type, electron-doped cuprates show a metallic resistivity along the c-axis, even when an energy gap is opened, for this reason the transfer matrix element is unaffected at the region in the Fermi

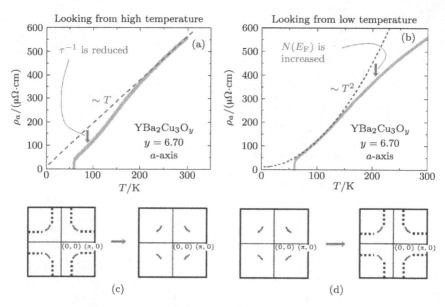

Fig. 9.15. Two scenarios about the ab-plane resistivity.

surface where an energy gap is opened. Pseudogap also affects resistivity in the ab-plane by reducing the density of the states $N(E_F)$, which diminishes τ^{-1} and then causes the decrease of ρ_{ab}. If the scattering comes from phonons or impurities, ρ_{ab} will increase as $N(E_F)$ decreases, which is inconsistent with the experimental facts.

For the underdoped samples, as Fig. 9.15 shows, the resistivity shows a T-linear behaviour above the gap-opening temperature, while a T^2 behavior can perfectly describe the low-temperature resistivity after the gap is opened. Thus, it could be considered in two ways: looking downward from the high-temperature side, the Fermi surface turns into Fermi-arc, and looking upward from low-temperature side, the Fermi-arc changes to a large Fermi surface. In conclusion, the k-dependence of transfer matrix element in c-direction makes it a good probe for detecting the pseudogap.

9.5. Transport Properties under High Magnetic Field

Transport properties of cuprates exhibit abundant anomalies under high magnetic field. The pioneering work was performed in the LSCO system

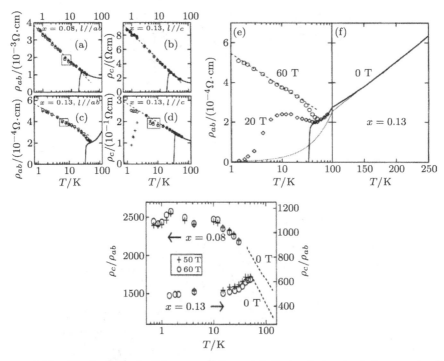

Fig. 9.16. The logarithmic divergence of the resistivity in ab-plane and along c-axis of $La_{2-x}Sr_xCuO_4$. The panels are from Ref. [10].

by Ando's group. Under a 60 T pulse magnetic field, an insulating state with a $\lg(1/T)$ temperature dependence of the low-temperature resistivity was revealed after the superconductivity was suppressed, as is shown in Fig. 9.16. The anisotropy between c-direction and in-plane resistivity is almost independent of temperature after the suppression of the superconducting order.

A crossover from insulating to metallic state, as shown in Fig. 9.17, was discovered in a series of following experiments on this material, which were carried out by Ando's group in 1996. In extremely underdoped samples, the temperature dependence of resistivity shows a strongly divergent $\lg(1/T)$ relation after the superconductivity was suppressed. However, a metallic state emerges upon reaching optimal doping between $x = 0.15$ and $x = 0.17$. Therefore, from the phase diagram, it can be concluded that the ground states of underdoped and overdoped region are insulating

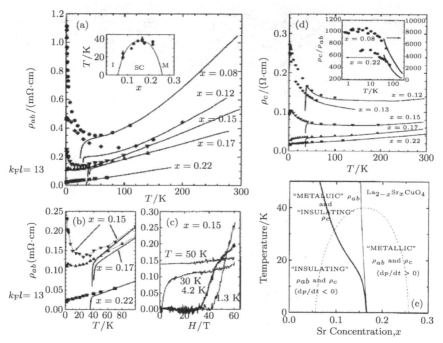

Fig. 9.17. The metal–insulator transition in the resistivity under high magnetic field in $La_{2-x}Sr_xCuO_4$ system. The panels are from Ref. [11].

and metallic, respectively. These earlier thoughts may not be totally correct. Subsequently, two different groups came up the neutron diffraction results showing that a spin density wave (SDW) or charge density wave (CDW) order was induced by a magnetic field. As a consequence, the low-temperature resistivity $\lg(1/T)$ behavior may be related to the magnetic or charge order, which is induced by an external magnetic field. Therefore, the previous localization is related to a field-induced magnetic order.

A conclusion can be made about the localization: typically, if the Fermi wave vector (k_F) multiplied by the free path length is equal to 0.1, it should be a rather bad metal or, in others words, an insulating state, but actually it behaves as an abnormal metal. Correspondingly, it should be in the normal metal state when the product is approximately equal to 10. However, the field-induced spin ordering may play an important role in this material, as will be discussed later. The insulating state referred here should also be induced by the magnetic field rather than being an intrinsic ground

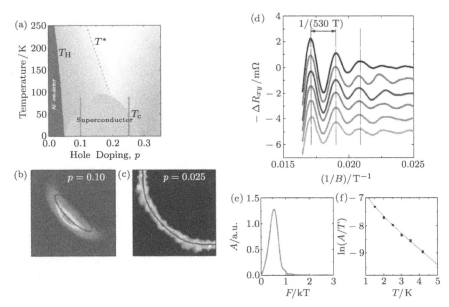

Fig. 9.18. The quantum oscillations and Fermi surface in underdoped HTSC. The panels are from Ref. [14].

state. Certainly, further experiments regarding this material have been carried out, and the neutron diffraction experiments demonstrated that the magnetic field could indeed induce SDW and CDW [12, 13].

In the previous sections, we have discussed the transport property under a high magnetic field. In 2007, L. Taillefer's group had carried out two transport experiments under high field. As shown in Fig. 9.18, they measured the Shubnikov de Haas effect under a field of approximately 50 T. They found that the previous concept of Fermi arc might be wrong since oscillation was observed. Such a quantum oscillation should be understood by using the Fermi pocket, instead of the Fermi arc. The experimental results were accurate, but the explanation may still be controversial. For instance, questions such as whether the Fermi surface would be modified by such a strong magnetic field could be presented. It might be Fermi arc under zero magnetic field, but the high field changes the Fermi surface topology, and as a result this possibility could not be ruled out. A further discussion on the comprehension of those experimental results is necessary.

In the same year, another article from L. Taillefer's group, which was published on *Nature*, reported a result on the electron pocket on the basis of observing negative Hall resistance at high magnetic field below a certain temperature, as shown in Fig. 9.19. They measured the Hall resistance

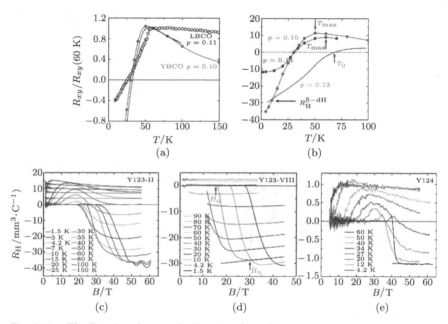

Fig. 9.19. The Fermi pockets in the Fermi surface of the hole-doped cuprate superconductor (YBCO). The panels are from Ref. [15].

in a high magnetic field (45–55 T). As is well known, many groups had already reported the Hall results of such materials in a low field, but they never discovered a negative Hall coefficient. However, a negative Hall coefficient was revealed under high magnetic field. In LBCO, a positive low-temperature Hall coefficient was discovered in previous experiments, even at temperatures which are close to T_C, but the intensity of the magnetic field of those experiments was only several to a dozen Tesla. There are several advantages when the measurement is done in high magnetic field. For example, the high field could suppress the superconductivity. Besides, a CDW order (which may be induced by magnetic field, for the samples are underdoped with $x = 0.1$) could be observed in high magnetic field. At the same time, one can find that the electron pocket reported by L. Taillefer's group is in good accordance with the earlier results of quantum oscillation experiments. However, we should pay attention to whether the electron pocket is actually induced by the magnetic field. The answer is especially complicated. Perhaps, it is not enough to determine the existence of an

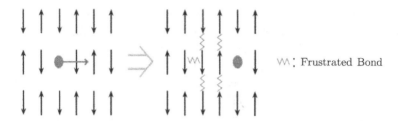

Fig. 9.20. Antiferromagnetic frustrated bond in single hole motion.

electron pocket only from the sign of the Hall coefficient, and more theoretical analysis is required for understanding this phenomenon. However, as a transport result measured in high magnetic field, it is worth our attention.

9.6. The Transport Properties of the Underdoped Samples

In the final section, we discuss the transport properties of the underdoped samples. In the underdoped case, the spin ordering still exists. For a single hole doping, there should be dynamic frustration: when the doped hole moves in the system, if the spin structure on the periphery of the hole changes completely, as shown in Fig. 9.20, long range ordering is still expected to exist, whereas there could be large interactions, which is called the dynamic frustration.

Figure 9.21 shows the experimental results reported by Ando's group with regard to the abnormal metallic transport behavior. Detailed doping research had been carried out on LSCO samples. It shows a perfect metallic behavior with $x = 0.01$, and no anomaly appears at the Néel temperature. Afterward, Ando carried out a series of doping experiments, and a result of $k_F l = 0.1$ was estimated in samples with $x = 0.01$, which corresponded to an insulating state within the classical theory of Ioffe–Regel limit. However, the experimental results display a good metallic behavior. Subsequently, Ando *et al.* measured the Hall coefficient of the same samples, and the results also showed a metallic behavior rather than an insulating or semiconducting behavior. The temperature dependence of mobility, as shown in Fig. 9.22, was then measured by the same authors. It is interesting that the relationship between mobility and carrier density is in good accordance with that of AFM correlation length and carrier density. It seems strange that the metallic state appears when there is a strong AFM correlation and a small $k_F l$. One possible explanation is the phase separation, which was suggested to have a strong effect in these samples, and a scenario called nematicity was proposed. As shown in Fig. 9.23, because of phase separation, electrons could always find one channel which was metallic with the neighboring area as AFM insulating state. This means the average value of $k_F l$ through the whole region is 0.1, but in some special regions the $k_F l$ exceeds the M-I limit and a metallic behavior appears, while other areas remain as an AFM insulating state. Thus, a satisfactory explanation for the abnormal metallic behavior based on the phase separation could be obtained.

Furthermore, for the hole-type superconductors, the CuO_2 planes have a strong distortion as shown in Fig. 9.24. For the octahedra or tetrahedra, if there exists a relative tilting, the spin will cant, consistent with the buckling of CuO_2 plane. Then, the AFM spin arrangement in-plane deviates from strict 180 degree. The overall magnetic order is AFM, but there will be an excess component along the c-direction. This component arranges in an AFM order between the layers under zero magnetic field, but if a large magnetic field is applied along the c-direction, this AFM order can be broken and a weak ferromagnetic state is induced in the c-direction.

All the issues discussed above should be taken into consideration when studying the properties of underdoped hole-type superconductors under applied magnetic field, especially in the c-direction. Figure 9.25 presents the results reported by Ando *et al.* When applying a magnetic field along the c-axis, the magnetic structure changes from AFM to a weak FM along c-direction, and the resistivity simultaneously decreases by 50%. Therefore, it should be carefully analyzed when discussing the transport data in a magnetic field along c-axis, such as the resistivity decreasing. The variation in susceptibility also proved the metamagnetic transformation from AFM to FM in applied field $H//c$. The data presented here were measured in underdoped samples, and for all the hole-type samples which have AFM order, the situations are also the same.

But, the situation is different for the n-type superconductors. Since the CuO_2 plane is flat with Cu–O–Cu angle equal to $180°$ in these materials, there is no spin component along c-direction. Thus, in extremely underdoped samples, we can discuss the responses in c-direction resistivity caused by the transformation of spin reorientation. Figure 9.26 shows the data measured under zero field and at 14 T for heavily underdoped NCCO crystal. There is a low-temperature resistivity anomaly between 70–30 K,

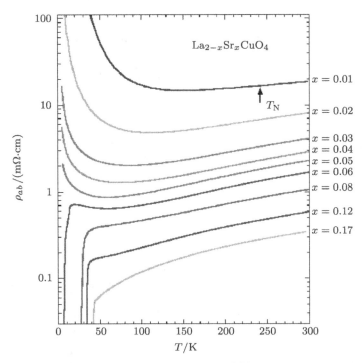

Fig. 9.21. The abnormal metallic resistivity in LSCO system. The panel is from Ref. [16].

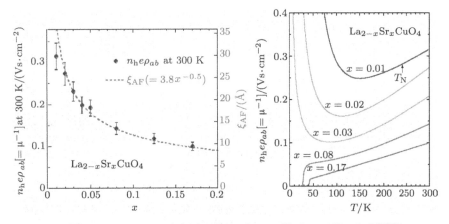

Fig. 9.22. The temperature and doping dependence of hole mobility in LSCO system. The panels are from Ref. [16].

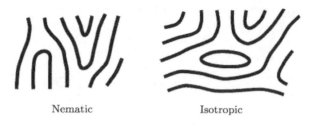

Fig. 9.23. The schematic diagram of nematic stripe state [17].

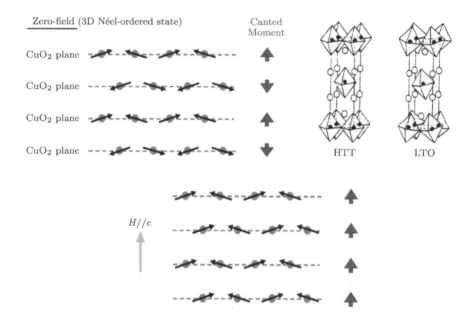

Fig. 9.24. The schematic diagram of weak ferromagnetic transition.

corresponding to a phase transition. Prof. P. C. Dai from ORNL and his collaborators have carried out work in this field. They revealed that for NCCO system, there is an AFM ordering taking place at 270 K and two spin reorientation transitions at 70 K and 30 K, respectively. Those results indicated that spin reorientation has an obvious effect on transport properties. In underdoped samples, the spin–charge coupling is rather strong.

Figure 9.27 shows the magnetoresistance data of NCCO at 20 K and 40 K published by our group. We found that at 40 K the behaviors are completely different when magnetic field is applied in the Cu–Cu direction

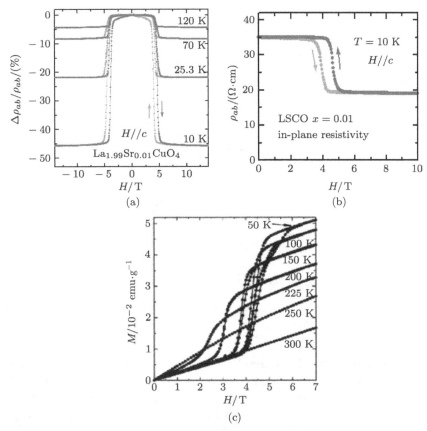

Fig. 9.25. The responses of the weak ferromagnetic transition in magnetoresistance and susceptibility. The panel is from Ref. [18].

and the Cu–O–Cu direction. That is, the transport properties are related to the orientation of the applied magnetic field.

The magnetoresistance data taken in heavily underdoped NCCO crystals at 5 K are summarized in Fig. 9.28. When a magnetic field is applied in the Cu–Cu direction, the magnetoresistance in c-direction basically remains unchanged, and the spin configuration is shown in Figs. 9.28 and 9.29. But if the field is applied in the Cu–O–Cu direction, the c-axis resistivity will change by about 100% (for heavily doped samples the variation can be as large as 200%). The angle dependence of in-plane and out-of-plane resistivity have been discussed thus far. At a lower temperature, the c-axis

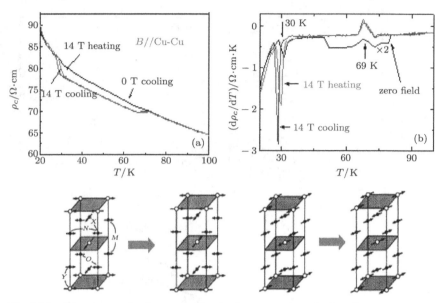

Fig. 9.26. The strong spin-charge coupling in the extremely underdoped n-type cuprates Ref. [19].

resistivity can change by an order of magnitude by just rotating the magnetic field while all of the other parameters remain unchanged. In this case, the electron spins have a strong effect on transport properties, and it is related to the spin–charge coupling. In the extremely underdoped, we also discovered this four-fold symmetry in spin susceptibility, which is perfectly coincidental with the four-fold behavior of magnetic-field-angle dependence of magneto- resistance.

Figure 9.30 shows the results from a work done in collaboration between P. C. Dai and Ando *et al.* The extremely underdoped $Pr_{1.29}La_{0.7}Ce_{0.01}CuO_4$ sample is non-superconducting. When the magnetic field is applied in the Cu-Cu direction and the Cu–O–Cu direction, completely different magnetoresistance behaviors were observed. Magnetoresistance for fields applied along Cu–Cu direction will saturate and even show a small decrease of the magnetoresistance above a not too high magnetic field, whereas a monotonical increase can be observed for magnetoresistance for the field applied along the Cu–O–Cu direction.

Work on the samples offered by P. C. Dai has been carried out in our group. However, those samples with a T_C of approximately 27 K that

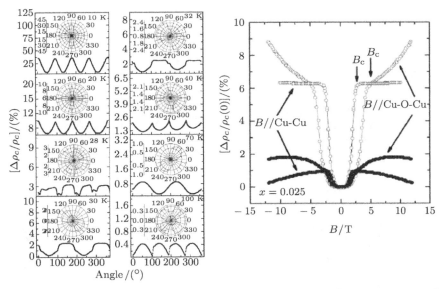

Fig. 9.27. The angle dependence of the magnetoresistance at a fixed temperature of $Nd_{1.975}Ce_{0.025}CuO_4$ Ref. [19].

had been achieved by P. C. Dai's group, do not show superconductivity before annealing, as shown in Fig. 9.31. We have measured the magnetoresistance of these unannealed samples with the applied magnetic field along both the Cu–Cu and the Cu–O–Cu direction. Similar behaviors as the annealed superconducting samples are observed. Although the samples above $x = 0.01$ and 0.12 are also both non-superconducting, the spin behaviors are completely different from the data discussed here. We have also carried out the angle dependence measurement, and found a weaker oscillation, which suggests a relatively weak spin correlation. The behavior is consistent with the earlier results.

There is a linear relationship between the c-axis and in-plane magnetoresistance. Thus, it is not appropriate to ignore the spin correlation in the underdoped samples when discussing transport properties. The detailed information of the correlation varies with the type of materials. As for hole-type superconductors, considering the existence of a weak ferromagnetic, which has a major influence on transport, the resistivity measured under magnetic field would show a decrease of approximately 50%.

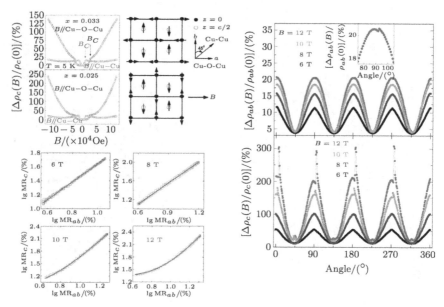

Fig. 9.28. The angle dependence of the low-temperature magnetoresistance of $Nd_{1.975}Ce_{0.025}CuO_4$. The panels are from Ref. [20].

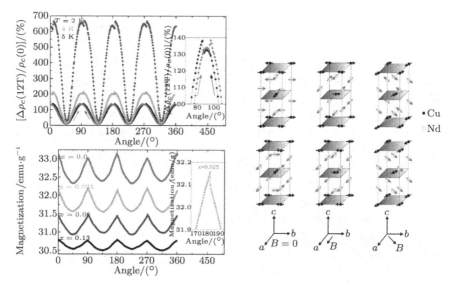

Fig. 9.29. The giant anisotropic magnetoresistance and the "spin valve" effect in $Nd_{1.975}Ce_{0.025}CuO_4$. The panels are from Ref. [20].

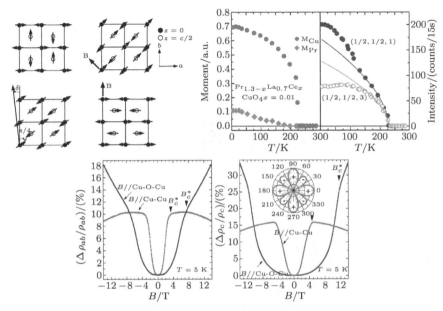

Fig. 9.30. The magnetic structure, the magnetoresistance at a fixed temperature of different angles and the neutron scattering result of $Pr_{1.29}La_{0.7}Ce_{0.01}CuO_4$. The panels are from Ref. [21].

References

[1] S. Martin et al., Phys. Rev B **41**, 846 (1990).
[2] T. R. Chien et al., Phys. Rev. B **43**, 6242 (1991).
[3] T. R. Chien et al., Phys. Rev. Lett. **67**, 2088 (1991).
[4] B. Wuyts et al., Phys. Rev. B **47**, 5512 (1993).
[5] C. H. Wang et al., Phys. Rev. B **72**, 132506 (2005).
[6] F. F. Balakirev et al., Nature **424**, 912 (2003).
[7] T. Tsukada et al., Phys. Rev. B **74**, 134508 (2006).
[8] S. Komiya et al., Phys. Rev. Lett. **94**, 207004 (2005).
[9] K. Takennaka et al., Phys. Rev. B **50**, 6534 (1994).
[10] Y. Ando et al., Phys. Rev. Lett. **75**, 4662 (1995).
[11] G. S. Boebinger et al., Phys. Rev. Lett. **77**, 5417 (1996).
[12] B. Lake et al., Nature **415**, 299 (2002).
[13] J. E. Hoffman et al., Science **295**, 466 (2002).
[14] N. Doiron-Leyraud et al., Nature **447**, 565 (2007).
[15] D. LeBoeuf et al., Nature **450**, 533 (2007).
[16] Y. Ando et al., Phys. Rev. Lett. **87**, 017001 (2001).
[17] S. A. Kivelson et al., Nature **393**, 550 (1998).
[18] T. Thio et al., Phys. Rev. B **38**, 905 (1988).

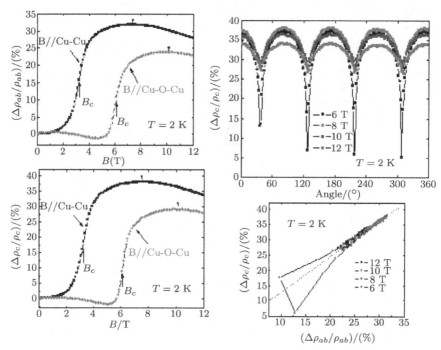

Fig. 9.31. The angle dependence of the magnetoresistance of non-superconducting $Pr_{0.88}La_{1.0}Ce_{0.12}CuO_4$.

[19] X. H. Chen et al., Phys. Rev. B **72**, 64517 (2005).
[20] T. Wu et al., J. Phys.: Conden Matt. **20**, 275226 (2008).
[21] A. N. Lavrov et al., Phys. Rev. Lett. **92**, 227003 (2004).
[22] S. Ono et al., Phys. Rev. B **67**, 104512 (2003).
[23] Seiki Komiya et al., Phys. Rev. B **65**, 214535 (2002).

10
Nernst Effect and Phase Fluctuation Picture of High-T_C Superconductors

Zhu-An Xu

Department of Physics, Zhejiang University, Hangzhou 310027, China

10.1. Introduction to Nernst Effect

Metals or semiconductor materials usually present a thermoelectric effect when there is a temperature gradient across the sample. An electric field is induced by the temperature gradient, and this effect is known as the Seebeck effect; the ratio of electric field and thermal gradient is often called the thermopower. Nernst effect is a corresponding magnetic thermoelectric effect. As shown in Fig. 10.1, the generation of a transverse electric field (y direction) can be observed, when the sample is applied a longitudinal thermal gradient (x direction) in the presence of a finite vertical magnetic field (z direction). This effect is called Nernst effect, and the Nernst coefficient ν is defined as:

$$\nu \equiv \frac{1}{B_z} \frac{E_y}{-(\nabla T_x)}. \tag{10.1}$$

We also define the Nernst signal as $e_y(H,T) = E_y/|\nabla T|$. Here, B is the magnetic induction intensity of external field, E_y is the transverse electric field (y direction), ∇T is the thermal gradient (x direction).

The external magnetic field generates Lorentz force on the charge carriers driven by thermal gradient, resulting in transverse deflection (y direction) of carriers; this transverse current is called Peltier current $\alpha_{yx}(-\nabla T)$. On the other hand, the thermoelectric field E_x and the external magnetic field H_z will produce another current called Hall current $\sigma_{yx} E_x$. Here, α_{yx}

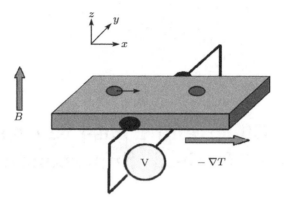

Fig. 10.1. Nernst effect.

is the Peltier conductivity and $\sigma_H = \sigma_{xy}$ is the Hall conductivity. Nernst signal originates from the difference between the two currents. However, for the usual single-band metal these two contributions are almost equal with opposite sign, and therefore the Nernst signal is usually very small.

For the layered structure high-T_C cuprates, electrical conductivity and other physical parameters in the ab plane can usually be considered to be isotropic, that is, the electrical conductivity $\sigma_{xx} = \sigma_{yy} = \sigma$, and Peltier conductivity meets $\alpha_{xx} = \alpha_{yy} = \alpha$. In the Nernst effect measurements, based on the boundary conditions, the absence of electric current loops makes the current zero in both x and y direction. It can be obtained from the definition of electric conductivity and Peltier conductivity α that the current density in x and y satisfies the following equations:

$$J_x = \sigma E_x - \sigma_H E_y + \alpha(-\partial_x T) = 0, \qquad (10.2a)$$

$$J_y = \sigma_H E_x + \sigma E_y + \alpha_{yx}(-\partial_x T) = 0. \qquad (10.2b)$$

Here, $\sigma_H = \sigma_{yx}$ is the Hall conductivity, and α, α_{yx} are diagonal and non-diagonal components of Peltier conductivity. From Eq. (10.2b), we obtain:

$$E_y = \frac{\alpha_{yx}\partial_x T - \sigma_H E_x}{\sigma}. \qquad (10.3)$$

Since the two items in Eq. (10.3) cancel each other, the Nernst effect of usual carriers is very small. Ignoring the small term $\sigma_H E_y$, the relation between Seebeck coefficient S and Peltier conductivity α is:

$$S = \frac{E_x}{-\partial_x T} = \frac{\alpha}{\sigma}. \tag{10.4}$$

Substituting Eq. (10.4) into Eq. (10.3), finally we get the Nernst coefficient:

$$\nu = \frac{E_y}{-(\partial_x T)B} = \left(\frac{\alpha_{xy}}{\sigma} - S\tan\theta\right)\frac{1}{B}, \tag{10.5}$$

where $\tan\theta = \sigma_H/\sigma$ is the Hall angle. For the usual single-band metal, because the Hall current and Peltier current cancel each other, Nernst coefficient is very small. Sondheimer proved in 1948 that if the Hall angle θ does not depend on energy, then these two parts can cancel each other completely [1]. This is called Sondheimer cancellation. For normal state of high-T_C cuparates, the magnitude of the Nernst coefficient ν is about 10^{-8} V/KT, while $S\tan\theta/B$ is about 10^{-6} to 10^{-7} V/KT [2], the Nernst coefficient is very small. But for the multi-band metals with opposite sign carriers, the Nernst coefficient could be very large [3].

For the mixed state of type-II superconductors, magnetic field can penetrate into the sample in the form of flux vortex. As shown in Fig. 10.2, the depinned vortices move toward to the cooler end driven by the thermal driving force $F_{th} = -S_\phi \nabla_x T$, where S_ϕ is the transport entropy of vortex per unit length. The vortex motion generates a transverse electric field

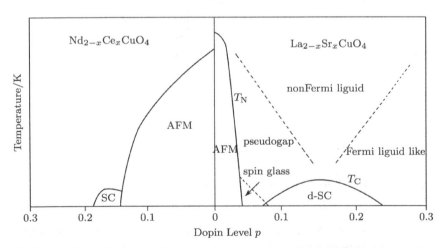

Fig. 10.2. Phase diagrams of electron-doped and hole-doped high-T_C superconductors (electron-doped superconductor $Nd_{2-x}Ce_xCuO_4$ and hole-doped superconductor $La_{2-x}Sr_xCuO_4$).

$E = B \times v$, which is perpendicular to both the vortex velocity v and B, called Josephson electric field [4]. Generally, the appearance of a transverse E_y by the motion of flux vortex (x direction) in the presence of a thermal gradient and magnetic field is known as the vortex Nernst effect. However, this E_y is not the same as the one in usual Nernst effect. The vortex Nernst signal is proportional to the number of vortices and its velocity v, much larger than the Nernst signal of usual charge carriers. The large vortex Nernst effect [7, 8] has been observed in both conventional type-II [5, 6] and high-T_C superconductors. Thus, the Nernst effect provides a highly sensitive probe for detecting vortices. Generally speaking, when $T > T_C$ the magnetic flux vortex will disappear with the external field penetrating sample uniformly, and the transition between superconductor and normal metal will make the vortex Nernst effect disappear; only the normal charge current contributes to the Nernst effect. However, for high-T_C cuprates, the large vortex Nernst signal can be detected in a wide temperature range above T_C, which is different from conventional superconductor. Though the Meissner effect is destroyed by strong phase fluctuation above T_C, there are still fluctuating vortices, which can contribute to the Nernst signal [2].

The Nernst effect measurement is nearly same as the thermopower measurement. As a temperature gradient ∇T is produced by a resistance heater at one end of the sample, a longitudinal voltage appears which corresponds to the thermopower. If a perpendicular magnetic field H (z direction) is also applied, we can obtain the transverse Nernst voltage and the Nernst coefficient can be calculated. The Nernst signal was measured at positive and negative field polarities, and the difference of the two polarities was taken to remove any thermopower contributions due to the electrode asymmetry.

10.2. High-T_C Superconductor Phase Diagram and Pseudogap State

10.2.1. *Electronic Phase Diagram*

The very short coherence length (about 1 nm), extremely low carrier concentration and strong anisotropy make the high-temperature superconducting cuprates very different from the conventional superconductors, and its mechanism could not be explained by the BCS theory. In particular, due to the strong correlation effect between Cu 3d electrons, the Fermi liquid theory cannot describe the normal state properties of the high-T_C superconductors. Furthermore, antiferromagnetic fluctuations, charge density wave and/or stripe phase are competing or coexisting with the superconducting order in high-T_C superconductors. In early 2000, Orenstein and Millis [9] pointed out that the spin and charge non-homogeneous nature, i.e. stripes state; low-temperature properties of superconducting state; phase coherent and the origin of pseudogap; Fermi surface and its anisotropy in non-superconducting or normal state will be the four key issues in high-temperature superconducting.

High-T_C cuprates have perovskite or derivative perovskite structure, and the tetragonal CuO_2 plane is essential to superconductivity [10]. Its parent compound La_2CuO_4 is an antiferromagnetic insulator. Through inducing the charge carriers into the CuO_2 plane by chemical doping, the system changes from an antiferromagnetic insulator into a metal and superconducting state. Therefore, understanding the phase diagram of high-T_C superconductors helps to better understand the mechanism of high-T_C cuprates.

Figure 10.2 shows the phase diagrams of the typical electron-doped high-T_C superconductor $Nd_{2-x}Ce_xCuO_4$ (NCCO) and hole-doped superconductor $La_{2-x}Sr_xCuO_4$ (LSCO). The x-axis (p) represents the induced carrier concentration by doping (defined as doping concentration), and the y-axis is temperature. Now, we will take LSCO as an example to introduce the hole-doped high-T_C cuprate phase diagram. For $La_{2-x}Sr_xCuO_4$, the parent compound La_2CuO_4 ($p = 0$) shows three-dimensional long-range antiferromagnetic order at $T_N = 300$ K, which means the parent La_2CuO_4 is a Mott insulator below Neel temperature T_N. Because of Cu $3d^9$, each copper atom can provides an unpaired electron in the CuO_2 plane; therefore, the parent compound should be metallic based on the band theory. Actually, the strong short-range Coulomb interaction between d-electrons localizes the electrons and makes the parent a Mott insulator. With increasing hole doping (p, for in $La_{2-x}Sr_xCuO_4$, $p = x$), the Neel temperature T_N decreases rapidly. At about $p = 0.02$, long-range antiferromagnetic order is destroyed, while the short-range antiferromagnetic fluctuations still exist and the system is turned into spin-glass state. With further doping, the system becomes metallic and superconducting at $p = 0.05$. The optimal doping level is about $p = 0.15$ at which T_C reaches a maximum. When $p > 0.15$, T_C drops as p increases, at $p = 0.25$, T_C drops to zero and the system becomes Fermiliquid-like.

The bell-shaped T_C-p curve means an important empirical relationship between T_C and the doping level exists. Below the curve, it is the superconducting state. Now scientists have reached a consensus that the hole-doped high-T_C cuprates have d-wave pairing symmetry. The doping level where T_C reaches its maximum is defined as the optimal doping (p_{opt}). For $p < p_{\text{opt}}$, it is the underdoped regime and for $p > p_{\text{opt}}$ the overdoped regime. In the underdoped regime, there is another curve called T^* above T_C corresponding to the opening of the pseudogap. However, the origin of the pseudogap is still under debate.

For electron-doped cuprates $Re_{2-x}Ce_xCuO_4$ (Re = Nd, Pr and Sm), we can induce the electron-type charge carriers by partially substituting Ce^{4+} for Ln^{3+} ions. However, many features of the electron-doped high-T_C cuprates are inconsistent with the hole-type counterparts, and even the superconducting pairing symmetry is debatable.

10.2.2. *Pseudogap*

The pseudogap was first revealed in the NMR experiment of underdoped $YBa_2Cu_3O_{7-\delta}$ (YBCO). In 1989, a group in Bell Lab found that the spin-lattice relaxation rate $1/T_1T$ of underdoped YBCO is gradually suppressed at temperatures far above T_C in the NMR experiments [11]. For the optimally doped system, the spin–lattice relaxation rate first increased with temperature decrease and dropped rapidly at T_C. But for the underdoped sample, the relaxation rate decreased far above T_C. Similar to spin-lattice relaxation rate, the Knight shift K, which is only weakly temperature dependent in normal state and drops rapidly in superconducting state, also decreases far above T_C in underdoped cuprates. This energy gap was first named as spin gap since the NMR experiment probes the spin excitation spectrum. Following experiments found that the energy gap also opens in the charge channel, and it was then called pseudogap (PG). The pseudogap is an incomplete gap. In the pseudogap region $T_C < T < T^*$, where T^* is defined as the temperature of pseudogap opening, some areas of Fermi surface become gapped (this part does not contribute to electronic conductivity) while the other Fermi surface is still preserved, and therefore the DOS at the Fermi level is partly suppressed. Now, pseudogap is widely recognized as a common feature in underdoped high T_C cuperates. A further description of the pseudogap can be found in review papers [12].

Despite the diverse T^* value obtained in different experiments, two distinct T^* scales are found in the underdoped cuprates above T_C, one is believed to be related to the energy gap on the freedom degree of spin, and the other on the freedom degree of charge. However, researchers argue that one may be associated to the stripe order and the other spin energy gap. So far, two consensus have been reached on T^*: (i) In the underdoped region, T^* is much higher than T_C, and it drops monotonically with increasing hole doping level and gradually reaches close to T_C. (ii) Similar to superconducting gap, the pseudogap has d-wave symmetry and is anisotropic with node or nodal arc.

Currently, there are a variety of theoretical works to explain the pseudogap. According to the relations between pseudogap and superconductivity, the theoretical scenarios can be divided into two categories. In the first category, the pseudogap phase is a precursor phase of superconductivity. For example, in the phase fluctuation model proposed by Emery and Kivelson [13], the Cooper pair should form at T^*, a temperature much higher than T_C. But its long-range phase coherence cannot be established until T_C due to the vortex excitations. The zero resistivity and Meissner effect can only be reached when the long-range phase coherence between Cooper pairs is established. Considering the low superfluid density and the quasi-two dimensionality of high T_C superconductor, Emery and Kivelson proposed that the phase fluctuation of superconducting order parameter is particularly essential for superconductivity and the long-range coherence determines the superconducting transition temperature T_C. However, the theoretical models of the second category argue that there is little relationship between the pseudogap and superconductivity, and it is claimed that the pseuodgap may be related to the competing ground state or fluctuations of some other orders such as AFM order, charge density wave (CDW), stripes phase, etc. These orders can compete or coexist with the superconducting ground state. In the second category, a quantum critical point (QCP) is proposed at a certain doping level where the pseudogap disappears [14].

These two categories of theories predict similar doping dependence of T^* in the underdoped regime where T^* is much higher than T_C, drops monotonically and gradually approaches T_C with increasing doping level. However, the different theories predict different behavior in the overdoped regime. If the pseudogap is a precursor to the superconducting state, as proposed in the theory of Emery and Kivelson [13], the two curves of T and T^* should decrease monotonically with increasing doping level and finally merge together. Meanwhile, other theories [14] of the second category argue that the T^* curve will drop into the superconducting state (i.e. $T^* < T_\mathrm{C}$) and reach zero at a critical doping level (i.e. $p \sim 0.19$), at which a quantum critical point is proposed.

Emery and Kivelson [13] proposed that the superconducting energy gap Δ and phase stiffness are two important energy scales to determine the superconducting order, as the former represents the ability to form Cooper pair bound state and the latter reflects the capacity of superconducting state to carry superfluid. For conventional superconductors, the superfluid density is very large, i.e. the phase stiffness is very large. Superconducting transition temperature depends only on the temperature of the Copper pair formation ($T_\mathrm{C}^{\mathrm{MF}} \sim \Delta$). The phase coherence between Cooper pairs is established simultaneously when the Cooper pairs form. However, in the high-T_C superconductors, the superfluid density is very low, so is the corresponding phase stiffness. For 2D superconducting systems, Emery and Kivelson [13] show that superconducting phase coherence disappears at a temperature $T_\mathrm{C}^\theta = \hbar^2 n_\mathrm{s}(0) d / m^*$, where d is the distance between conducting layers, $n_\mathrm{s}(0)$ is the superfluid density at $T = 0$, m^* is the effective mass of Cooper pairs. T_C^θ represents the phase stiffness quantitatively. In the underdoped regime, due to the low superfluid density, T_C^θ is less than the mean field transition temperature $T_\mathrm{C}^{\mathrm{MF}}$, and thus the actual superconducting transition temperature T_C depends on T_C^θ. It is believed that in the pseudogap state, that is, $T < T^*$, the localized Cooper pairs start to form; however, the long-range phase coherence is not established until $T < T_\mathrm{C}^\theta$ due to the vortex excitations. Obviously $T_\mathrm{C} = T_\mathrm{C}^\theta$ for the underdoped regime, that is, T_C is proportional to the doping level (p). For the overdoped regime, because T_C^θ is higher than the mean-field transition temperature $T_\mathrm{C}^{\mathrm{MF}}$, T_C depends on $T_\mathrm{C}^{\mathrm{MF}}$. $T_\mathrm{C}^{\mathrm{MF}}$ drops monotonically with the doping level, so does T_C. This picture is consistent with the doping dependence of T_C shown in Fig. 10.2. Uemura's group [15] has used the μSR measurement to probe the superfluid density of high-T_C cuprates, and they have revealed a proportional relationship between T_C and $n_\mathrm{s}(0)/m^*$ in the underdoped regime, so-called Uemura scaling plot. It is an important experimental evidence to support the pre-pairing theory suggested by Emery and Kivelson.

According to the above pre-pairing superconductor theory suggested by Emery and Kivelson, the local dynamic Cooper pairs should exist in the pseudogap state, that is some kind of residual superconductivity, the pseudogap is actually caused by the superconducting pairing. This scenario has aroused great interest in this community, and experimental physicists have tried their best to discover experimental evidences to support the theoretical model. Orenstein's group [16] studied high-frequency (100–600 GHz) conductivity of Bi2212 superconductor and found that there indeed exist short-lived vortex excitations above T_C in underdoped regime. They also indicated that the phase stiffness can still be detected in very high frequency (600 GHz) though the phase coherence time drops rapidly above T_C. Their result confirms that the dynamic vortex excitations indeed exist above T_C, and it is an experimental evidence to support Emery and Kivelson's pre-pairing theory. Nernst effect measurement, which will be introduced below, is another important experiment to confirm the existence of vortex excitation above T_C.

10.3. Nernst Effect of Hole-Doped High-T_C Superconductor

Nernst effect has already been studied in the normal metals, magnetic materials, strongly correlated electron systems and the type-II superconductors. Shortly after the discovery of high-T_C superconductor, the Nernst effect in the mixed state was reported. However, the early reports mainly focus on the Nernst signal caused by flux motion below T_C. Compared to conventional type-II superconductors, the liquid phase of vortices in high-T_C cuprates is very large, and thus large vortex Nernst signals can be observed in very large regime of the mixed state.

10.3.1. Vortex Nernst Signal Above T_C

In early 2000, Ong's group [2] performed Nernst effect measurements on the $La_{2-x}Sr_xCuO_x$ single crystals for the first time, and a surprising result was discovered: the large vortex Nernst signal still exists at the temperature range from 50 K to 100 K above T_C. The Nernst effect in LSCO single crystal with Sr doping level from 0.05 to 0.17 was systematically investigated, [2] and the report attracted great interest in this community.

The field dependence of Nernst signal $E_y/\nabla_x T$ in a typical underdoped LSCO sample ($x = 0.10$) is shown in Fig. 10.3. The numbers beside the curves represent the temperature. The inset shows the temperature dependence of Nernst coefficient. The superconducting critical temperature of this sample is 28 K measured from the resistivity. The non-linear field dependence of the Nernst signal $e_y = E_y/\nabla T$ below T_C is the feature of vortex

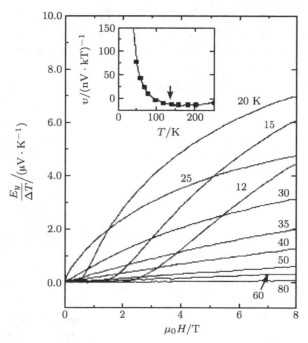

Fig. 10.3. Field dependence of Nernst voltage $E_y/\nabla_x T$ in $La_{2-x}Sr_xCuO_4$ ($x = 0.10$); The inset shows the temperature dependence of Nernst coefficient.

in mixed state. For low magnetic field, since the flux is pinned and cannot move freely, the Nernst voltage is nearly zero; for the field H higher than the melting field H_m, the vortices begin to move and e_y increases dramatically as the number of vortices increases proportionally. But at high magnetic field, the transport entropy of vortex drops as the field further increases; thus, the vortex velocity becomes small and e_y first saturates and then finally decreases according to Eq. (10.6). These two effects compete with each other, accounting for the nonlinear field dependence of e_y which becomes maximum at H^* then gradually drops as $H > H^*$. The inset in Fig. 10.3 also indicates that even in a wide range of $T > T_C$, the Nernst signal is still large. For the normal state far above T_{C0}, the background of Nernst signal changes little with temperature, and the Nernst signal of vortex-like excitation appears around 140 K, it is enhanced gradually with cooling temperature and exhibits a sharp increase as T is close to T_C.

The anomalous large Nernst signal above T_C suggests that vortex-like excitations exist in the range from T_{C0} to 140 K. The temperature at which

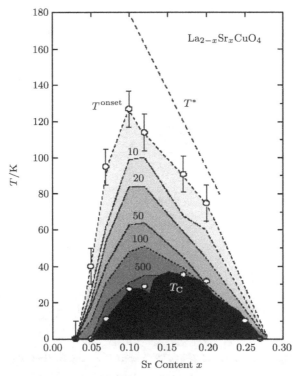

Fig. 10.4. Contour plot of residual Nernst signal $v-v_n$ and doping dependence of T^{onset} in LSCO; the numbers denote the value of residual Nernst signal in the unit of nV/KT.

the Nernst coefficient starts to deviate from the normal state background is defined as T_v (also known as T^{onset}), which corresponds to the occurrence of vortex-like excitations. T_v is usually lower than T^*, but proportional to it in a certain rage of doping.

Figure 10.4 shows the Sr doping dependence (x) of T^{onset} in LSCO from underdoped regime to overdoped regime and the contour of residual Nernst signal ($v - v_n$), where v_n is the Nernst coefficient of normal state background. Especially since the Nernst coefficient v can be written as $v = S/B(\tan\theta_\alpha - \tan\theta)$, $S\tan\theta/B$ is usually from the contribution of normal charge carriers, but the other term $\tan\theta_\alpha = \dfrac{\alpha_{xy}}{\alpha}$ (called Peltier current) would consist of two contributions: one from the normal charge carriers (written as α_{xy}^n/α), the other, known as the vortex excitation item (α_{xy}^s/α), i.e. $\tan\theta_\alpha = \alpha_{xy}^n/\alpha + \alpha_{xy}^s/\alpha$. T_v, defined from the temperature at which the vortex excitation item goes to zero.

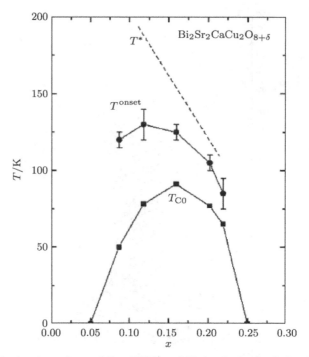

Fig. 10.5. Doping dependence of T_{C0}, T^{onset} and T^* for the $Bi_2Sr_2CaCu_2O_{8+\delta}$ system.

The anomalous vortex Nernst signal above T_C exists not only in LSCO system but is also observed in other hole-type high-temperature superconducting cuprates, such as YBCO [17, 18], Bi2201 [17] and Bi2212 [19]. Thus, people believe that the existence of vortex excitations in a certain temperature range above T_C is a universal feature in all the hole-type cuprate superconductors. Figure 10.5 shows the doping dependence of T_{C0} and T^{onset} and T^* for Bi2212 system. There is a larger vortex Nernst effect in a certain temperature range between T_{C0} and T^*. The result is similar to the case of LSCO.

According to the result of the Nernst effect, as shown in Fig. 10.6, a general phase diagram of hole-doped cuprate superconductors can be established. It can be seen that there is a regime of vortex-like excitation in the pseudogap regime (area B in Fig. 10.6), strongly implying that there is a relationship between the pseudogap phase and the vortex-like excitation. In the theoretical model [15] of superconducting phase fluctuation, the opening of the pseudogap corresponds to the emergence of preformed Cooper pairs,

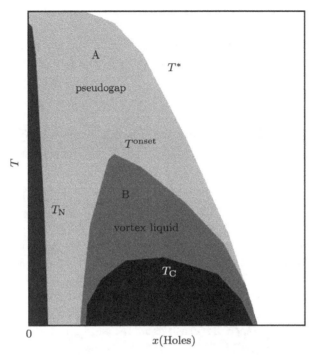

Fig. 10.6. Phase diagram of hole-doped high T_C cuprates. Area A is the pseudogap state; area B is the phase fluctuation region; Area C is the d-wave superconducting state.

that is, the amplitude of the superconducting order parameter is non-zero $|\psi| \neq 0$ in the pseudogap phase. But there is no Meissner effect and zero resistance in the range of $T_C < T < T^*$, because the vortex excitations destroy the phase coherence of Cooper pairs and there are no long-range phase coherence. Since Nernst effect is sensitive to the vortex excitations, it is an excellent way to detect these excitations and provides a strong experimental evidence to support the scenario of superconducting phase fluctuation. Note that the vortex-like excitations do not immediately appear at T^* but in a lower-temperature T^{onset}, which indicates that the pseudogap opening may not result from the pre-formation of Cooper pairs directly, and might just be a precursor phase or even another order which is not related to the pre-formed Cooper pairs such as the spin gap. The Nernst effect measurements can only sketch a general picture of superconducting phase fluctuation, and it is difficult to reveal the intrinsic relation between the pseudogap and superconducting phase fluctuation.

10.3.2. Upper Critical Field H_{C2}

According to time-dependent Ginzburg–Landau theory, the diamagnetic magnetization $|M|$ of mixed state goes to zero linearly with the magnetic field H as H is close to H_{C2}, i.e.:

$$M(T, H) = -[H_{C2}(T) - H]/2.32\kappa^2. \tag{10.6}$$

According to Caroli–Maki's expression [20–22], the entropy S_ϕ of magnetic flux is:

$$S_\phi(T, H) = \frac{\Phi_0}{T} |M(T, H)| L_D(T). \tag{10.7}$$

Here, Φ_0 is the magnetic flux quanta and $L_D(T)$ is a numerical function which decreases gradually with T from 1 to T_{C0}. According to Eqs. (10.6) and (10.7), with the external magnetic field close to H_{C2}, the entropy S_ϕ carried by the magnetic flux linearly decreases to 0, and the thermal dragging force $F_{th} = -S_\phi \nabla T$ on the magnetic flux driven by thermal gradient also drops gradually to 0, so does the vortex Nernst signal e_y. Therefore, the upper critical field H_{C2} can be obtained by the zero position of e_y.

The field dependence of Nernst signal measured in the optimally doped single crystal $Bi_2Sr_{2-y}La_yCuO_6$ ($y = 0.4$) is shown in Fig. 10.7. The high field Nernst effect was performed in the National High Magnetic Field Lab in Florida. The critical temperature of the sample is $T_C = 28$ K. However, as can be seen from Fig. 10.7, the Nernst signal e_y does not fall rapidly to zero even around T_C. The extrapolation of the Nernst signal to zero at 30 K indicates that the upper critical field H_{C2} is much larger than 45 T. The H_{C2} phase line does not drop to zero at T_{C0}, but retains a large non-zero value. The temperature dependence of H_{C2} both in this sample and another overdoped Bi2201 sample ($y = 0.2$, $T_C = 22$ K) is shown in Fig. 10.8. This result is in contrast to the conventional superconductor behavior, and also inconsistent with the H_{C2} values obtained by the usual methods. The H_{C2} obtained by magneto-resistance measurement drops to zero at T_C.

This anomalous "$H_{C2}(T)$" behavior can easily be understood in the scenario of superconducting phase fluctuations theory (SPF). According to the SPF model, the superconducting critical temperature obtained by the usual methods is related to the temperature where long-range phase coherence is established, instead of the real Cooper pair forming temperature. Therefore, the H_{C2} obtained via magneto-resistance is the field where the phase coherence, instead of the Cooper pairs, is destroyed. On the contrary, the

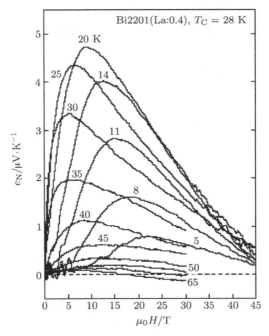

Fig. 10.7. The field dependence of Nernst signal measured in the optimally doped single crystal $Bi_2Sr_{2-y}La_yCuO_6$ ($y = 0.4$).

H_{C2} obtained from the Nernst effect is the real depairing field. Since Cooper pair phase fluctuations appear above T_{C0}, H_{C2} does not drop to zero at T_{C0} until T^{onset} is reached.

The upper critical field obtained from the Nernst effect truly reflects the size of Cooper pair and the strength of the pairing potential. Figure 10.9 shows the doping (x) dependence of H_{C2} in Bi-2201 and Bi-2212 systems determined by the Nernst effect, and the inset shows the doping dependence of the coherence length calculated using the relation $H_{C2} = \Phi_0/2\pi\xi^2$. It can be seen that H_{C2} increases as the doping level x decreases, similar to the x dependence of superconducting gap, indicating the increase of Cooper pair attracting potential as the system is less doped. The obtained ξ_0 is in good agreement with the ARPES [23] and STM results [24, 25].

10.3.3. 2D Nature of Vortex Excitations Above T_C

In 2003, Wen and Xu et al. [26] performed further angle-dependent measurements of Nernst effect to reveal the relationship between Nernst signal and

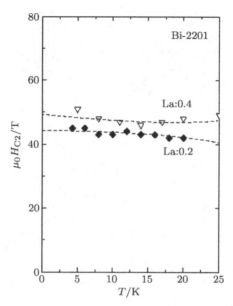

Fig. 10.8. The temperature dependence of H_{C2} obtained from Nernst effect for two $Bi_2Sr_{2-y}La_yCuO_6$ samples (y =0.2, 0.4). T_{C0} is 28 K ($y = 0.4$, optimally doped) and 22 K ($y = 0.2$, overdoped), respectively.

the angle between magnetic field H and the c-axis for the underdoped and optimally doped single-crystal LSCO. Their experiments confirmed that the vortex-like excitations above T_C are strictly in-plane excitations. When the angle between the direction of H and the c-axis is θ, the Nernst electric field $E_y = \alpha|\nabla T|H\cos\theta$. Generally speaking, the coefficient α is θ dependent; therefore, the Nernst electric field is not strictly linear to $H\cos\theta$ unless the system is isotropic or the vortex-like excitations is strictly two dimensional. They discovered that, when $T > T_{C0}$, the Nernst electric field E_y was strictly linear to $H\cos\theta$, although the high T_C superconductor should be strongly anisotropic, which implies that the vortex-like excitations was strictly of a 2D nature. As shown in Fig. 10.10, there were the data of three different angles, $\theta = 0°$ (square), $25°$ (circle), and $50°$ (triangle). It was clear that the in-plane Nernst signals are well scaled together. The inset shows the experiment setup, where a, b and c represent the crystalline axes.

Figure 10.11 shows the temperature dependence of Nernst signal for $H//ab$ and $H//c$ in the underdoped and optimally doped single-crystal $La_{2-x}Sr_xCuO_4$ ($x = 0.0, 0.15$). We found that, when $H//ab$, the Nernst signal induced by Josephson vortex below T_C is significant, while it drops

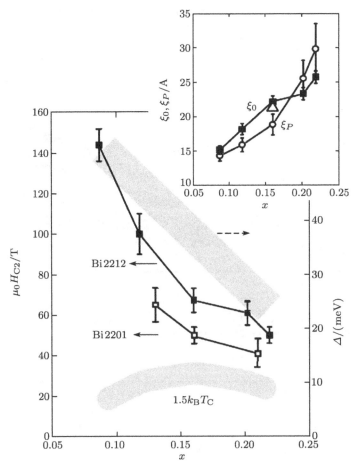

Fig. 10.9. The doping (x) dependence of H_{C2} in Bi2201 and Bi2212 system determined by the Nernst effect, and the inset shows the doping dependence of the coherence length calculated using the relation $H_{C2} = \Phi_0/2\pi\xi^2$.

quickly to the background signal level above T_C. This result, from another point of view, indicates that the vortex-like excitations Nernst signal has 2D characteristic above T_C.

10.3.4. *Superconducting Phase Fluctuations Picture*

Generally speaking, when $T > T_C$, the external field will penetrate the sample uniformly, leading to the disappearance of vortex and the vortex Nernst

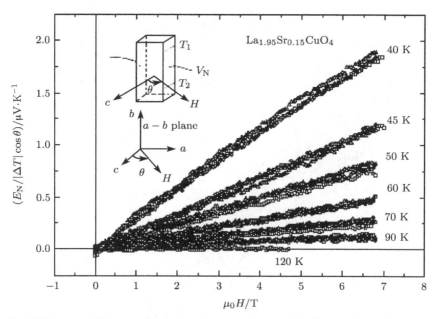

Fig. 10.10. Angle (between H and c axis) dependence of the Nernst signal (here were three angles, $\theta = 0°$ (square), $25°$ (circle) and $50°$ (triangle).)

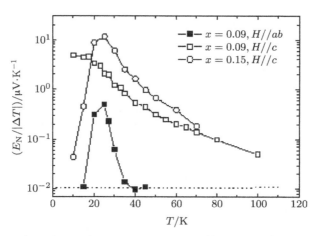

Fig. 10.11. T dependence of Nernst signal when $H//ab$ and $H//c$ in the underdoped and optimally doped single-crystal $La_{2-x}Sr_xCuO_4$ ($x = 0.0, 0.15$).

effect as well. However, the experiments showed that, in cuprate superconductors, the anomalously enhanced Nernst signal persists at temperatures even up to T^{onset}, much higher than T_{C0}, especially in the underdoped region. The origin of the anomalous Nernst signal has attracted more attention. Moreover, the onset temperature T^{onset} at which the Nernst signal drops to background level lies in the pseudogap region ($T_{\text{C0}} < T^{\text{onset}} < T^*$); therefore, it has been proposed that the anomalous Nernst signal has some connections with the pseudogap or there is a kind of interplay between pseudogap state and superconducting state. So far, the theoretical explanations of the anomalous Nernst effect above T_{C} can be classified into three categories: (i) conventional superconducting fluctuations theory; (ii) superconducting phase fluctuations theory (or superconducting pre-pairing model) and (iii) novel excitations derived from Resonance Valence Band (RVB) theory.

10.3.4.1. *Conventional fluctuations theory*

Ussishkin *et al.* considered that the general Gaussian type superconducting fluctuations (amplitude fluctuations) can account sufficiently for the vortex Nernst signal above T_{C} in the overdoped and optimally doped region. In the underdoped region, they suggested that the strong fluctuations are non-Gaussian fluctuations, which led to a decrease in the mean field transition temperature T_{C}^{MF}; therefore, the real superconducting transition temperature T_{C} should be replaced by T_{C}^{MF}. Using the relevant parameters of LSCO, Wang *et al.* found that this theory can fit the experimental Nernst effect results of the optimally doped and overdoped cuprates very well, but it cannot account for the stronger fluctuations of the severely underdoped sample. From the phase diagram of high-T_{C} superconductors, it can be found that T_{C} decreases with decreasing doping level (x) in the underdoped region, but both T^* (at which the pseudogap opens), and T_{v} (at which the vortex Nernst signal appears) increase instead. This doping dependence of T_{v} cannot be understood in the conventional superconducting fluctuation picture. Moreover, according to the theory, due to the longer coherent length, there should be stronger fluctuations in the electron-doped superconductor $Nd_{2-x}Ce_xCuO_4$ (NCCO). Actually, the vortex-like excitation region is much narrower in NCCO, only about 2 K above T_{C} (see the following Section 10.4). Carlson *et al.* predicted that, for hole-doped cuprates, order parameter phase fluctuations will lead to anomalous Nernst signal in a certain temperature region above T_C. At high temperature, both the phase and amplitude fluctuations contribute to the anomalous Nernst signal. In superconducting state, the order parameter can be expressed by the wave function $\Psi = |\Psi|e^{i\theta}$, and it is long-range phase coherent. In conventional superconductors, the amplitude of Cooper pair disappears at T_C, where T_C is the same as the mean field transition temperature T^{MF} in BCS theory. The fluctuations of order parameter Ψ are just the fluctuations of the amplitude $|\Psi|$. The Gaussian type approximation can describe the conventional superconductor fluctuation effect well. In the systems of low dimension and low superfluid density, e.g. high-T_C superconductors, owing to the much smaller phase stiffness, phase fluctuations become very important. Thus, it is not surprising that Gaussian type fluctuations picture does not apply to the high-T_C superconductors any more.

10.3.4.2. *Strong phase fluctuation theory (preformed pair theory)*

Shortly after the discovery of high-T_C superconductors, Baskaran et al. [30], Doniach and Inui et al. [31] noticed that T_C was determined by the disappearance of Cooper pair phase coherence in the underdoped region. Owing to the much lower superfluid density in the high-T_C superconductors, and the quasi-two-dimensional nature of the CuO plane, Emery and Kivelson [13] pointed out the importance of phase fluctuations of superconducting parameter in 1995, and that it is the long-range phase coherence that determines the superconducting transition temperature T_{C0}. When temperature is lower than T^*, superconducting pairs begins to form, and the pseudogap is the same as the superconducting gap. Due to the existence of a large number of vortex excitations, or destruction by thermal fluctuations, the phase of the superconducting order parameter is fluctuating. Only when temperature falls below T_C, the phase coherence can be established between the Cooper pairs, and then the Meissner state appears, and the zero resistance and diamagnetism (Meissner effect) can be observed. T^* corresponds to the temperature of superconducting pair formation, and T_{C0} corresponds to the temperature at which the long-range phase coherence is established. For the temperature range between $T_C < T < T^*$, though the Cooper pairs are not long-range phase coherent, the amplitude of order parameter is non-zero, and short range or dynamic phase correlation may exist. The short-range coherent Cooper pairs may account for the vortex-like excitations detected by Nernst effect. When temperature is raised relatively high, e.g. up to T_v, dynamic phase coherence disappears, and so does the Nernst signal.

10.3.4.3. *Resonance valence bond (RVB) theory*

Based on the spin-charge separation RVB picture, Weng *et al.* proposed another explanation for the residual Nernst signal above T_{C0}. In the phase string theory of a $t - J$ model, thermally excited spinons destroy the phase coherence, leading to a new phase characterized by the free spinon vortex–spontaneous vortex phase in the temperature range of $T_C < T < T_v$, where the characteristic temperature T_v corresponds to holon condensation. The Nernst effect is an intrinsic feature of this spontaneous vortex phase. Below T_C, superconducting phase coherence is established and superconductivity appears, the spinon vortex turns into the usual vortex. In this picture, the spinon vortex is not same as the conventional vortex excitation, and there are no fluctuating Cooper pairs in the pseudogap state. However, the pseudogap state and the superconducting state are closely related — the former is the precursor of the latter.

In summary, the phase fluctuations theory based on the scenario of preformed pairs can account for the vortex-like excitations detected by the Nernst effect measurements reasonably and comprehensively, and this viewpoint has been supported by more and more experiments. As mentioned above, the high-frequency conductivity experiment by Orenstein's group indicated that there are indeed short-lived vortex excitations above T_C. Though the supercurrent cannot exist for $T > T_C$ due to the destruction of the phase coherence by vortex excitations, it displays many features of superconducting state, implying that there is close intrinsic connection between the Nernst signal of vortex-like excitations and superconductivity. The Nernst effect results strongly support the theoretical model of preformed pairs. Recently, Wang *et al.* have measured the magnetic susceptibility above T_{C0} in LSCO using a torque technology with high accuracy and confirmed that there is indeed weak diamagnetism at the temperature range where vortex-like Nernst signal exists, and the diamagnetism strength is scaled well with the Nernst signal. The torque measurement results will be presented in detail below.

10.4. Nernst Effect in Electron-Doped High-T_C Superconductors

The Nernst effect in hole-doped high-temperature superconductors were presented in the previous section. No matter if it is underdoped or overdoped hole type cuprates, the anomalous vortex Nernst signal is detectable in a wide temperature range above T_C. Especially in the underdoped regime, the vortex Nernst signal persists at the temperature 100–150 K above T_C, and this temperature remarkably exceeds the conventional superconducting fluctuation temperature range. In this section, the Nernst effect in the electron-doped high-T_C superconductors will be presented.

The Nernst effect experiments have revealed that there is a significant difference between the electron-doped cuprates and the hole-doped cuprates. In electron type cuprates, the vortex-like excitations is confined to a narrow range above T_C, i.e. the superconducting phase fluctuations exist in a narrow temperature range. Moreover, the normal state Nernst effect is remarkable compared to the hole type cuprates. Reference [19] reported the Nernst effect in $Nd_{2-x}Ce_xCuO_4$ ($x = 0.15$, $T_{C0} = 24.5$ K). Figure 10.12 shows the field dependence of Nernst signal after subtracting the normal-state background signal in NCCO.

As seen from Fig. 10.12, the vortex Nernst signal in electron-doped NCCO was different from the hole doped cuprates. Namely, the vortex Nernst signal disappeares immediately once the temperature exceeded T_{C0} (e.g. the curve at 26 K), indicating that the fluctuation range was only about 2 K. Futhermore, the Nernst signal curves exhibits a very sharp peak below T_{C0}, other than the dome-like shape in the hole-doped cuprates. The phase diagram derived from Nernst effect is shown in Fig. 10.13. The upper critical field is small, only about 10 T, and the H_{C2} curve ends at T_{C0}. When T_{C0} is approached, H_{C2} decreases linearly with increasing temperature. These features should be usually observed in conventional superconductors, and the hole-doped cuprates exhibit quite different features.

Afterward the Nernst effect in another electron-doped cuprate $Pr_{2-x}Ce_xCuO_4$ was reported, and the results are similar to $Nd_{2-x}Ce_xCuO_4$. With the doping level increasing from underdoped regime to overdoped regime, the upper critical field H_{C2} decreases. Even in the underdoped regime, H_{C2} is no more than 8 T, similar to the case of NCCO. For the optimally doped and overdoped PCCO and NCCO, there is no anomalous vortex Nernst signal above T_C; meanwhile, the Nernst signal e_y^N produced by

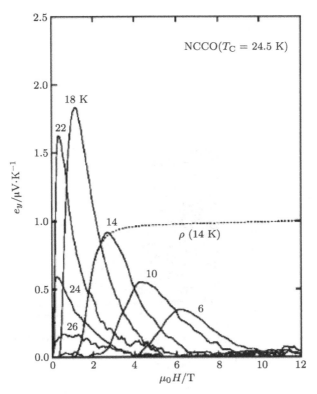

Fig. 10.12. The field dependence of Nernst signal after subtracting the normal state background signal in NCCO. The panel is from Ref. [19].

the normal state quasi-particle is very large, unlike the hole doped cuprates such as Bi2201 and Bi2212, where e_y^N is negligibly small. Fournier and Hamza *et al.* explained this large normal state Nernst signal in terms of two-band charge carriers, and the two-band electronic structure has been supported by the ARPES experiment (the electron pocket appeares in the hole like Fermi surface). There are two type of charge carriers in the Ce-doped NCCO or PCCO, leading to the invalidity of Sondheimer cancellation, which could be the origin of the large Nernst signal in the normal state.

Ong *et al.* have briefly summarized the vortex phase diagrams of the hole doped, electron-doped high-T_C superconductors and conventional type-II superconductors. Figure 10.14 shows the comparison of vortex melting field H_m and upper critical field H_{C2} between the conventional

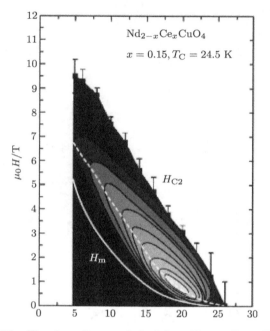

Fig. 10.13. The phase diagram derived from Nernst effect of NCCO.

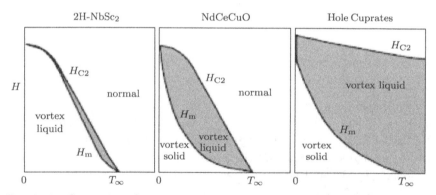

Fig. 10.14. Comparison of vortex melting field H_m and upper critical field H_{C2} between conventional superconductor NbSe$_2$, electron-doped high-T_C superconductor NCCO and hole doped high-T_C superconductor.

superconductor NbSe$_2$, electron-doped high-T_C superconductor NCCO and the hole-doped high-T_C superconductor. Obviously, the former two superconductors have the same H_{C2} behavior, a characteristic of BCS superconductors: H_{C2} is almost temperature independent at low temperature, and decreases linearly with increasing temperature, and finally ends up at the transition temperature T_{C0}. However, the vortex fluid region is quite different, i.e. the vortex fluid region in high-T_C superconductors is much larger than the conventional superconductor NbSe$_2$. For the hole-doped cuprates, H_{C2} decreases slowly with increasing temperature, and remains large even at the superconducting transition temperature. Estimated from the trend, H_{C2} will reach zero at a very high temperature. The vortex fluid state continues to exist even at temperature much above T_{C0}, which should be closely associated with the existence of the pseudogap above T_{C0} in the hole-doped cuprates. On the other hand, there is no experimental evidence confirming the existence of the pseudogap state above T_{C0} in the electron-doped NCCO, implying the different superconducting mechanism from the hole-doped cuprates.

10.5. Further Experimental Evidence of Superconducting Phase Fluctuation Picture: Magnetic Susceptibility Measured by Torque Technique

The phase fluctuation picture of high-T_C superconductors is very similar to the well-known 2D Kosterlitz–Thoulesee (KT) phase transition. In the KT phase transition theory, the KT phase transition temperature T_{KT} lies below the mean field phase transition temperature T^{MF}. Once temperature becomes higher than T_{KT}, vortex–anti-vortex pair will be destroyed by thermal excitations. Though these spontaneous vortices destroy the superfluid long-range phase coherence, the order parameter amplitude is finite. Owing to the disorder of phase, the average of total order parameter is zero, i.e.:

$$\langle \widehat{\psi} \rangle = |\psi| \langle e^{i\theta(r)} \rangle = 0.$$

Generally speaking, the phase transition temperature T_{KT} is proportional to the superfluid stiffness and the superfluid density. Only when the mean field phase transition temperature T^{MF} is higher than T_{KT} does the order parameter amplitude $|\psi|$ go toward zero.

The phase fluctuation in the cuprate high-T_C superconductors is a kind of 3D version of the KT phase transition. The observed superconducting transition temperature is not the mean field phase transition temperature T^{MF}, but the temperature at which the long-range phase coherence is destroyed, i.e. just like T_{KT} in the KT phase transition. When $T_C < T < T^{MF}$, the order parameter amplitude is finite, but the phase is totally disordered. According to the recent theories, the phase fluctuation is different from the normal amplitude fluctuation (Gaussian type fluctuation). There should still exist supercurrents surrounding the vortices; therefore, weak diamagnetism is presented. Owing to the large depairing field, the diamagnetism of the phase fluctuations state will persist up to large magnetic fields. Previous Nernst experiments have confirmed that these vortex excitations are strictly two dimensional, i.e. the diamagnetic magnetization is strictly along the c axis; therefore, the diamagnetism could be detected by a torque technique.

If the angle between the magnetic moment of the sample m and external field H is φ, then the sample will be applied a torque $\tau = m \times B$. The sample is glued to the tip of a Si cantilever with its c axis at a small angle to the external field. The torque will lead to a small change between the Si cantilever and the substrate. The small distance change can be detected precisely by the change in the capacitance between the Si cantilever and the substrate, and then the sample moment m can be calculated based on the capacitance change. Compared with SQUID, the torque technique can be applied at high fields with a resolution of 5×10^{-9} emu at 10 T, which is comparable to SQUID.

Naughton et al. have discovered the weak diamagnetism above T_C in Hg2212 single crystal by the torque technique, which cannot be interpreted by the usual Gaussian type fluctuations. Whereafter, Ong's group measured the weak diamagnetism above T_C in more high-T_C superconductors using torque technique and revealed that the diamagnetism is well consistent with the superconducting phase fluctuations picture. Combined with the Nernst effect results, these experiments strongly support the superconducting phase fluctuations above T_C. Figure 10.15 shows the M vs. H curves measured on the underdoped single crystal Bi-2212 ($T_C = 50$ K). The almost temperature-independent paramagnetic magnetization has been subtracted. It can be seen that the diamagnetism is obvious even 70 K above T_C, and it is non-linear to the external magnetic field, consistent with the analysis based on the phase fluctuation model. Figure 10.16 compares the diamagnetism M and the Nernst signal e_y, for the underdoped and the optimally doped Bi2212. Both M and e_y are measured under a magnetic field $\mu_0 H = 14$ T. Obviously, the temperature dependence of M and e_y above T_C match each other both in the underdoped and optimally doped

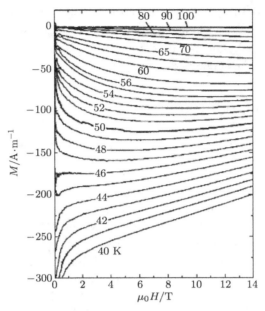

Fig. 10.15. The M vs. H measured in the underdoped single crystal Bi2212 ($T_C = 50$ K).

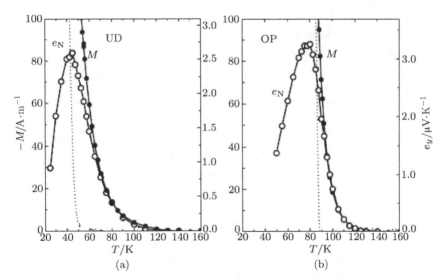

Fig. 10.16. Comparison of the diamagnetism M and the Nernst signal e_y, (a) in the underdoped Bi2212, (b) in the optimally doped Bi2212. Both M and e_y were measured with $B = 14$ T.

samples. This result clearly indicates that besides the existence of vortex excitations above T_C, there is also weak diamagnetism, which is different from the usual Gaussian type amplitude fluctuations, but in excellent agreement with the superconducting phase fluctuation picture.

10.6. Conclusions

The Nernst effect results in high-T_C superconductors are summarized as follows:

(1) There are large vortex Nernst signals in the mixed state in high-T_C superconductors, and the vortex Nernst signals persist to a characteristic temperature T^{onset} well above T_C, where T^{onset} is smaller than T^* (the pseudogap opening temperature).
(2) The upper critical field H_{C2} can be obtained from the Nernst effect measurement. H_{C2} is weakly temperature dependent and does not drop to zero at T_{C0} in hole-doped cuprates, in contrast to the conventional superconductors. Moreover, H_{C2} increases with decreasing doping level, indicating the enhanced pairing potential in the underdoped regime. This critical field H_{C2} corresponds to the Cooper pair depairing field, whereas the critical field obtained from the magnetoresistance measurement actually corresponds to the destruction field of long-range phase coherence between Cooper pairs.
(3) The Nernst effect results support the superconducting phase fluctuations picture. The superconducting transition critical temperature T_C corresponds to the destruction temperature of the long-range phase coherence. There is finite order parameter amplitude in the temprature range between $T_C < T < T_C^{\text{MF}}$, but the phase is disordered in space owing to the thermally activated vortices; therefore, the average superconducting order parameter is zero.
(4) KT phase transition theory indicated that there is weak diamagnetism above the KT phase transition temperature, and superconducting phase fluctuations model is a 3D version of the KT phase transition theory. The weak diamagnetic signal obtained by the torque technique is consistent with the Nernst results, and it again supports the superconducting phase fluctuations picture.
(5) For the electron-doped cuprates, vortex Nernst signal disappears immediately above T_C, and $H_{C2} - T$ phase diagram shows the same behavior

as the conventional superconductors, but with much larger vortex fluid region.

Acknowledgments

The author would like to thank Ya-Yu Wang, Zheng-Yu Weng and Hai-Hu Wen for the helpful discussions, and also Jing-Qin Shen, Zeng-Wei Zhu and Chen-Yi Shen for their assistance in preparing the chapter. The work is partially supported by the NSFC (Natural Science Foundation of China).

References

[1] E. H. Sondheimer, *Proc. R. Soc. London, Ser. A* **193**, 484 (1948).
[2] Z. A. Xu, N. P. Ong, Y. Wang *et al.*, *Nature (London)* **406**, 486 (2000).
[3] R. Bel, K. Behnia, H. Berger, *Phys. Rev. Lett.* **91**, 066602 (2003).
[4] B. D. Josephson, *Phys. Lett.* **16**, 242 (1965).
[5] F. Vidal, *Phys. Rev. B* **8**, 1982 (1973).
[6] R. P. Huebener, A. Seher, *Phys. Rev.* **181**, 701 (1969).
[7] S. J. Hagen, C. J. Lobb, R. L. Greene, *Phys. Rev. B* **42** (10), 6777 (1990).
[8] M. Zeh, H. C. Ri, F. Kober, *Phys. Rev. Lett.* **64** (26), 3195 (1990).
[9] J. Orenstein, A. J. Millis, *Science* **288**, 468 (2000).
[10] J. K. Liang, G. C. Che, X. N. Chen, *Phase Relationship and Crystal Structures of High-T_C Superconducting Oxides* (Science Press, Beijing, 1994).
[11] W. Warren, R. E. Walstedt, G. F. Brennert *et al.*, *Phys. Rev. Lett.* **62**, 1193 (1989).
[12] T. Timusk, B. Statt, *Rep. Prog. Phys.* **62**, 61 (1999).
[13] V. J. Emery, S. A. Kivelson, *Nature (London)* **374**, 434 (1995).
[14] J. L. Tallon, J. W. Loram, G. V. M. Williams, *Phys. Stat. Sol.* **215** (1), 531 (1999).
[15] Y. J. Uemura, G. M. Luke, B. J. Sternlieb, *Phys. Rev. Lett.* **62**, 2317 (1989).
[16] J. Corson, R. Mallozzi, J. Orenstein, *Nature (London)* **398**, 221 (1999).
[17] Y. Wang, Z. A. Xu, T. Kakeshita, *Phys. Rev. B* **64** (22), 224519 (2001).
[18] Y. Y. Wang, N. P. Ong, Z. A. Xu, T. Kakeshita, *Phys. Rev. Lett.* **88**, 257003 (2002).
[19] Y. Y. Wang, S. Ono, Y. Onose *et al.*, *Science* **299** (5603), 86 (2003).
[20] C. Caroli, K. Maki, *Phys. Rev.* **164**, 591 (1967).
[21] A. Houghton, K. Maki, *Phys. Rev. B* **3**, 1625 (1971).
[22] C. R. Hu, *Phys. Rev. B* **13**, 4780 (1976).
[23] H. Ding, J. R. Engelbrecht, Z. Wang *et al.*, *Phys. Rev. Lett.* **87**, 227001 (2001).
[24] S. H. Pan, E. W. Hudson, A. K. Gupta *et al.*, *Phys. Rev. Lett.* **85**, 1536 (2000).

[25] J. E. Hoffman, E. W. Hudson, K. M. Lang et al., *Science* **295** (5554), 466 (2002).
[26] H. H. Wen, Z. Y. Liu, Z. A. Xu et al., *Europhys. Lett.* **63**, 583 (2003).
[27] I. Ussishkin, S. L. Sondhi, D. A. Huse, *Phys. Rev. Lett.* **89**, 287001 (2002).
[28] Y. Wang, Princeton University, 2003 (unpublished).
[29] E. W. Carlson, V. J. Emery, S. A. Kivelson et al., cond-mat/0206217 (2002).
[30] G. Baskaran, Z. Zou, P. W. Anderson, *Solid State Commun.* **63**, 973 (1987).
[31] S. Doniach, M. Inui, *Phys. Rev. B* **41**, 6668 (1990).
[32] Z. Y. Weng, V. N. Muthukumar, *Phys. Rev. B* **66**, 094509 (2002).
[33] Y. Y. Wang, L. Li, M. J. Naughton et al., *Phys. Rev. Lett.* **95**, 247002 (2005).
[34] X. G.Jiang, W. Jiang, S. N. Mao et al., *Physica B* **2305**, 194 (1994).
[35] P. Fournier, X. Jiang, W. Jiang et al., *Phys. Rev. B* **56**, 14149 (1997).
[36] F. Gollnik, M. Naito, *Phys. Rev. B* **58**, 11734 (1998).
[37] H. Balci, C. P. Hill, M. M. Qazilbash et al., *Phys. Rev. B* **68**, 054520 (2003).
[38] N. P. Armitage, D. H. Lu, D. L. Feng et al., *Phys. Rev. Lett.* **86**, 1126 (2001).
[39] N. P. Ong, Y.Wang, *Physica C* **408**, 11 (2004).

11
Very Low-Temperature Heat Transport Properties of High-Temperature Superconductors

Xue-Feng Sun

University of Science and Technology of China, Heifei, Anhui 230026, China

It has been more than 20 years since the cuprate high-temperature superconductors (HTSC) were discovered. The various physical properties of the normal state and the superconducting state have been studied in great detail. The nature and transport behaviors of the low-energy quasi-particle (QP) in the superconducting state are undoubtedly important to explore the unconventional superconducting electronic states and the mechanism of superconductivity. In addition to the common techniques, such as the angle-resolved photoemission spectroscopy (ARPES), the penetration depth, the AC electrical conductivity and so on, the very low-temperature thermal conductivity κ also plays a very important role. Many experimental developments on heat transport have been made in the past ten years, but meanwhile more and more problems emerged in this field. In this chapter, we mainly introduce the developments and controversies of the very low-temperature thermal conductivity in HTSC.

11.1. Introduction of Heat Transport

Figure 11.1 shows the schematic plot of the very low-temperature thermal conductivity measurement, in which the "one heater, two thermometers" steady-state technique is used. As shown in Fig. 11.1, one end of the sample

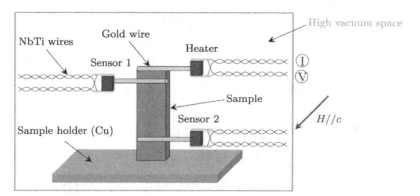

Fig. 11.1. Schematic plot of the steady-state measurement of low-temperature heat transport.

is fixed on the sample holder with epoxy to achieve good thermal contact. A resistive heater is mounted at the other end of the sample, and another two resistance thermometers are mounted in the middle to measure the temperature gradient. When the temperature of sample holder is stabilized, applying a certain current I through the heater, which generates Joule heat $P = IV$, a stable heat flow and temperature gradient are then formed in the sample. The temperature difference ΔT can be measured using two thermometers, and then the thermal conductivity κ can be defined as:

$$\kappa = \frac{P}{\Delta T}\frac{l}{S}, \qquad (11.1)$$

where l is the distance between two thermometers, S is the cross-sectional area of the sample. Typically, the samples have length of 1–2 mm, width of ∼0.5 mm, and thickness of ∼0.1 mm. The thermal conductivity measurements of HTSC described in this chapter, if without any special mention, were done with the heat flowing along the ab plane (CuO_2 plane) and the magnetic field along the c direction.

In principle, the very low-temperature thermal conductivity measurement is a very simple technique in the HTSC experimental studies, but there are still some difficulties. For example, the temperature measurement needs very high accuracy and stability. Moreover, in consideration of the measurements in the magnetic field, both the resistance thermometers on the sample and the standard thermometer on the holder exhibit magnetic resistance effects, therefore the magnetic field dependence of their resistance should be calibrated precisely before the experiment. In addition, special attention should be paid to the error due to the heat leakage (using very fine NbTi superconducting lines as the test leads of resistance heater and thermometers can effectively reduce heat leakage; besides, the high vacuum environment is also necessary). On the other hand, high-quality single-crystal samples are needed to guarantee the reproducibility of the results. At present, the experimental measurement error can be less than 10% in absolute value.

As we know, thermal conductivity is related to the ability of energy transport of elementary excitations in solid, including phonons, electrons or magnons, etc. The classical kinetic theory gives a simple formula [1]:

$$\kappa = \frac{1}{3}cvl, \qquad (11.2)$$

where c is the heat capacity, v is the mean velocity, and l is the mean free path of heat carriers. Thermal conductivity can provide the physical information of a variety of elementary excitations. When there is more than one kind of heat carrier in the system, the behavior of thermal conductivity usually becomes complicated and difficult to analyze.

Phonon thermal conductivity exists in all solid materials, of which the simplest case is the non-magnetic insulators. When the temperature is gradually decreased from room temperature or higher, the phonon–phonon scattering becomes weakened and κ is gradually increased; at nearly 20 K, where κ achieves a maximum, the magnitude and location of the peak are strongly dependent on the extrinsic scattering mechanism such as point defects, dislocations and so on, and therefore are good characterizations for the crystal quality. With further decreasing temperature, the reduction of the phonon population dominates the decrease of κ and at the limit of absolute zero the κ heads to zero. Therefore, the behavior of phonon thermal conductivity throughout the whole temperature range is very complicated. Only at very low temperatures (usually below 10 K in the insulating materials, and at even lower temperatures in the cuprates) does the phonon thermal conductivity show a simple T^3 relation. This is because at very low temperatures, the microscopic scattering mechanisms like phonon–phonon scattering and defect scattering are smeared out; the mean free path of phonon is gradually increased and finally determined by the geometric size of the sample and is independent of the temperature. This is the so-called boundary scattering limit. The T^3 relation of $\kappa(T)$ results from the temperature dependence of the phonon specific heat at low temperatures. Such a kind of temperature dependence is helpful for separating the electronic thermal conductivity from the phononic one in metals or unconventional superconductors at very low temperatures. It is worth noting that in the HTSC materials, due to the electron–phonon scattering and magnetic scattering, the behavior of phonon thermal conductivity at very low temperatures is much more complicated than that of usual insulators, which makes the experimental results difficult to analyze.

In normal metals, the heat transport is dominated by electrons [1]. Thermal conductivity at high temperatures depends on the electron–phonon scattering. When the temperature is higher than or equal to the Debye temperature, κ is a constant; when the temperature is lower than the Debye temperature, κ is proportional to $1/T^2$. At very low temperatures, electrons are scattered elastically only by the defects and impurities, so κ shows a linear dependence on temperature. The well-known Wiedemann–Franz (WF) law points out that the scattering mechanism of electronic heat transport and electrical transport are the same at very low temperature, and they meet the following relation:

$$\frac{\kappa}{\sigma T} = \frac{\pi^2}{3} \left(\frac{k_B}{e}\right)^2 \equiv L_0, \qquad (11.3)$$

where the Lorentz number $L_0 = 2.44 \times 10^{-8}$ W/K^2. WF law is a fundamental property in Fermi liquid theory. Therefore, it is a basic method to judge whether a correlated electron system is Fermi liquid by comparing with the WF law from the measurements of thermal conductivity and electronic conductivity.

For the conventional s-wave superconductors, the study of low-temperature heat transport is not very informative. The reason is very simple, being that the s-wave superconducting gap in the momentum space is isotropic, so the number of quasi-particle (QP) excitations decrease exponentially with decreasing temperature in the superconducting state [2]. Therefore, the contribution of QPs to the thermal conductivity at very low temperature is negligibly small and can provide us little information.

For a d-wave cuprate superconductor, the superconducting gap along some particular directions in the momentum space is zero, the so-called gap nodes. The low-energy QPs can be excited and form a finite density near the nodes as long as the impurity scattering works. These QPs can

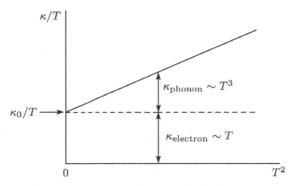

Fig. 11.2. At very low-temperature region, in which the boundary scattering limit establishes, κ/T is proportional to T^2. The slope is dependent on the phonon thermal conductivity and the residual thermal conductivity κ_0/T is the contribution of electrons.

exist at very low temperatures and contribute to the heat transport, so the low-temperature thermal conductivity of HTSC could include a significant contribution from QPs [3]. But there is a problem: how to separate the contribution of electrons from the phonon conductivity? It is usually easy to achieve for the data at very low temperatures. As mentioned above, κ_{phonon} is proportional to T^3 and $\kappa_{electron}$ is linear with T at very low temperatures, we can get:

$$\frac{\kappa}{T} = a + bT^2. \qquad (11.4)$$

As shown in Fig. 11.2, the plot of κ/T vs. T^2 is expected to be well fitted linearly at very low temperatures, with the slope b characterizing the phonon contribution and the intercept a characterizing the contribution of electrons, which is named as the residual thermal conductivity (often written as κ_0/T). The first important issue of the very low-temperature thermal conductivity is to study whether there is non-zero κ_0/T, because it is one of the most effective ways to judge whether the order parameter is the traditional s-wave or the unconventional d-wave (or the p-wave with nodes). Second, the dependences of κ_0/T on the concentration of heat carriers and the magnetic field also provide physics about the QP transport, the ground-state properties, and the quantum phase transitions, etc. Since Eq. (11.4) is usually established only at very low temperatures, the very low-temperature measurement becomes an indispensable tool to study the QP heat transport in HTSC.

11.2. Universal Thermal Conductivity

Universal thermal conductivity is a prediction by the Fermi liquid theory on the classical d-wave superconductor [4, 5]. For the quasi-2D superconductors with $d_{x^2-y^2}$ symmetry, the anisotropic energy gap can be expressed as

$$\Delta(k) = \Delta_0 \cos 2\phi, \tag{11.5}$$

where Δ_0 is the maximum of the energy gap, ϕ is the in-plane azimuth angle in the momentum space. Due to the impurity scattering, there could always be low-energy QP excitations near the nodes. Although the impurities can create the elementary excitations, simultaneously they also scatter QPs. In some cases, the effects of creation and scattering balance, causing the thermal conductivity of QPs to be independent of the strength of impurity scattering. This phenomenon is called the universal thermal conductivity. By using the self-consistent T-matrix approximation (SCTMA) theory for the classical d-wave superconductors, one can get that, when the impurity scattering strength meets $k_B T \ll \gamma \ll \Delta_0$ (γ is the width of impurity energy band), the residual thermal conductivity is independent of the strength of QP scattering, and satisfies the following relation [5]:

$$\frac{\kappa_0}{T} = \frac{k_B^2}{3\hbar} \frac{n}{d} \left(\frac{v_F}{v_\Delta} + \frac{v_\Delta}{v_F} \right) \approx \frac{k_B^2}{3\hbar} \frac{n}{d} \frac{v_F}{v_\Delta}. \tag{11.6}$$

It is clearly seen that the κ_0/T relies on the number of CuO_2 planes in the unit cell, n, the lattice constant along the c axis, d, the Fermi velocity, v_F, and the tangential velocity of the Fermi surface at the node, v_Δ, and is independent of the impurity scattering strength. Since v_Δ is related to the slope of the energy gap at the nodes:

$$v_\Delta = \frac{1}{\hbar k_F} \left. \frac{d\Delta}{d\phi} \right|_{node}, \tag{11.7}$$

the κ_0/T is sometimes expressed as [6]:

$$\frac{\kappa_0}{T} \approx \frac{k_B^2}{6\hbar} \frac{n}{d} \frac{k_F v_F}{\Delta_0}. \tag{11.8}$$

From the experimental results of ARPES, we know that k_F and v_F are the parameters that have little to do with the materials and the carrier concentrations, so κ_0/T is directly related to superconducting gap Δ_0. The universal thermal conductivity is a criterion whether the HTSC can be described by the classical d-wave theory. If the scenario is correct, it would be very helpful to provide a true bulk measurement of superconducting gap, which is essentially more effective than the ARPES and tunneling spectroscopy experiments.

It is known that, although most experiments have shown that HTSC may have $d_{x^2-y^2}$ symmetry of the superconducting order parameter [7], many complicated physical behaviors, like strong electron correlation, electronic inhomogeneity, charge/spin ordering and competition, make HTSC exhibit many non-Fermi liquid behaviors and even the quantum phase transitions associated with the change of superconducting symmetry [8–11]. Therefore, it is questionable that the universal thermal conductivity, a prediction of the classical d-wave superconductor, can describe accurately the QP transport behavior in HTSC. In particular, the progress of many experiments carried out by the scanning tunneling microscope (STM) in recent years shows that the electronic state of HTSC is strongly inhomogeneous [12–18]; some theories have shown that taking the spatial inhomogeneity of the superconducting order parameter into account, the low-temperature transport of QPs may localize and deviate from the universal thermal conductivity [19, 20]. Then, what about the experimental results?

Shortly after the theoretical prediction of the universal thermal conductivity, Taillefer et al. gave the first experimental evidence [21]. As shown in Fig. 11.3, they observed the non-zero $\kappa_0/T = 0.019$ W · K^{-2} · m

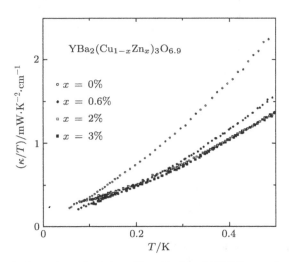

Fig. 11.3. *A*-axis thermal conductivity of four Zn-doped YBCO crystals, plotted as κ/T vs. T. The panel is from Ref. [21].

in the optimally doped $YBa_2Cu_3O_y$ (YBCO, $y = 6.9$). By doping different amounts of Zn to change the impurity scattering strength and measuring the thermal conductivity, the authors pointed out that the κ_0/T of the samples with Zn content ranging from 0% to 3% is nearly constant ($\kappa_0/T = 0.017 \sim 0.025$ W $\cdot$ K^{-2} $\cdot$ m). They concluded that the residual thermal conductivity is independent of impurity scattering strength, which directly confirmed the picture of the universal thermal conductivity. However, the errors of data in this work were rather large, and thereby the effect of the Zn-doping on the κ_0/T of YBCO was not very clear. In fact, this problem also showed up in their subsequent work [22], in which the κ_0/T of Zn-doped YBCO samples (Zn content, $x = 0\%$–3%; O content, $y=6.9$) were found to be about 0.012–0.014 W$\cdot$K$^{-2}\cdot$m and differ much from the previous results. Obviously, in these experiments the absolute value of the κ_0/T and its relation with the impurity concentration were in fact far from the universal behavior.

Chiao et al. [23] reported the second experimental evidence for the universal thermal conductivity. As shown in Fig. 11.4, an optimally doped $Bi_2Sr_2CaCu_2O_{8+\delta}$ (Bi2212) single crystal ($T_C = 89$ K) showed $\kappa_0/T = 0.015$ W $\cdot$ K^{-2} $\cdot$ m. Using Eq. (11.6) they got the superconducting gap parameter $v_F/v_\Delta = 19$, which was in good agreement with the result

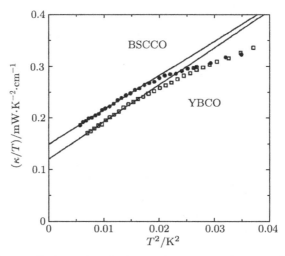

Fig. 11.4. κ/T vs. T^2 of $YBa_2Cu_3O_7$ (squares) and $Bi_2Sr_2CaCu_2O_8$ (circles) at optimum doping. The lines are linear fits to the data below 130 mK, using Eq. (11.4). The panel is from Ref. [23].

by ARPES [24]. The authors therefore believed that this result strongly supported the scenario of universal thermal conductivity. However, this work did not explore the relation between the κ_0/T and the impurity scattering strength and was not able to directly prove the validity of the universal behavior. Moreover, since the Bi2212 single crystal used in this work was grown by the flux method, the content of impurity in the sample cannot be ignored. It was also questionable to calculate v_F/v_Δ using the observed κ_0/T as a universal quantity. Therefore, it is still a question whether this result reflected the intrinsic property of Bi2212 system or not.

Nakamae et al. [25] modified the impurity scattering strength by using electron irradiation to introduce point defects in Bi2212 single crystals (see Fig. 11.5). They found that although there were changes in the low-temperature thermal conductivity of the samples before and after the irradiation, the κ_0/T were almost constant. This work made up for the shortcoming of Chiao et al.'s [23] work, and strongly supported the validity of universal thermal conductivity. Note that before and after irradiation, the low-temperature thermal conductivity were greatly different down to 100 mK, which indicated that even at such low temperatures, the impurity scattering still had significant effect on the QP transport, and the electron term κ_e/T was not independent of the temperature as the universal behavior expected, but was proportional to T, approximately. Additionally, there were some other problems in this work, for example, the impurity scattering of these samples are significant, judging by the relatively large residual resistivity, and the experimental noise of the thermal conductivity is so high that great error exists in the heat transport measurement.

The above three works had been considered as the experimental evidence to support the picture of the universal thermal conductivity. Based on these, it was believed that the universal thermal conductivity is a common phenomenon for HTSC, and more information might be obtained from the thermal conductivity data. For example, the results in Figs. 11.6 and 11.7 were given by Sutherland et al. [26] and Hawthorn et al. [6] in 2003 and 2007, respectively. They calculated the superconducting gap of YBCO, Bi2212, $Tl_2Ba_2CuO_{6+\delta}$ (Tl2201) samples from the residual thermal conductivity using Eq. (11.8) and when compared with the values obtained by other experimental techniques, it was found that all the results are consistent. The authors therefore believed that the nature of QPs can be described by the simple d-wave superconductivity theory in a wide doping region of HTSC. (Doping dependence of the superconducting gap Δ_0 obtained from

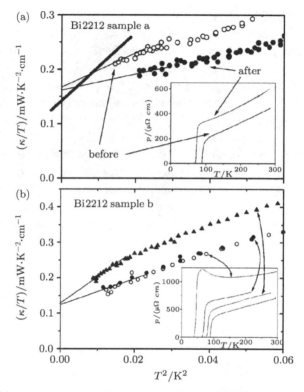

Fig. 11.5. (a) Low-temperature thermal conductivity of Bi2212 crystal (sample a) plotted as a function of T^2. The thin lines are guides to the eye. The thick line represents the expected asymptotic lattice conductivity at the ballistic regime. The inset shows the resistivity data of the same sample. (b) Same for sample b. Note that the thermal conductivity of this sample in the pristine state was not measured. The panel is from Ref. [25].

the quasi-particle velocity v_2 defined in Eq. (11.3) (filled symbols). Here we assume $\Delta = \Delta_0 \cos 2\phi$, so that $\Delta_0 = \hbar k_F v_2/2$, and we plot data for YBCO alongside Bi2212 [7] and Tl2201 [8]. For comparison, a BCS gap of the form $\Delta_{\mathrm{BCS}} = 2.14 k_B T_C$ is also plotted, with T_C taken from Eq. (11.1) (and $T_C^{\max} = 90$ K). The value of the energy gap in Bi2212, as determined by ARPES, is shown as measured in the superconducting state [29] and the normal state [30–32] (open symbols). The thick dashed line is a guide to the eye.)

At almost the same time, however, there appeared more and more experimental results challenging the universal thermal conductivity. Early

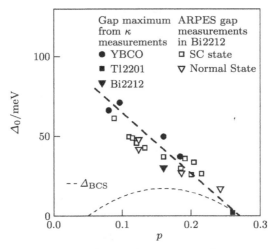

Fig. 11.6. Comparison between the calculated superconducting energy gap Δ_0 from the residual thermal conductivity of YBCO, Bi2212, and Tl2201 single crystals and the results from ARPES with different carrier concentration [26]. The thin dotted line shows the expected BCS gap $\Delta_{BCS} = 2.14 k_B T_C$. The panel is from Ref. [26].

in 2000, Hussey et al. [27] first found in the underdoped $YBa_2Cu_4O_8$ (Y124) some phenomenon inconsistent with the universal thermal conductivity. They carried out a detailed study on a group of Y124 single crystals with $T_C = 80$ K and the measurement temperature down to 100 mK. They found that the low-temperature thermal conductivity can be fitted well using Eq. (11.4), as shown in Fig. 11.8. The fitting parameter b is in good correspondence with the coefficient of the lattice specific heat, and it was interesting to find that all the obtained κ_0/T are nearly zero. This phenomenon was considered to result from some kind of QP localization behavior. Note that the universal thermal conductivity predicts that there is always finite κ_0/T in the d-wave superconductors, which is incompatible with Hussey et al.'s work [27].

Second, although Sutherland et al. [26] and Hawthorn et al. [6] found that the superconducting energy gaps derived from the κ_0/T were consistent with those from the tunneling spectroscopy and ARPES (shown in Figs. 11.6 and 11.7), recent ARPES [28, 29] and Raman spectroscopy [30] measurements have revealed a double-gap feature in HTSC. The larger one may be the superconducting gap, whose dependence on the carrier concentration is similar to the behavior of the gap obtained from the κ_0/T. However, this gap is along the anti-nodal direction, and should be less relevant

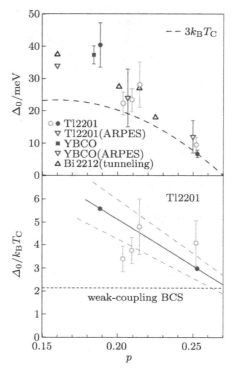

Fig. 11.7. Comparison between the calculated superconducting energy gap Δ_0 from the residual thermal conductivity of Tl2201 and YBCO single crystals and the results from ARPES and tunneling spectroscopy with different carrier concentrations. The panel is from Ref. [6].

to the low-energy QP excitations. The other gap is located at the nodal direction and directly relevant to the QP excitations. It is nearly invariant with the doping concentration in the higher doping region but decreases gradually in the underdoped region, which is, however, inconsistent with the behavior of the gap from the κ_0/T. From this point of view, the dependences of superconducting gap on the doping concentration from ARPES and Raman spectroscopy are actually different from those from the thermal conductivity.

In fact, the early studies found that for a variety of HTSC, there was a "universal" dependence of κ_0/T on the carrier concentration. Takeya et al. [31] systematically studied the relationship between the κ_0/T and the carrier concentration for the first time. They measured the thermal conductivity of a series of $La_{2-x}Sr_xCuO_4$ (LSCO) single crystals with $x = 0 - 0.22$

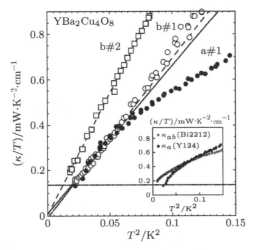

Fig. 11.8. Thermal conductivity of Y124 single crystals along the a axis and b axis at very low temperatures and the linear fit using Eq. (11.4). The panel is from Ref. [27].

down to tens of mK, as shown in Fig. 11.9. Using Eq. (11.4) to separate the κ_0/T term from the experimental data, they found that κ_0/T in the non-superconducting samples with $x \leq 0.05$ are nearly zero. The κ_0/T becomes finite near $x = 0.06$, accompanied by the presence of superconductivity; then, except for the singularity at $x = 1/8$, κ_0/T in all the other samples increased with x, as shown in Fig. 11.10. According to Eq. (11.8), it is easy to know that the dependence of κ_0/T on x means a decrease of the superconducting gap with the increase of x. It was inconsistent with the results from ARPES and Raman spectroscopy, which indicated that the validity of Eq. (11.8) and even the scenario of universal thermal conductivity are questionable.

The doping dependence of κ_0/T has also been studied in YBCO, Bi2201, Tl2201, and Bi2212. Figures 11.11 and 11.12 are the YBCO results reported by Sutherland et al. [26] and Sun et al. [32], respectively. For the sake of avoiding the contribution of electron thermal conductivity along the Cu–O chains in the b axis, all these measurements were carried out along the a axis of the untwined single crystals. The obtained electron thermal conductivity was completely the QP conduction behavior in the CuO_2 superconducting layers. These data indicated that when the carrier concentration increased with increasing the oxygen content, κ_0/T was also gradually increased. Similarly, for the Bi2201 system, there were two experimental results. Figure 11.13 is for the La-doped $Bi_2Sr_{2-x}La_xCuO_{6+\delta}$

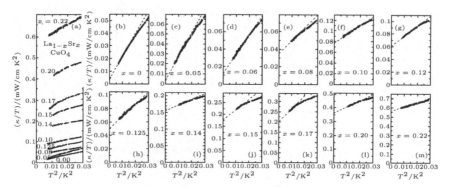

Fig. 11.9. The very low-temperature thermal conductivity of LSCO ($x = 0 - 0.22$) single crystals and the linear fit using Eq. (11.4). The panel is from Ref. [31].

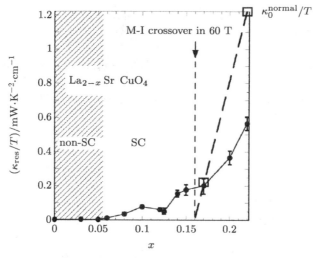

Fig. 11.10. The dependence of residual thermal conductivity on the carrier concentration in LSCO single crystal. The panel is from Ref. [31].

(BSLCO) [33], in which the highest superconducting transition temperature could achieve above 36 K; Fig. 11.14 shows the result of $Bi_{2+x}Sr_{2-x}CuO_{6+\delta}$ system [34], in which the carrier concentration can be modified by controlling the ratio of Bi and Sr and the highest superconducting transition temperature was about 10 K. Obviously, the relation between the κ_0/T and doping concentration in Bi2201 system was qualitatively the same as those in LSCO and YBCO. In addition, the aforementioned thermal conductivity

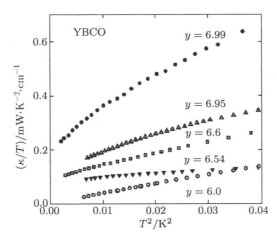

Fig. 11.11. The a-axis thermal conductivity of YBCO single crystal for different oxygen content (y) at low temperatures. The panel is from Ref. [26].

results of Tl2201 by Hawthorn et al. [6] (Fig. 11.7) also exhibited the same behavior.

For Bi2212 system, the early works by Chiao et al. [23] and Nakamae et al. [25] only showed the results of the optimally doped samples. Later on, Sun et al. [35] systematically studied the Bi2212 samples with different carrier concentrations, including adjusting the oxygen content and doping rare-earth element Dy in Ca site. The behavior of low-temperature thermal conductivity was found to be qualitatively consistent with LSCO and YBCO, as shown in Fig. 11.15. The authors also used Eq. (11.6) to obtain the relation between the gap parameter v_F/v_2 (v_2 is v_Δ mentioned before) and the carrier concentration, and found that it differed from the results by ARPES [24].

In summary, almost all the HTSC display similar relation between the κ_0/T and the carrier concentration. If one simply uses Eqs. (11.6) or (11.8) to obtain the dependence of gap parameter on the carrier concentration, it will be found to be inconsistent with that from ARPES and other measurements. The reason may be that there is a large discrepancy between the experimental κ_0/T and the universal thermal conductivity; or that the universal thermal conductivity could not accurately describe the transport properties of QPs in HTSC.

The experimental results on Bi2212 by Sun et al. [35] also showed that the κ_0/T for the optimally doped Bi2212 single crystals ($T_C = 94\,\mathrm{K}$,

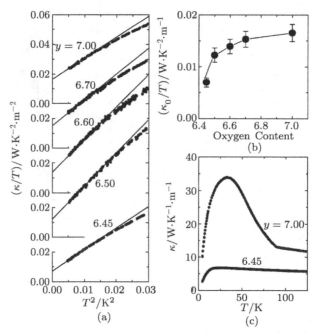

Fig. 11.12. (a) The a-axis thermal conductivity of YBCO single crystal for different oxygen content (y) at low temperatures; (b) The y dependence of residual thermal conductivity; (c) The high-temperature thermal conductivity data indicating the high quality of these single-crystal samples Ref. [32].

$p \approx 0.155$) could achieve nearly 0.05 W/K^{2}m; even in the underdoped samples ($T_C = 70$ K, $p \approx 0.13$), the κ_0/T can be as large as 0.03 W/K^{2}m. In comparison, the earlier work reported the κ_0/T about 0.015 W/K^{2}m, which may be because those samples ($T_C \approx 90$ K) had relatively high content of impurities [23,25]. It should be noted that, different from the Chiao et al.'s [22] and Nakamae et al.'s [24] works, the single crystals used by Sun et al. were grown by optical floating-zone method, and the quality of their samples was very high, which could be seen from the resistance data of these samples [36,37]. Therefore, the above experimental results actually demonstrated that the κ_0/T was closely related to the quality of single crystals or the impurity concentration. Apparently, further investigations on the relation between the κ_0/T and the impurity scattering strength are called for.

For this purpose, Sun et al. [35] carried out a detailed study on a series of superconducting systems like YBCO, LSCO, Bi2212, etc., by doping

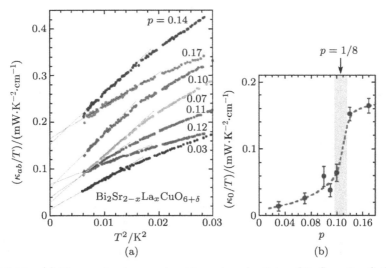

Fig. 11.13. (a) The very low-temperature thermal conductivity of $Bi_2Sr_{2-x}La_xCuO_{6+\delta}$ single crystals for different carrier concentrations p; (b) The p dependence of residual thermal conductivity Ref. [33].

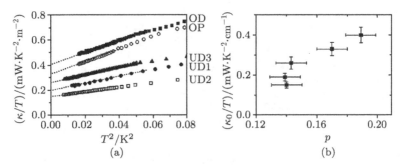

Fig. 11.14. (a) The very low-temperature thermal conductivity of $Bi_{2+x}Sr_{2-x}CuO_{6+\delta}$ single crystals for different carrier concentrations p; (b) The p dependence of residual thermal conductivity. The panel is from Ref. [34].

small amounts of Zn to change the impurity scattering strength. They found that in all systems, a distinct suppression of the κ_0/T (magnitude reduction by 20%–50%) was detected in the optimally doped and underdoped samples, as shown in Fig. 11.16. This is a direct challenge to the picture of universal thermal conductivity. It should be noted that the results in YBCO by Taillefer et al. [21] indicated that, when the Zn-doped concentration

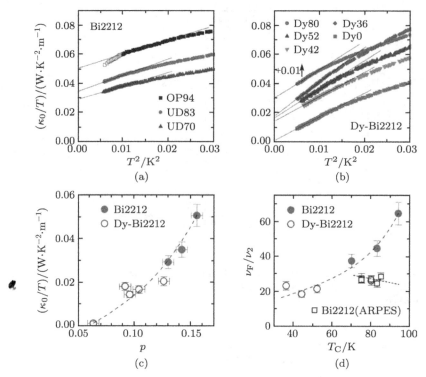

Fig. 11.15. The low-temperature thermal conductivity of Bi2212 single crystals with different carrier concentrations [35]. Carrier concentration can be adjusted by changing oxygen content (a) or Dy doping (b). (c) Shows the dependence of residual thermal conductivity on the carrier concentration. (d) Shows the relation between the gap parameter v_F/v_2 (v_2 is v_Δ in this article) obtained by Eq. (11.6) and the transition temperature, compared with the measured v_F/v_2 by ARPES.

increased from 0% to 3%, the impurity scattering strength became about 10 times larger, but the corresponding change in the κ_0/T was much smaller. They pointed out that this was just the meaning of the "universal" thermal conductivity. However, the theory of SCTMA [4, 38] predicted that if the impurity scattering strength was too big to deviate from the prediction of the universal thermal conductivity behavior $k_B T \ll \gamma \ll \Delta_0$, the κ_0/T should be larger than the universal thermal conductivity, which was opposite to the experiment results showed in Fig. 11.16. In fact, a slight Zn-doping could induce the increase of QP thermal conductivity in the overdoped samples, which is consistent with the prediction from the classical d-wave theory.

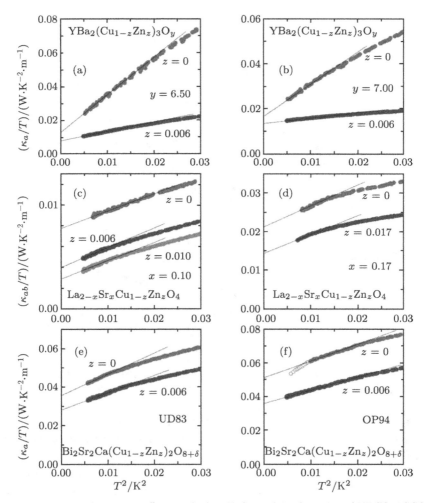

Fig. 11.16. Zn-substitution effect on the low-T thermal conductivity of YBCO, LSCO, and Bi2212 in the underdoped and optimally doped regimes. The doping levels of Bi2212 samples are indicated by the T_C values of pure ones (Zn-substituted ones are annealed to achieve the same oxygen contents as the pure ones). The solid lines are linear fits to the lowest-T data to extract κ_0/T, except for the $z = 0$ sample in (f) Ref. [35].

There were also some other experimental results which were different from the behavior of the universal thermal conductivity. For example, Sun et al. [35] found that the optimally and underdoped Bi2212 samples showed strongly anisotropic QP transport behavior, that is, the thermal conductivity along the a axis and b axis in the CuO_2 plane are not the same.

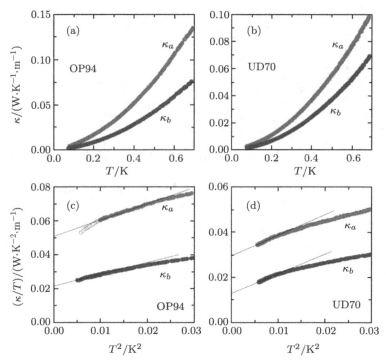

Fig. 11.17. Thermal conductivity of underdoped and optimally doped Bi2212 crystals along the a and b axis [35]. The solid lines are the fit to the data using Eq. (11.4).

As shown in Fig. 11.17, κ along the a axis is nearly twice the one along the b direction, which was also not explained by the classical d-wave theory, only if the superconducting gap is not of fourfold symmetry. Sutherland et al. [26] noted that there was a minimum in Eq. (11.6) predicted theoretically, but the obtained κ_0/T from the underdoped LSCO was smaller than the theoretical minimum, which was another phenomenon inconsistent with the universal thermal conductivity.

As we mentioned before, the experimental phenomena in HTSC which are inconsistent with the universal thermal conductivity are not difficult to understand. The universal thermal conductivity is established under a simple theoretical framework, the object is a classical d-wave superconductor. From the viewpoint of experiments, recent STM experiments showed that the electronic states of HTSC are strongly inhomogeneous [12–18], which was not taken into account in early theoretical works [4, 5]. From the viewpoint of theory, in 2002 Atkinson and Hirschfeld [19] had found the influence of disorder or inhomogeneity on the heat transport of QPs. They reported that the density of state of the low-energy QPs could be suppressed or even a small pseudogap at the Fermi level having formed due to the disorder or inhomogeneity in the superconducting state, which could strongly suppress the conduction of QPs. On the other hand, Andersen and Hirschfeld [20] recently considered that the introduction of non-magnetic impurities would affect the magnetic properties of the underdoped samples and meanwhile strongly suppress the transport of QPs.

In brief, the validity of the universal thermal conductivity is still questionable. A majority of experimental results actually could not support this scenario. In particular, the superconducting gap calculated from the thermal conductivity combined with Eq. (11.6) or (11.8) in the underdoped region is quite unreliable. The more important point is that the low-energy QP heat transport is definitely observed in HTSC and most of them showed that the residual thermal conductivity κ_0/T is finite, and κ_0/T becomes larger with the increasing carrier concentration. The existence of these low-energy QPs is undoubtedly the strongest evidence to support the unconventional superconducting symmetry with gap nodes, and the presence of nodal QPs calls for the new theory of the electronic states of HTSC. Particularly in the underdoped region, whether there are nodal excitations in a strongly correlated system, which has the spin/charge stripe phase and singular normal-state properties like the pseudogap, is an essential issue in theory.

11.3. Very Low-Temperature Thermal Conductivity and Metal–Insulator Crossover

The properties of the ground state of HTSC in the normal state has attracted a lot of attention [40]. Since the upper critical field is too high, it is a challenging task to detect the nature of the ground state experimentally. For LSCO and Bi2201 crystals with relatively low T_C and H_{C2}, one can probe the ground state properties directly using pulsed magnetic field up to 60 T to destroy the superconductivity completely and measure the DC resistivity. Ando et al. [41, 42] measured the ab-plane resistivity of LSCO single crystal in 60 T field and found that there was a metal–insulator crossover (MIC) of the ground state near the optimal carrier concentration ($x = 0.16 - 0.17$ hole/Cu). In the high carrier concentration region, the resistivity at zero-temperature limit showed to a finite value, implying the metallic ground state; in the low carrier concentration region, the resistivity at low temperatures behaved as a $\lg(1/T)$ divergence, which is a special insulating ground state and for which there is no appropriate theoretical explanation at present. For the La-doped Bi2201 system, the highest T_C is about 36 K. Similar to LSCO, 60 T pulsed magnetic field can completely destroy the superconductivity of BLSCO too. Ono et al. [43] had done a similar electrical measurement under the pulsed magnetic field and found that the MIC point of the Bi2201 system lies near the carrier concentration of 0.12 hole/Cu, different from the case of LSCO. For other cuprate superconductors like YBCO and Bi2212, the H_{C2} is too high to achieve in the laboratory, it is therefore not practical to measure the resistivity without destroying the superconductivity completely. It is therefore necessary to find other experimental methods to detect the properties of the ground state.

Recently it has been found that one can characterize the properties of the normal state in HTSC from the magnetic-field dependence of thermal conductivity at very low temperatures. Here, we first introduce briefly the process of cognition on the field dependence of low-temperature thermal conductivity.

An early interesting finding on the thermal conductivity of HTSC is the "plateau" behavior of $\kappa(H)$ discovered by Krishana et al. [44] in Bi2212 single crystals. It was found that with the increasing magnetic field, the κ first decreased gradually and then no longer changed above a characteristic magnetic field H_k. This phenomenon emerged in a temperature range from about 20 K down to 4 K, and the characterized magnetic field H_k decreased with temperature decreasing. At the beginning, this plateau behavior was considered to have originated from some kind of magnetic-field induced phase transition [44, 45]. But soon after Aubin et al. [46] found that the plateau behavior was related to the history of applying magnetic field, excluding the possibility of phase transition at H_k. Then, it was gradually realized that magnetic field could induce two different effects on the QP transport. One was the dominant vortex scattering on QPs at higher temperature, which weakened the QP heat conductivity [37, 47, 48]; the other one was the shift of QP energy spectrum with the increasing field due to the presence of the superconducting current, resulting in an increase in the number of QPs, which is the so-called "Volovik" effect [49]. The later effect was dominant at low temperatures, which lead to the increase of QP thermal conductivity with the increasing magnetic field [48]. The competition between these two effects therefore gave rise to the $\kappa(H)$ plateau phenomenon. This interpretation was quickly proved by the thermal conductivity at very low temperatures [22, 39, 50]. These evidences indicated that as the temperature went down to the very low temperature region, the dependence of κ on magnetic field can reflect the intrinsic properties of superconductors because of the disappearance of the vortex scattering.

Sun et al. [51] studied systematically the change of thermal conductivity in the magnetic field for a series of LSCO single crystals with various doping from the underdoped to the overdoped region ($x = 0.08, 0.10, 0.14, 0.17$, and 0.22). Note that in these samples there was finite residual thermal conductivity or the metallic transport behavior in zero field, as already mentioned before. For the optimally and overdoped samples, their κ increased with the increasing magnetic field at very low temperatures, as shown in Fig. 11.18. At relatively high temperatures, the scattering of vortex on QPs was dominant so that the κ decreased with the increasing magnetic field. That is to say, the nature of the ground state can be detected only at very low temperatures. The behavior shown in Fig. 11.18 was firstly discovered by Chiao et al. [22] in optimally doped YBCO, the κ_0/T was increased with increasing the applied magnetic field, which resulted from the "Volovik" effect [49]. Therefore, in the metallic phase of

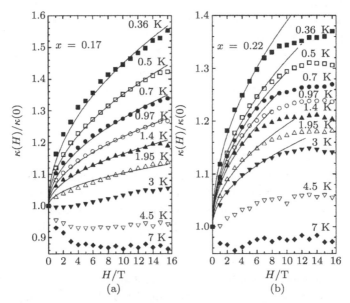

Fig. 11.18. The magnetic-field-induced thermal conductivity of optimally doped and overdoped LSCO single crystals at low temperatures Ref. [51].

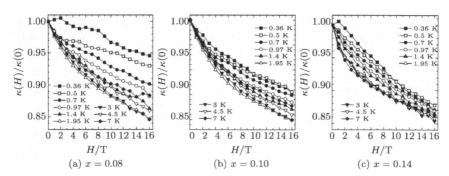

Fig. 11.19. The magnetic-field-induced thermal conductivity of underdoped LSCO single crystals at low temperatures Ref. [51].

superconducting materials, the number of QP excitations can be increased by increasing the magnetic field which leads to the enhancement of κ, a classical behavior of d-wave superconductors.

However, in the underdoped LSCO crystals, although the increase of the magnetic field could cause an increase in the number of QPs, the low-temperature thermal conductivity was reduced [51], as shown in Fig. 11.19. It is notable that the "Volovik" effect is the classical behavior of d-wave superconductors and is independent of the carrier concentration, which had been supported by low-temperature heat-specific results [52]. The reduction of κ in the underdoped LSCO with the increasing magnetic field indicated that the QPs were localized. Considering the spin/charge stripe phase in HTSC revealed by neutron scattering [53–55], and STM [56], Sun et al. [51] first pointed out that this peculiar field-induced QP localization was closely related to the static spin/charge order caused by the magnetic field in the underdoped HTSC, and this point of view was soon verified by some theoretical works [57, 58]. Clearly, the MIC behavior from the dependence of low-temperature κ on magnetic field is the same as the one by resistivity measurement; both techniques revealed the critical carrier concentration at the optimal doping. This work indicated that the ground-state transition from the metallic behavior in the high doping region to the insulating state in the low doping region can be detected by low-temperature heat transport, as shown in Fig. 11.20.

Hawthorn et al. [59] reported essentially the same results, and they measured the thermal conductivity of LSCO crystals in zero field and high field at lower temperatures, as shown in Fig. 11.21. In the highly doped LSCO, the field-induced enhancement of QPs resulted in the increase of κ.

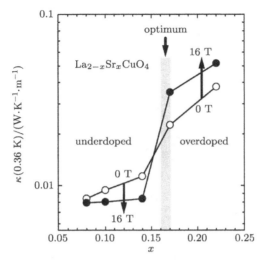

Fig. 11.20. The thermal conductivity of LSCO single crystal with different carrier concentrations at 0.36 K in 0 T and 16 T. The effect of magnetic field on the thermal conductivity demonstrated the cross-over of the QP transport properties in the vicinity of the optimal carrier concentration Ref. [51].

If the doping was lowered, the result was opposite to the former one, which was consistent with the results by Sun et al. [51]. For the non-superconducting LSCO samples ($x \leq 0.05$), the κ did not vary with magnetic field indicating no electron contribution to thermal conductivity at low temperatures, which was the same as the ground-state properties of non-superconducting LSCO insulators [40]. In addition, since the measurement was carried out down to tens of mK, one can obtain the dependence of κ_0/T on the field directly. The authors considered that their results reflected a kind of thermal metal–insulator crossover.

For BSLCO system with $T_C = 30$ K, Ando et al. [33] proceeded with a similar experimental exploration. They measured the thermal conductivity of a series of single crystals with different carrier concentrations (p) and then compared the magnetic-field dependences of κ at very low temperatures, as shown in Fig. 11.22. The result showed that at 0.36 K, the κ of the samples with $p = 0.13$ and 0.15 increased with the magnetic field; while for $p = 0.10$ and 0.11 samples, the κ decreased with increasing the magnetic field. Similar to the LSCO system, the dependence of κ on the magnetic field at low temperatures indicated that MIC occurred at $p = 0.12$, the same as the previous resistivity measurement under 60 T.

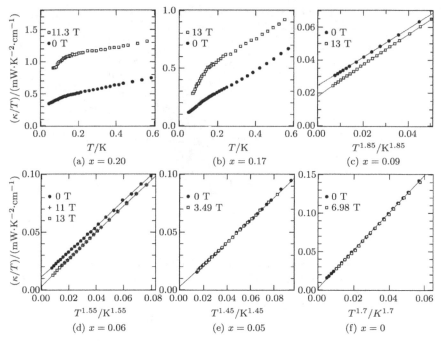

Fig. 11.21. Thermal conductivity of the LSCO single crystals with different carrier concentrations in zero and external magnetic field at very low temperature. The panel is from Ref. [59].

The results of LSCO and BSLCO suggested that the dependence of κ on the magnetic field at low temperatures can reflect the MIC of ground state of HTSC, and that the magnetic field required is not very high for experiment. It is not unexpected that the transport properties of QPs in the low field should gradually transfer to the normal state behavior with increasing the magnetic field. Although no direct experiments verified whether the transport properties of QPs change in this process, there is indeed reasonable correspondence between those existing phenomena.

So, the low-temperature thermal conductivity can be used to detect the MIC of the ground state of those HTSC in which the upper critical fields are too high to measure the resistivity, such as YBCO and Bi2212, etc. Sun et al. [32] systematically studied the low-temperature thermal conductivity of a series YBCO single crystals with oxygen content y from 6.45 to 7.00, as shown in Fig. 11.23. For the optimally doped samples, an early work by Chiao et al. [22] had proved that the QP transport was metallic at

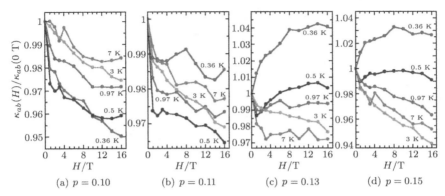

Fig. 11.22. The magnetic-field dependence of thermal conductivity of $Bi_2Sr_{2-x}La_x CuO_{6+\delta}$ single crystal with different carrier concentrations (p) at low temperatures [33].

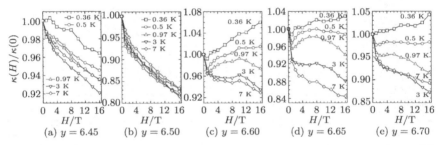

Fig. 11.23. The field dependence of low-temperature thermal conductivity for YBCO single crystals with different oxygen contents [32].

low temperatures. Results by Sun et al., showed that even if the oxygen content was gradually decreased to 6.60, the transport behavior of QP was still metallic at low temperatures, but the samples of oxygen content below 6.50 behaved as localization behavior at low temperatures. Thus, in the YBCO system, the MIC is located at about $y = 6.55$ (the carrier concentration is about 0.07 hole/Cu). The phase diagram of the ground state of YBCO is shown in Fig. 11.24, where the author also pointed out that $y = 6.55$ corresponded to a transition point of in-plane resistivity anisotropic behavior. For $y < 6.55$, the resistivity behavior indicated the formation of static charge stripes phase [60]. Therefore, the MIC is closely related to the charge self-organization behavior, which is similar to the LSCO system.

So far, the work by Sun et al. [32] is still the only one showing the ground-state properties of YBCO. Recently, Rullier-Albenque et al. [61]

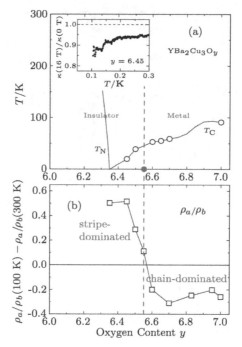

Fig. 11.24. (a) Phase diagram of YBCO, T_N and T_C indicate anti-ferromagnetic transition temperature and the superconducting transition temperature, respectively. The red solid dots show the metal–insulator transition point of the ground state, given by the low-thermal conductivity. (b) The resistivity anisotropy ρ_a/ρ_b at 100 K and 300 K, indicating the formation of charge-stripe phase in the low-doped samples [32].

measured the resistivity of YBCO single crystals with oxygen content of 6.60 in 60 T field and down to about 4 K. It was suggested that the sample exhibited the metallic behavior at low temperatures, which was consistent with the thermal conductivity results by Sun *et al.* [32]. But at present there is no resistivity data at lower temperatures and high field in the lower-doped YBCO crystals. Subsequently, the μSR experiments [62] also demonstrated that the MIC of the ground state in YBCO and LSCO crystals and the peculiar electron localization in the low-doped region originated from the competition between the field-induced static anti-ferromagnetism and superconductivity. On the other hand, neutron scattering [63] indicated that the ground-state MIC in YBCO was associated with the dramatic changes of spin-excitation spectrum. The neutron resonance peak and the apparent spin energy gap were likely the fundamental characteristics of

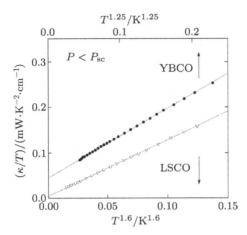

Fig. 11.25. The low-temperature thermal conductivity of underdoped YBCO ($y = 6.33$) and LSCO ($x = 0.05$) single crystals. The solid lines are the fit to Eq. (11.1). The panel is from Ref. [64].

the superconductors with metallic ground state, while the superconductors with insulating ground state behaved as gapless spin excitations.

However, the following experimental work shown in Fig. 11.24 by Sutherland et al. [64] questioned the ground-state property of YBCO. They measured the low-temperature thermal conductivity of non-superconducting underdoped YBCO single crystal ($y = 6.33$). Different from the analysis method in Eq. (11.4), they used the formula:

$$\frac{\kappa}{T} = a + bT^{\alpha-1}, \tag{11.9}$$

where $\alpha < 3$, to fit the low-temperature data and obtain a non-zero residual thermal conductivity, as shown in Fig. 11.25. (The origin of this formula would be introduced later.) This meant that in the non-superconducting underdoped YBCO a 'thermal metal' phase was present. The results were inconsistent with the insulating ground state in YBCO with $y < 6.55$ reported by Sun et al. It should be noted that Sutherland et al.'s result might be of importance in physics, which indicated that a "nodal metal" phase can exist in the non-superconducting underdoped HTSC [65].

To address this controversial issue, Sun et al. [66] used a more direct means to study the ground state of the non-superconducting underdoped YBCO ($y = 6.35$) single crystals. As this YBCO crystal is non-superconducting, its property of the ground state can be detected directly

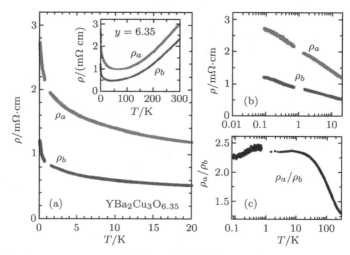

Fig. 11.26. Low-temperature resistivity of YBCO ($y = 6.35$) single crystal along the a and b directions below 20 K; inset shows the data up to 300 K [66]. (b) Semilog plot of the data to show the $\lg(1/T)$ divergence. (c) T dependence of the in-plane anisotropy, ρ_a/ρ_b.

from the DC resistivity measurement. In addition, the sample used by Sutherland et al. was the twin crystal, which may affect the low-temperature transport behavior of QPs [67]. Therefore, Sun et al. [66] carried out detailed measurements of low-temperature resistivity and thermal conductivity along the a axis (the direction perpendicular to the Cu–O chains) and the b axis (along the Cu–O chain direction) by using a large number of non-twinned underdoped YBCO single crystals ($y = 6.35$). The results are shown in Figs. 11.26 and 11.27. First, the a-axis and b-axis resistivity at low temperatures diverged as $\lg(1/T)$, suggesting the insulating ground state. Secondly, the low-temperature κ along the a axis and b axis cannot be fitted using Eq. (11.4), but it was obviously seen that κ/T was gradually heading to zero as $T \to 0$ K. When Eq. (11.4) cannot describe the low-temperature thermal transport, one possibility that was proposed by Sutherland et al. [25, 64], that is, the low-temperature κ might be fitted by Eq. (11.1). Sun et al. [66] also attempted to analyze data in this way as shown in Fig. 11.27(b). However, although Eq. (11.1) could indeed fit the low-temperature κ along the a axis and b axis well, the fitting parameter a is negative, indicating no real physical meaning. Therefore, from the results in Fig. 11.27 one can conclude that for the non-superconducting crystal of underdoped YBCO ($y = 6.35$), the residual thermal conductivity is zero

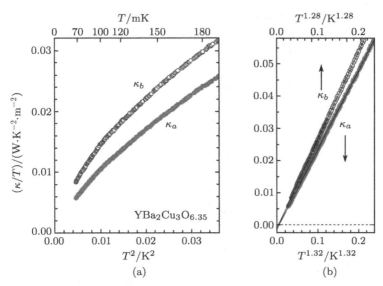

Fig. 11.27. Low-temperature thermal conductivity of YBCO ($y = 6.35$) along the a and b directions [66]. (a) The relation between κ/T and T^2 cannot be fitted linearly. (b) The low-temperature data can be well fitted by Eq. (11.1), but the fitting parameter a is negative and does not have any physical meaning.

and the ground state of this sample is insulator, which is actually the same as the resistivity measurement shown in Fig. 11.26.

Results by Sun et al. [66] on the non-superconducting underdoped YBCO ($y = 6.35$) single crystals were consistent with the previous ground-state phase diagram shown in Fig. 11.24. Afterward, Doiron-Leyraud et al. [68] gave further analysis of the low-temperature resistivity of non-superconducting underdoped YBCO ($y = 6.33$), which displayed insulator behavior like $\lg(1/T)$, confirming the result by Sun et al. But Doiron-Leyraud et al. [68] obtained a non-zero residual thermal conductivity using Eq. (11.1). The authors considered that there may be another kind of QP excitations in the underdoped copper oxides, which was the electronic insulator but was a thermal conductor. This interesting phenomenon needs to be explained and proved in theory.

It is worth mentioning that the insulating $\lg(1/T)$ behavior in the non-superconducting YBCO is similar to the behaviors of underdoped LSCO and BSLCO in high magnetic field up to 60 T, which reflects that the detected properties in 60 T field may be the real ground-state nature of the normal state.

In summary, the dependence of thermal conductivity on the field at low temperatures can probe the nature of the ground state in high field. However, there are still some remaining problems in analyzing the experimental data. In above discussions, the effect of the magnetic field was entirely attributed to the QPs, it is therefore necessary to study further whether the thermal conductivity of phonons is invariant with the magnetic field. Note that a recent paper reported that the phonon conductivity in the parent insulator of cuprates is strongly dependent on the magnetic field [69].

11.4. Wiedemann–Franz Law

Another importance of studying the low-temperature thermal conductivity is that one can examine whether some kind of HTSC could be described by the Fermi liquid theory through testing the Wiedemann-Franz (WF) law.

Hill *et al.* [70] first examined the WF law for the optimally doped $Pr_{2-x}Ce_xCuO_{4-y}$ (PCCO) single crystal. Due to the relatively low H_{C2} in PCCO, it is easy to obtain the normal-state properties. On the one hand, the authors measured the resistivity in strong field (up to 14 T), finding that the optimally doped PCCO crystal displayed the metallic electric transport behavior down to the zero-temperature limit, and gave residual resistivity. On the other hand, they measured the low-temperature thermal conductivity in zero-field and external magnetic field. A surprising finding was that although the κ changed with the field significantly, the residual thermal conductivity κ_0/T was nearly zero and invariant with the magnetic field. It meant that the Lorentz number from Eq. (11.3) was zero instead of $L_0 = 2.44 \times 10^{-8}$ W/K^2 required by the WF law.

This experimental result soon attracted intensive attention because it revealed that the electronic state of the cuprate superconductors cannot be described by the Fermi liquid theory. However, as we will mention in the following section, this result was an extrinsic behavior, therefore many theoretical works on this issue will not be introduced here.

Although the above phenomenon in PCCO was verified to be extrinsic, it stimulated people to study further on the validity of WF law in HTSC. Proust *et al.* [71] examined the overdoped Tl2201 single crystal with $T_C = 13$ K, in which the carrier concentration was about 0.26 hole/Cu. As shown in Fig. 11.28, from the resistivity and thermal conductivity at low temperatures and high field, they obtained the residual resistivity and residual thermal conductivity of the ground state in the

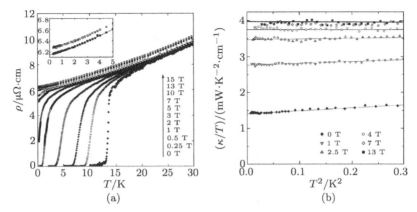

Fig. 11.28. (a) The in-plane resistivity of Tl2201 single crystal at low temperatures, the upper critical magnetic field is about 13 T. (b) Low-temperature thermal conductivity of Tl2201, the solid lines are the fit by $\kappa_0/T = a + bT^2$. The right panel is from Ref. [71].

normal state (13 T and $T \to 0$ K), and calculated the Lorentz ratio $L = \rho_0 \kappa_0/T = (0.99 \pm 0.01)L_0$. It indicated that in this material, the WF law was strictly obeyed. Nakamae et al. [72] studied the heavily overdoped non-superconducting LSCO ($x = 0.30$) samples, which also indicated the validity of WF law. Therefore, the conventional Fermi liquid theory may describe the ground state of the overdoped HTSC.

As we know, the cuprate superconductors display pseudogap, quantum critical behavior, spin/charge order and other non-Fermi liquid behaviors when the carrier concentration gradually decreases through the optimal doping to the underdoping, in which case the invalidity of WF law is naturally expected. It is notable that if the ground state is insulating (as mentioned before, it is the case for most materials in the underdoped region), it is not meaningful to discuss the WF law. However, materials with metallic ground state in the underdoped region are helpful for studying the WF law in the strongly correlated systems.

Bel et al. [73] and Proust et al. [34] studied a series of $Bi_{2+x}Sr_{2-x}CuO_{6+\delta}$ single crystals from overdoped to underdoped region by changing the relative content of Bi and Sr (carrier concentration is $0.19 - 0.14$ hole/Cu). The magnetoresistance showed that the magnetic field of 25 T was high enough to completely destroy the superconductivity of these samples and get the normal-state properties. The results in Fig. 11.29 were given from the resistivity and thermal conductivity at low temperature and in 25 T field. The transport properties of the overdoped samples satisfied

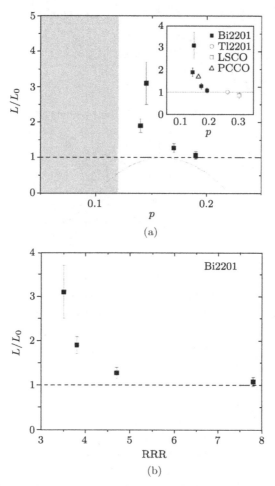

Fig. 11.29. (a) The relation between the normalized Lorentz number (L/L_0) and the carrier concentration p, and (b) the residual resistivity ratio RRR, where RRR was the ratio of room-temperature resistivity and the residual resistivity in 25 T, which characterized the degree of disorder in the sample. The shaded area ($p < 0.12$) in (a) shows the insulating ground state. The panel is from Ref. [34].

the WF law, further confirming the above overdoped behaviors on Tl2201 and LSCO. For the optimally doped samples, the transport behavior started to deviate from the WF law. Upon further decreasing the carrier concentration, the deviation of experimental value of L/L_0 from the expected one became significant, which suggested that the Fermi-liquid theory cannot be used to describe the electronic states of HTSC with the decreasing carrier concentration. Actually, a lot of theories have been proposed to describe the HTSC. One picture is the Luttinger liquid, that is, in the quasi-1D correlated electronic system, the picture of QP fails while the spin and charge excitations can be separated from each other [74, 75]. For another example, when the quantum critical behavior becomes dominant due to the order competition, the Fermi liquid picture breaks down and various non-Fermi liquid behaviors emerge [76, 78]. Proust et al. also compared L/L_0 with the residual resistivity ratio (RRR), and considered that the deviation from the WF law near the MIC point originated from disorder [79–81].

11.5. Acquirement and Analysis of the Intrinsic Low-Temperature Thermal Conductivity Data

It is known from above that in HTSC there are a lot of consistent results in the low-temperature heat transport, but meanwhile much dispute still exists on some issues. In this section, how to get and analyze the experimental intrinsic data will be introduced, which is a key in the study of thermal conductivity.

It is actually very difficult to get the intrinsic data of the very low-temperature heat transport. The question is whether one can get the transport behavior of QPs accurately from the thermal conductivity at the temperature down to tens of mK. Since at such low temperatures the electron temperature may not be the same as that of phonons, it is then very difficult to measure the electron temperature accurately. This problem exists in many materials, and the theoretical discussion has been proposed for HTSC. Due to the easy coupling of phonons with the surroundings, the electron thermal conductivity can only be determined accurately in the following two cases. One is the situation that the thermal resistance of contacts is small enough to be negligible, compared to the thermal resistance of QPs in the sample. The other one is that the electron–phonon coupling is so strong that the electron temperature is the same as that of phonon. In HTSC, especially those with large carrier concentration, where the thermal conductivity of QPs is dominant (which is increased with increasing the carrier concentration as mentioned before), the first condition is not met. On the other hand, it is known from the ARPES results that the electron–phonon coupling weakens apparently with the increasing carrier concentration in HTSC, so the second condition is also not easy to achieve.

For these reasons, the extrinsic phenomena are often observed in the thermal conductivity data at very low temperatures, especially the sudden decrease of κ at low temperatures in the optimally doped and overdoped LSCO and Bi2212 samples and the electron-type superconductor PCCO [31, 35, 36, 59, 70]. For example, there are some problems in the data of overdoped ($x = 0.20$) and optimally doped ($x = 0.17$) LSCO shown in Figs. 11.21(a) and 11.21(b). In the zero field, there is an obvious kink in the data below 0.15 K. In the high field, due to the increase of QP thermal conductivity, the kink shifts to higher temperature (0.18 K) and becomes more distinct. A similar phenomenon was also observed in Takeya et al.'s results of LSCO (Fig. 11.9) and Sun et al.'s Bi2212 data (Figs. 11.15–11.17). In addition, the experiment by Hill et al. [70] showing that the residual thermal conductivity in the optimally doped PCCO is almost zero and is invariant with the applied magnetic field, can be ascribed to the decoupling between electrons and phonons [82], which is similar to the case in LSCO and Bi2212.

Smith et al. [82] carried out a deep theoretical study on the issue. When a heat current Q flows through the sample, the route of heat flux in the sample can be equivalent to a heat circuit shown in Fig. 11.30, including the thermal contact resistance $R_{ph(c)}$ and $R_{el(c)}$, the thermal resistance of QPs and phonons, R_{el} and R_{ph}, and the thermal resistance of electron–phonon coupling R_{el-ph}. They calculated the relation between the κ_{el}/T and the parameter r at very low temperature, where r is defined as $R_{el(c)}/R_{el}$, the ratio of the thermal contact resistance and electron thermal resistance. In Fig. 11.31, when r is not very small, the measured κ_{el}/T is much smaller

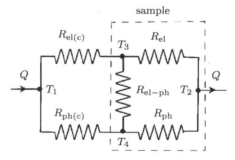

Fig. 11.30. The equivalent heat flux diagram in the thermal conductivity measurement. $R_{ph(c)}$ and $R_{el(c)}$ are the thermal resistances of phonon and electron heat flux before entering into the sample, R_{ph} and R_{el} are the thermal resistances of phonon and electron heat flux in the sample and R_{el-ph} is the thermal exchange between electron and phonon. T_1, T_2, T_3, and T_4 are the temperatures of different positions. The panel is from Ref. [82].

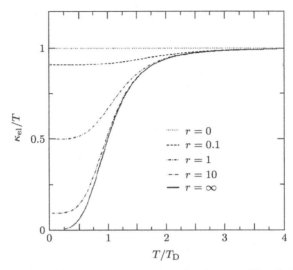

Fig. 11.31. The relation between the thermal conductivity $\kappa_{\rm el}/T$ and the temperature with different strength of thermal contact resistance (r). The panel is from Ref. [82].

than the real value, especially since when the temperature is below the characteristic temperature $T_{\rm D}$, the measured $\kappa_{\rm el}/T$ will decrease rapidly as the temperature decreases. $T_{\rm D}$ is a characteristic temperature related to the electron–phonon coupling and heat exchange. When $T < T_{\rm D}$, the temperature and the thermal conductivity of electrons cannot be properly measured due to the electron–phonon decoupling, and the observed thermal conductivity decreases rapidly. Their results explained well the origin of abnormal low-temperature thermal conductivity in PCCO, LSCO and Bi2212 systems.

The electron–phonon decoupling could lead to the extrinsic experimental results from the low-temperature thermal conductivity, which requires us to pay attention to the truth of the data. It is an effective method to compare the slope from the κ/T vs. T^2 relation with the lattice specific heat. It is known that the phonon thermal conductivity at low temperatures is proportional to T^3 and the coefficient depends on the lattice heat capacity and the sample size (the boundary scattering limit). However, it has been found recently that the phonon thermal conductivity is so complicated that it is difficult to separate the electronic term from the phonon conductivity, even if the data are intrinsic.

In the earliest experiments, Taillefer et al. [21] measured the thermal conductivity of YBCO ($y = 6.0$) single crystal without charge carriers, the

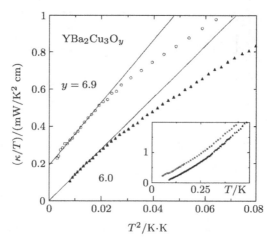

Fig. 11.32. Low-temperature thermal conductivity of YBCO single crystal ($y = 6.0$ and $y = 6.9$) along the a direction. The panel is from Ref. [21].

low-temperature data could be fitted well using Eq. (11.4) below 130 mK, as shown in Fig. 11.32. It is important that the fitting parameter b is consistent with the phonon specific heat coefficient and the sample size, meanwhile the residual thermal conductivity a was zero, which was reasonable for this insulator. Based on this, Taillefer et al. [21] and many other research groups usually used this method to fit the κ at low temperatures in order to separate the electron thermal conductivity from the phonon conductivity. It is necessary to note that the equation $\kappa/T = a + bT^2$ actually requires a phonon term proportional to T^3 and an electronic term linear with T. However, these two preconditions cannot be met easily.

As is already known, at very low temperature the boundary scattering limit is established, and the mean free path of phonon l_ph is determined by the sample surface or boundary scattering, that is [83]:

$$l_\mathrm{ph} = W \equiv \frac{2}{\sqrt{\pi}} \bar{w}, \qquad (11.10)$$

where $\bar{w}$ is the geometric mean width of the sample. The T^3 temperature dependence of phonon thermal conductivity in boundary scattering region is therefore the same as that of the lattice heat capacity. It has already been found that when the sample surface is very smooth, phonons can be reflected specularly at the boundaries, and the mean free path of phonons can exceed the geometry size and strengthen with decreasing the temperature, resulting

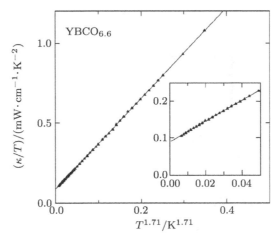

Fig. 11.33. Low-temperature thermal conductivity of YBCO single crystal ($y = 6.6$) along the a direction. The solid line is the fit using Eq. (11.9). The panel is from Ref. [26].

in a reduction of phonon thermal conductivity slower than T^3, showing a $T^\alpha (\alpha < 3)$ behavior. For example, Pohl and Stritzker [84] reported a $T^{2.77}$ ($\alpha = 2.77$) dependence of phonon thermal conductivity in a polished Al_2O_3 sample at low temperatures.

Sutherland et al. [26] pointed out that for the general materials, the T^3 relation of phonon thermal conductivity below 1 K was applicable, but for HTSC, the T^3 relation could be fitted only in a narrow temperature region (usually below 130 mK). They attributed this to the specular reflection, so one could apply the formula (11.9) to fit the low-temperature κ of YBCO and LSCO in a wider temperature region. As shown in Fig. 11.33, the κ of YBCO ($y = 6.6$) can be fitted well using Eq. (11.9) in a wide temperature region (from tens of mK to 0.6 K). The fitting parameter α was found to be 2.71, meaning that the specular reflection may play an important role in phonon heat conduction for HTSC. Subsequently, Sutherland et al. [26] and Hawthorn et al. [59] also used Eq. (11.9) to analyze thermal conductivity data of YBCO and LSCO and obtained residual thermal conductivity, where the parameter α mostly lay between 2.45 and 2.85.

It should be noted that although the data analysis by Sutherland et al. [26] can be used in a wide temperature region, the obtained residual thermal conductivity using this method by Taillefer et al. was negative in YBCO ($y = 6.0$), which was obviously unreasonable. This may be due to either the inaccurate experimental data in early times or the problematic data

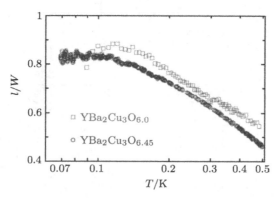

Fig. 11.34. The temperature dependence of the ratios of phonon mean free path l to the geometric size W of YBCO single crystals ($y = 6.0$ and 6.45) [85].

analysis. Ando et al. [85] pointed out that whether the specular reflection takes effect can be judged by calculating the phonon mean free path and comparing it to the geometric size of the sample. If the specular reflection worked, the mean free path would be larger than the geometric size and keep increasing with the decreasing temperature, which has already been evidenced in Al_2O_3. Ando et al. analyzed the data similarly of YBCO with the oxygen content of 6.0 by Taillefer et al. [21] and YBCO with the oxygen content of 6.45 by Sun et al. [32]. It was found that the ratio of the mean free path l to the sample geometry size W, was less than 1 until 100 mK, as shown in Fig. 11.34. This result was opposite to the prediction of the specular reflection. If there was specular reflection, the ratio should be significantly larger than 1 at such low temperatures. Therefore, whether the method proposed by Sutherland et al. [26] is valid and whether the specular reflection is working are quite questionable and needs further investigations.

Electron–phonon scattering could also make the thermal conductivity deviate from T^3 relation. Hawthorn et al. [6] reported that in the overdoped Tl2201 system, due to the significant electron–phonon scattering at low temperatures, the phonon thermal conductivity showed a T^2 relation with temperature. The low-temperature thermal conductivity can be described as:

$$\frac{\kappa}{T} = a + bT. \tag{11.11}$$

A similar case was observed in the Bi2212 system [36]. Note that the effect of electron–phonon coupling in the overdoped Tl2201 at low temperatures

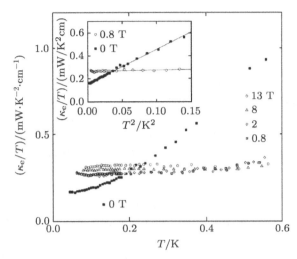

Fig. 11.35. Thermal conductivity of QPs of YBCO single crystal ($y = 6.99$) in magnetic field. The panel is from Ref. [87].

is contradictory to that expected in the overdoped LSCO, Bi2212 and PCCO. Therefore, the temperature dependence of thermal conductivity in HTSC at low temperatures is actually very complicated and lacks accurate understanding.

Except for the uncertainty of the temperature dependence of phonon thermal conductivity at low temperatures, whether the electron thermal conductivity κ_e is linear to the temperature is also not clear. Although the microwave electric conductivity measurement [86] suggested that in HTSC the QP scattering was elastic at low temperatures, the real case may not be so simple as seen from the recent low-temperature thermal conductivity results. Hill et al. [87] measured a very pure YBCO single crystal and found that even at very low temperatures, the QP thermal conductivity was not the simple linear relation with temperature, but proportional to T^3, as shown in Fig. 11.35. This result immediately told us that the fitting to the very low thermal conductivity of YBCO by Sutherland et al. [26] using Eq. (11.9) was problematic without including the T^3 dependence of QP thermal conductivity. On the other hand, it is more reliable to fit the κ of YBCO using Eq. (11.4) when the temperature dependence of phonon thermal conductivity is still controversial. Even if the electron conduction is proportional to T^3, one can still obtain the correct residual thermal conductivity using Eq. (11.4) (the parameter a).

11.6. Conclusions

The very low-temperature thermal conductivity is a powerful tool to study the transport behavior of QPs and the superconducting gap symmetry in HTSC. In previous works, many important results, especially the observed finite residual thermal conductivity, directly indicated the unconventional gap symmetry and the existence of nodes. In addition, the magnetic-field-induced dependence of QP thermal conductivity is an efficient technique to detect the nature of the ground state. For many other unconventional superconductors, the relations between the low-temperature thermal conductivity and the temperature/magnetic field, and the anisotropic behavior in the magnetic field are often used to characterize the superconducting gap symmetry, and a great deal of achievement has been made [88], which has not been introduced here.

However, as mentioned in this chapter, there are some difficulties in measuring and analyzing the thermal conductivity data. At present, there are many controversial issues on the experimental results calling for further study and discussions.

References

[1] R. Berman, *Thermal Conduction in Solids* (Oxford University Press, Oxford, 1976).
[2] C. B. Satterthwaite, *Phys. Rev.* **125**, 873 (1962).
[3] N. E. Hussey, *Adv. Phys.* **51**, 1685 (2002).
[4] M. J. Graf, S. K. Yip, J. A. Sauls et al., *Phys. Rev. B* **53**, 15147 (1996).
[5] A. C. Durst, P. A. Lee, *Phys. Rev. B* **62**, 1270 (2000).
[6] D. G. Hawthorn, S. Y. Li, M. Sutherland et al., *Phys. Rev. B* **75**, 104518 (2007).
[7] C. C. Tsuei, J. R. Keitley, *Rev. Mod. Phys.* **72**, 969 (2000).
[8] P. A. Lee, N. Nagaosa, X. G. Wen, *Rev. Mod. Phys.* **78**, 17 (2006).
[9] S. A. Kivelson, I. P. Bindloss, E. Fradkin et al., *Rev. Mod. Phys.* **75**, 1201 (2003).
[10] S. Sachdev, *Rev. Mod. Phys.* **75**, 913 (2003).
[11] M. Vojta, *Rep. Prog. Phys.* **66**, 2069 (2003).
[12] S. H. Pan, E. W. Hudson, K. M. Lang et al., *Nature* **403**, 746 (2000).
[13] S. H. Pan, J. P. O'Neal, R. L. Badzey et al., *Nature* **413**, 282 (2001).
[14] C. Howald, P. Fournier, A. Kapitulnik, *Phys. Rev. B* **64**, R100504 (2001).
[15] K. M. Lang, V. Madhavan, J. E. Hoffman et al., *Nature* **415**, 412 (2002).
[16] K. McElroy, Lee Jinho, J. A. Slezak et al., *Science* **309**, 1048 (2005).
[17] K. K. Gomez, A. N. Pasupathy, A. Pushp et al., *Nature* **447**, 569 (2007).

[18] A. N. Pasupathy, A. Pushp, K. K. Gomes et al., *Science* **320**, 196 (2008).
[19] W. A. Atkinson, P. J. Hirschfeld, *Phys. Rev. Lett.* **88**, 187003 (2002).
[20] B. M. Anderson, P. J. Hirschfeld, *Phys. Rev. Lett.* **100**, 257003 (2008).
[21] L. Taillefer, B. Lussier, R. Gagnon et al., *Phys. Rev. Lett.* **79**, 483 (1997).
[22] M. Chiao, R. W. Hill, C. Lupien et al., *Phys. Rev. Lett.* **82**, 2943 (1999).
[23] M. Chiao, R. W. Hill, C. Lupien et al., *Phys. Rev. B* **62**, 3554 (2000).
[24] J. Mesot, M. R. Norman, H. Ding et al., *Phys. Rev. Lett.* **83**, 840 (1999).
[25] S. Nakamae, K. Behnia, L. Balicas et al., *Phys. Rev. B* **63**, 184509 (2001).
[26] M. Sutherland, D. G. Hawthorn, R. W. Hill et al., *Phys. Rev. B* **67**, 174520 (2003).
[27] N. E. Hussey, S. Nakamae, K. Behnia et al., *Phys. Rev. Lett.* **85**, 4140 (2000).
[28] K. Tanaka, W. S. Lee, D. H. Lu et al., *Science* **314**, 1910 (2006).
[29] T. Kondo, T. Takeuchi, A. Kaminski et al., *Phys. Rev. Lett.* **98**, 267004 (2007).
[30] M. Le Tacon, A. Sacuto, A. Georges et al., *Nature Phys.* **2**, 537 (2006).
[31] J. Takeya, Y. Ando, S. Komiya et al., *Phys. Rev. Lett.* **88**, 077001 (2002).
[32] X. F. Sun, K. Segawa, Y. Ando, *Phys. Rev. Lett.* **93**, 107001 (2004).
[33] Y. Ando, S. Ono, X. F. Sun et al., *Phys. Rev. Lett.* **92**, 247004 (2004).
[34] C. Proust, K. Behnia, R. Bel et al., *Phys. Rev. B* **72**, 214511 (2005).
[35] X. F. Sun, S. Ono, Y. Abe et al., *Phys. Rev. Lett.* **96**, 017008 (2006).
[36] X. F. Sun, S. Ono, X. Zhao et al., *Phys. Rev. B* **77**, 094515 (2008).
[37] Y. Ando, J. Takeya, Y. Abe et al., *Phys. Rev. B* **62**, 626 (2000).
[38] Y. Sun, K. Maki, *Europhys Lett.* **32**, 355 (1995).
[39] Y. Ando, J. Takeya, Y. Abe et al., *Phys. Rev. Lett.* **88**, 147004 (2002).
[40] M. Imada, A. Fujimori, Y. Tokura, *Rev. Mod. Phys.* **70**, 1039 (1998).
[41] Y. Ando, G. S. Boebinger, A. Passner et al., *Phys. Rev. Lett.* **75**, 4662 (1995).
[42] G. S. Boebinger, Y. Ando, A. Passner et al., *Phys. Rev. Lett.* **77**, 5417 (1996).
[43] S. Ono, Y. Ando, T. Murayama et al., *Phys. Rev. Lett.* **85**, 638 (2000).
[44] K. Krishana, N. P. Ong, Q. Li et al., *Science* **277**, 83 (1997).
[45] R. B. Laughlin, *Phys. Rev. Lett.* **80**, 5188 (1998).
[46] H. Aubin, K. Behnia, S. Ooi et al., *Science* **280**, 9a (1998).
[47] M. Franz, *Phys. Rev. Lett.* **82**, 1760 (1999).
[48] I. Vekhter, A. Houghton, *Phys. Rev. Lett.* **83**, 4626 (1999).
[49] G. E. Volovik, *JETP Lett.* **58**, 469 (1993).
[50] H. Aubin, K. Behnia, S. Ooi et al., *Phys. Rev. Lett.* **82**, 624 (1999).
[51] X. F. Sun, S. Komiya, J. Takeya et al., *Phys. Rev. Lett.* **90**, 117004 (2003).
[52] S. J. Chen, C. F. Chang, H. L. Tsay et al., *Phys. Rev. B* **58**, R14753 (1998).
[53] B. Lake, G. Aeppli, K. N. Clausen et al., *Science* **291**, 1759 (2001).
[54] B. Lake, H. M. Rùnnow, N. B. Christensen et al., *Nature* **415**, 299 (2002).
[55] B. Khaykovich, Y. S. Lee, R. W. Erwin et al., *Phys. Rev. B* **66**, 014528 (2002).
[56] J. E. Hoffman, E. W. Hudson, K. M. Lang et al., *Science* **295**, 466 (2002).
[57] V. P. Gusynin, V. A. Miransky, *Eur. Phys. J. B* **37**, 363 (2004).
[58] M. Takigawa, M. Ichioka, K. Machida, *Physica C* **404**, 375 (2004).

[59] D. G. Hawthorn, R. W. Hill, C. Proust et al., Phys. Rev. Lett. **90**, 197004 (2003).
[60] Y. Ando, K. Segawa, S. Komiya et al., Phys. Rev. Lett. **88**, 137005 (2002).
[61] F. Rullier-Albenque, H. Alloul, C. Proust et al., Phys. Rev. Lett. **99**, 027003 (2007).
[62] J. E. Sonier, F. D. Callaghan, Y. Ando et al., Phys. Rev. B **76**, 064522 (2007).
[63] S. L. Li, Z. Yamani, H. J. Kang et al., Phys. Rev. B **77**, 014523 (2008).
[64] M. Sutherland, S. Y. Li, D. G. Hawthorn et al., Phys. Rev. Lett. **94**, 147004 (2005).
[65] T. Timusk, Phys. World **18**, 31 (2005).
[66] X. F. Sun, K. Segawa, Y. Ando, Phys. Rev. B **72**, R100502 (2005).
[67] Y. Ando, Phys. Rev. Lett. **100**, 029702 (2008).
[68] N. Doiron-Leyraud, M. Sutherland, S. Y. Li et al., Phys. Rev. Lett. **97**, 207001 (2006).
[69] X. F. Sun, I. Tsukada, T. Suzuki et al., Phys. Rev. B **72**, 104501 (2005).
[70] R. W. Hill, C. Proust, L. Taillefer et al., Nature **414**, 711 (2001).
[71] C. Proust, E. Boaknin, R. W. Hill et al., Phys. Rev. Lett. **89**, 147003 (2002).
[72] S. Nakamae, K. Behnia, N. Mangkorntong et al., Phys. Rev. B **68**, R100502 (2003).
[73] R. Bel, K. Behnia, C. Proust et al., Phys. Rev. Lett. **92**, 177003 (2004).
[74] C. L. Kane, M. P. A. Fisher, Phys. Rev. Lett. **76**, 3192 (1996).
[75] T. Senthil, M. P. A. Fisher, Phys. Rev. Lett. **86**, 292 (2001).
[76] S. Charkravarty, R. B. Laughlin, D. K. Morr et al., Phys. Rev. B **63**, 094503 (2001).
[77] T. Senthil, M. Vojta, S. Sachdev, Phys. Rev. B **69**, 035111 (2004).
[78] D. Podolsky, A. Vishwanath, J. Moore et al., Phys. Rev. B **75**, 014520 (2007).
[79] R. Raimondi, G. Savona, P. Schwab et al., Phys. Rev. B **70**, 155109 (2004).
[80] D. R. Niven, R. A. Smith, Phys. Rev. B **71**, 035106 (2005).
[81] G. Catelani, I. L. Aleiner, JETP **100**, 331 (2005).
[82] M. F. Smith, J. Paglione, M. B. Walker et al., Phys. Rev. B **71**, 014506 (2005).
[83] P. D. Thacher, Phys. Rev. **156**, 975 (1967).
[84] R. O. Pohl, B. Stritzker, Phys. Rev. B **25**, 3608 (1982).
[85] Y. Ando, X. F. Sun, K. Segawa, J. Phys.: Conf. Ser. **108**, 012001 (2008).
[86] A. Hosseini, R. Harris, Kamal Saeid et al., Phys. Rev. B **60**, 1349 (1999).
[87] R. W. Hill, C. Lupien, M. Sutherland et al., Phys. Rev. Lett. **92**, 027001 (2004).
[88] Y. Matsuda, K. Izawa, I. Vekhter, J. Phys.: Condens Matter **18**, R705 (2006).

12
A Brief Overview of Raman Scattering in Cuprate Superconductors

Mei-Jia Wang, An-Min Zhang and Qing-Min Zhang

Department of Physics, Renmin University,
Beijing 100872, P. R. China

High-temperature superconductivity is one of the most fascinating and challenging areas of condensed matter physics. In the past two decades, a huge number of theoretical and experimental studies have been contributed to the field and many crucial progresses have been made. Raman scattering has been extensively applied to high-T_C superconductors owing to its unique abilities to probe multiple primary excitations and symmetry selection rules. In this chapter, we present a brief review on Raman scattering experimental achievements in cuprate superconductors, with emphasis on pairing mechanism-related aspects such as electron–phonon coupling, pair-breaking peak and two-magnon, etc. We will also address some open questions.

12.1. Introduction

When light propagates through a medium, it will interact with microscopic particles or elementary excitations in the medium, leading to a change of energy and momentum — this phenomenon is called light scattering.

If scattered light has an identical frequency with incident light, it is an elastic scattering, or the so-called Rayleigh scattering, otherwise it is called inelastic light scattering. There are a lot of excitations which can cause inelastic scattering in solids, such as optical phonons, acoustic phonons, electrons, magnons and many other kinds of elementary excitations. Generally, we call such inelastic scattering as Raman scattering. The scattering with a very small energy transition ($\sim 2 - 3$ cm^{-1}) is called Brillouin scattering. Raman scattering is different from infrared absorption, where a monochromatic light as incident light is required. After the incident light enters a solid, it will interact with elementary excitations and an energy transfer will happen. The Raman shift we usually see in a Raman spectrum is exactly the transfer energy between incident light and elementary excitations. Raman scattering can be classified into Stokes and anti-Stokes procedures, corresponding to energy transfer from incident light to elementary excitation and vice versa.

Raman scattering has been widely applied in fundamental research fields such as physics, chemistry, biography, etc., and many industrial areas. Basically, it can be used to characterize and measure rotating and vibrating modes, by which one can study composition, phase transition, stress etc., in condensed matter. In addition, Raman scattering can also be used to study the properties related to electrons, which originate from the scattering by electrons or holes near the Fermi surface. Raman scattering is also a good technique to detect magnetic excitation. Electronic Raman scattering (ERS) had very limited applications in the 1960s and 1970s because the penetration depth is very small in normal metals, so the effective intensities of incident light are low, and Coulomb screening effect is very strong due to high density of electrons in metals. ERS reflects charge density-density correlation, but the strong screening in metals will much weaken density correlation effect. High-temperature superconductors (HTSC) are suitable for electronic Raman scattering due to its lower density of charge carriers. Moreover, the Cu–O planes in HTSC have a basic tetragonal symmetry, which is a great advantage that can be used by Raman scattering. One can study the anisotropic excitations from nodes and antinodes of Brilliouin zone (BZ) by symmetry analysis, which is particularly useful in exploring d-wave pairing in cuprate superconductors. Today, ERS has become one of the most important experimental techniques in studying HTSC [1, 2].

12.2. Electron Raman Scattering in High-Temperature Superconductors

12.2.1. Phenomenological and Microscopic Theoretical Pictures

According to the classical theory of electromagnetic field, atoms (molecules) will be polarized by light radiation field entering into solid, inducing vibrating dipoles and instantaneous polarization. Instantaneous polarization can be described as a linear plus a non-linear term as the following:

$$P = \varepsilon_0(\chi E_I + \chi' X E_I).$$

The first term corresponds to elastic scattering (Rayleigh scattering) and the second term represents inelastic scattering that changes the frequency of incident light.

Phenomenologically, the intensity of inelastic scattering can be described by linear response theory. The response from a linear passive system can be decomposed to a number of damped harmonic oscillators. We define the generalized displacement as:

$$\overline{X}(q,\omega) = T(q,\omega)F(\omega)$$

using fluctuation–dissipation theorem:

$$\langle X(q)X^+(q)\rangle_\omega = \frac{\hbar}{\pi}\{n(\omega)+1\}\mathrm{Im}T(q,\omega)$$

we can see that Raman scattering is a measure of the imaginary part of the response function. For ERS, the effective Raman response function reflects the response of charge system to the perturbation by light radiation field. Electronic Raman response function is dominated by vertex function between light radiation field and electrons, which is determined by band structure and crystal symmetry. The theoretical details of ERS can be found in Refs. [1, 2].

One of the common features for HTSC is the same Cu–O squares. In general, it is treated as D_{4h} point group, though in some cases there may exist a small deviation from the ideal squares. The following discussions are based on this symmetry.

According to general phenomenological theory, the intensity of Raman scattering is proportional to the square of the matrix of second-order polarizability tensors. The Raman-active irreducible representations of D_{4h} point group are listed here, as well as the corresponding polarizability tensors, x, y and z representing the principal axes determined by second-order polarizability tensors.

So the modes with different symmetries are selected by polarization geometries. Usually, we use symbols like $z(x,y)z$ to represent the experimental geometry. The first and last letters give the propagation directions of incident and scattered light, respectively. The two letters in the bracket indicate the polarizations of incident and scattered light, respectively. We can also detect the combination of different modes by selecting polarizations of incident and scattered light if it is difficult to directly obtain a pure symmetry by polarization geometry. The polarization geometries for HTSC are shown in Fig. 12.1, in which polarizations of incident and scattered light are indicated by the arrows. The third $(A_{2g}+B_{1g})$ and fourth $(A_{2g}+B_{2g})$ cases are two kinds of vertical geometries which are used most frequently. Generally, Raman intensity is dominated by the contribution from B_{1g} channel in the third geometry because the contribution of A_{2g} is negligible below 1000 cm^{-1}. So the third geometry can be considered as B_{1g} symmetry. Similarly, the fourth case in Fig. 12.1 is called B_{2g} symmetry. Circular polarization geometries are illustrated in the fifth and sixth cases in Fig. 12.1, which is particularly helpful when exploring some circular excitations such as phonic or magnetic chiral modes. Using all of the six geometries one can make a complete symmetry analysis to precisely separate a pure spectrum for each symmetry.

The above example of cuprate superconductors introduces formal symmetry phenomenologically. In a microscopic view, the cross section of ERS is closely related to its structural factor, formally which is similar to neutron scattering. In principle, the cross section is proportional to the imaginary part of correlation function. When a solid is irradiated by an electromagnetic wave, oscillatory electric field $\boldsymbol{E}$ will cause a perturbation for charge system, leading to an effective charge density fluctuation, which reflects an interaction between charge and light radiation field. Effective fluctuation of charge system is determined by interaction vertex with light, which is a key of electronic Raman scattering. Theoretically, in the absence of interband transition the vertex is proportional to

$$A_{1g} \quad A_{2g} \quad B_{1g} \quad B_{2g} \quad E_g$$

$$\begin{bmatrix} a & & \\ & a & \\ & & b \end{bmatrix} \begin{bmatrix} & c & \\ -c & & \end{bmatrix} \begin{bmatrix} d & & \\ & -d & \end{bmatrix} \begin{bmatrix} & e & \\ e & & \end{bmatrix} \begin{bmatrix} & & f \\ & & \\ g & & \end{bmatrix}$$

$$x = \begin{pmatrix} 1 \\ 0 \\ 0 \end{pmatrix} \quad y = \begin{pmatrix} 0 \\ 1 \\ 0 \end{pmatrix} \quad z = \begin{pmatrix} 0 \\ 0 \\ 1 \end{pmatrix}$$

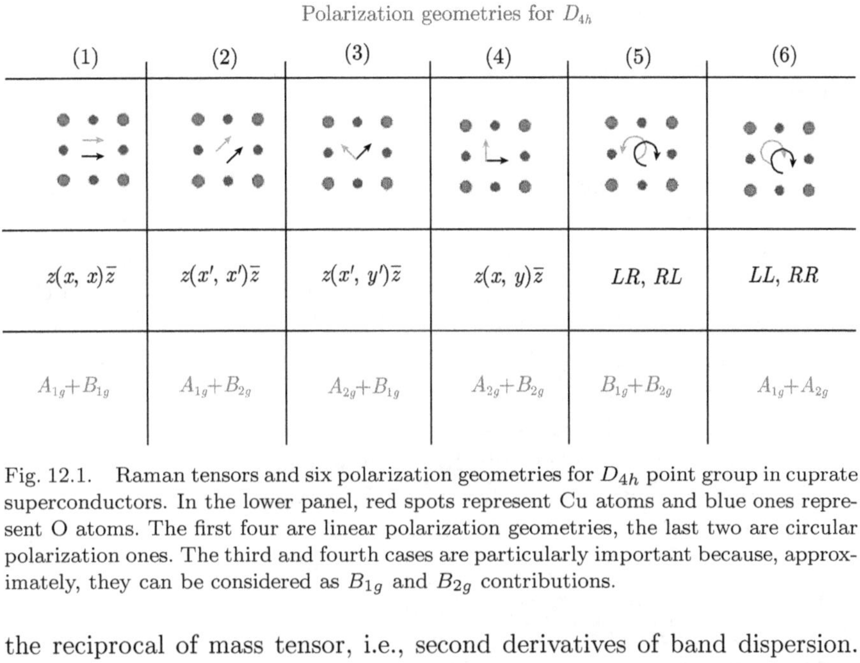

Polarization geometries for D_{4h}

(1)	(2)	(3)	(4)	(5)	(6)
$z(x,x)\bar{z}$	$z(x',x')\bar{z}$	$z(x',y')\bar{z}$	$z(x,y)\bar{z}$	LR, RL	LL, RR
$A_{1g}+B_{1g}$	$A_{1g}+B_{2g}$	$A_{2g}+B_{1g}$	$A_{2g}+B_{2g}$	$B_{1g}+B_{2g}$	$A_{1g}+A_{2g}$

Fig. 12.1. Raman tensors and six polarization geometries for D_{4h} point group in cuprate superconductors. In the lower panel, red spots represent Cu atoms and blue ones represent O atoms. The first four are linear polarization geometries, the last two are circular polarization ones. The third and fourth cases are particularly important because, approximately, they can be considered as B_{1g} and B_{2g} contributions.

the reciprocal of mass tensor, i.e., second derivatives of band dispersion. Considering HTSC as a tetragonal symmetry, we can expand the vertex in an analytical way. Taking the first-order term, the vertex for B_{1g} can be written as:

$$\gamma_k = \cos(k_x) - \cos(k_y),$$

and the vertex of B_{2g} can be written as:

$$\gamma_k = \sin(k_x)\sin(k_y),$$

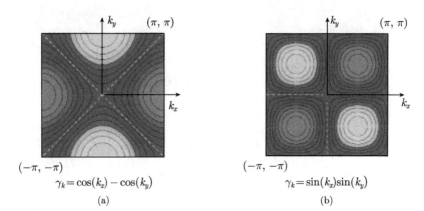

Fig. 12.2. The amplitudes of the vertices of B_{1g} (a) and B_{2g} (b) in the first BZ, which indicates B_{1g} and B_{2g} electronic Raman spectra detect antinodal and nodal regions, respectively.

which have the same symmetry as d-wave. Figure 12.2 shows the relative amplitudes of the two vertices in the first BZ. B_{1g} and B_{2g} channels pick up quasi-particles excitation at antinodes ($\pm\pi, 0$) and nodes ($\pm\pi/2, \pm\pi/2$), respectively. This is a unique advantage for ERS in the study of HTSC, corresponding to 45° angle-resolution anisotropy. Note that the spectra from two channels corresponding to low-energy excitations are not exactly at the nodes and antinodes in k space, but a convolution in the region around the nodes and antinodes.

In the normal state, electronic Raman scattering can be seen as a Drude process similar to infrared conductivity, whereas a pair-breaking peak is a prominent feature in superconducting state which is contributed by breaking cooper pairs with incident light. For an isotropic conventional BCS superconductor, a uniform pair-breaking peak is expected exactly at $2\Delta_0$, while it is quite different for a d-wave superconductor. A well-established ERS theory for d-wave superconductors predicts that a d-wave superconductor will show a pair-breaking peak at $2\Delta_0$ in B_{1g} channel and below $2\Delta_0$ ($\sim 1.7\Delta_0$) in B_{2g} channel. The d-wave anisotropy is exactly resolved by ERS, as we mentioned above. Furthermore, low-energy behavior is power law form for B_{1g} and linear for B_{2g} in the limit of $\Omega \to 0$, while it is an exponential decay at low energies for a conventional s-wave superconductor (see Fig. 12.3).

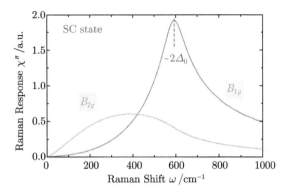

Fig. 12.3. Electronic Raman spectra below T_C in B_{1g} and B_{2g} channels based on d-wave superconducting theory. B_{1g} and B_{2g} pair-breaking peaks show an apparent anisotropy, which appears at $2\Delta_0$ and below $2\Delta_0$, and with linear and power law decays at low energies, respectively.

12.2.2. Superconducting State

12.2.2.1. Superconducting gap and anisotropic pairing

A good-quality surface can be obtained for Bi2212 system because it is a layered structure and easy to be cleaved. Many Raman scattering measurements were performed on this system and some consistent results have been reached. Figure 12.4 shows typical Raman scattering results of Bi2212 system from two groups. The sharp peaks are attributed to phonons, [4] and the results from two groups are basically the same after subtracting phonon peaks.

In Fig. 12.4(a), from up to down, doping level changes from overdoping to underdoping are shown. For electronic Raman spectra of B_{1g} symmetry, there is a clear pair-breaking peak in overdoped region. The shape of pair-breaking peak becomes much broader and its position shifts to a higher frequency with the evolution from overdoping to underdoping [3]. The theory of electronic Raman scattering for cuprate superconductors points out that B_{1g} peak position corresponds to twice of maximum superconducting energy gap. And the ratio of superconducting gap to $k_B T_C$ continuously increases from 4 to 9 when doping level evolves from overdoping to underdoping, showing an essential difference from conventional BCS superconductors. The significant deviation from 3.52 given by BCS

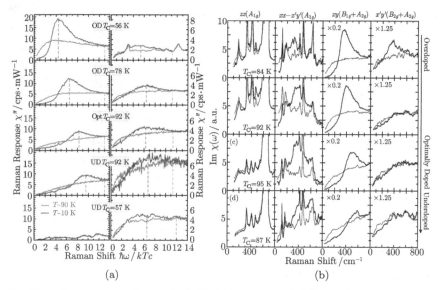

Fig. 12.4. Raman scattering results in Bi2212 system with different doping levels from two groups. Red and blue colors denote normal and superconducting state. The thin line represents normal state and the thick one, superconducting state. (a) From Ref. [3], (b) From Ref. [4].

in underdoped region implies an unconventional superconducting mechanism. B_{2g} peak position is hard to be determined precisely because the peak is much broader. Some reports suggested that B_{2g} peak position traces doping dependence of T_C. We will come back to this later when discussing the topic on two-gap structure. The obvious anisotropy between B_{1g} and B_{2g} is consistent with other experiments like angle-resolved photoemission, which strongly supports a scenario of d-wave pairing in cuprate superconductors.

The results from YBaCuO system are shown in Fig. 12.5. It is not easy to separate electronic contributions from the B_{1g} spectra because there exists a strong B_{1g} phonon around 340 cm^{-1} (so-called buckling mode). B_{2g} pair-breaking peak can still be observed, which is very broad, similar to that of Bi2212 system [5].

Figure 12.6 shows ERS from LaSrCuO, which is different from other copper oxide systems. It looks like there occurs an exchange in intensity between B_{1g} and B_{2g} spectra, i.e. B_{2g} pair-breaking peak is much stronger than that in B_{1g} channel [6, 7]. This is related to electronic stripe phase which we will discuss later.

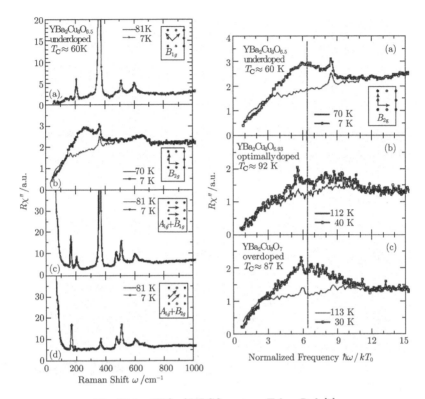

Fig. 12.5. ERS of YBCO system. Taken Ref. [5].

In Fig. 12.7, ERS of TlBaCuO and HgBaCaCuO are presented. We can see that pair-breaking peak in B_{1g} channel is clear and the magnitude of peak position over T_C is the same as Bi2212, while we can barely see pair-breaking peak in B_{2g} channel [8]. The clear peak in A_{1g} channel below T_C is considered to be related to spin resonance mode observed by neutron scattering [7].

Figure 12.8 is ERS of electron-doped cuprate superconductor NdCeCuO. Controversial conclusions on pairing symmetry are reported based on the existing ERS results. In Fig. 12.8(a), an earlier Raman result is shown. One can see that pair-breaking peak gets a rapid asymptotic form in low energy both in B_{1g} and B_{2g} channels, which implies an anisotropic s-wave pairing [10]. Different Raman results come from Blumberg's group [11], which are shown in Figs. 12.8(c) and 12.8(b). An obvious difference between B_{1g} and B_{2g} spectra below T_C indicates a strong anisotropic

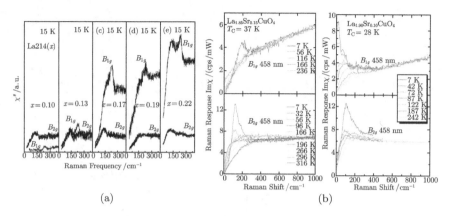

Fig. 12.6. ERS of LSCO system. (a) From Ref. [6], (b) From Ref. [7].

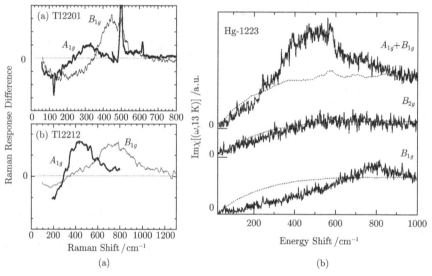

Fig. 12.7. ERS of TlBaCuO and HgBaCaCuO systems. (a) From Ref. [8], (b) From Ref. [9].

pairing. Interestingly, the B_{2g} peak position is higher than the B_{1g} one, which is opposite to the Bi2212 system, other hole-doped systems and theoretical predictions for a pure d-wave. This behavior is explained as an indication of non-monotonic d-wave gap. In the picture of non-monotonic d-wave gap, the maximum of d-wave gap appears between the nodal and antinodal regions, as shown in Fig. 12.8(d). This is different from the case of hole-doping where the maximum and zero are located exactly in the antinodal

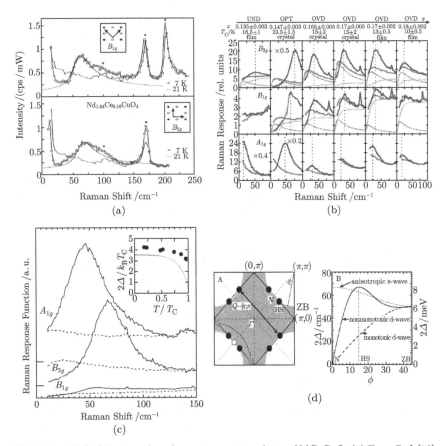

Fig. 12.8. ERS of electron-doped cuprate superconductor NdCeCuO. (a) From Ref. [10], (b) From Ref. [11], in which blue color denotes superconducting state. (c) and (d) From Ref. [12].

and nodal directions. The shift of the maximum gap in electron-doped systems gives rise to the different B_{1g} and B_{2g} spectra from hole-doped systems. In a microscopic view, electronic Raman spectra are determined by Raman vertex, and hence depend on band structures [12]. Actually, a two-band model has been proposed to present a good interpretation of the behavior of ERS in electron-doped cuprate superconductors [1].

12.2.2.2. *Two gaps*

Recently two-gap structure has attracted much attention in the study of cuprate superconductors. The two energy gaps were first proposed in a paper on electronic Raman scattering in 2006, in which Hg2201 crystals with different doping levels were studied [13]. The paper presents the evolution of pair-breaking peak with varying doping. A well-defined B_{1g} peak below T_C can be seen in overdoped region, while it gets more and more obscure and monotonically moves to higher energies with decreasing doping level from overdoped to underdoped. On the other hand, B_{2g} peak is quite broad and shows a non-monotonic dependence on doping level, which reaches a maximum in frequency near optimal doping. In B_{1g} and B_{2g} channels, two-gap features are clearly revealed. B_{2g} position follows the dome of T_C versus doping very well, while B_{1g} position shows a monotonic linear dependence on doping, and it intersects with T_C dome in overdoped region rather than optimal doping. So the different doping dependences of B_{1g} and B_{2g} peaks give rise to two energy scales: one is the gap from B_{2g} channel covering nodal regions, which looks like a superconducting gap in both gap size and doping dependence; the other one comes from B_{1g} channel detecting antinodal regions, which behaves as a pseudogap. The gap values are measured by Raman and other experiments such as ARPES, tunneling spectroscopy, etc. All of the gap values are classified into two categories: one follows B_{1g} behavior and the other B_{2g}. There are still some controversies about the relationship between the two gaps and the superconducting mechanism.

12.2.2.3. *Impurity effect*

Electronic Raman spectra with magnetic and non-magnetic impurities are shown in Fig. 12.9. The impurity effect in cuprate superconductors on electric Raman Scattering spectrum has been theoretically studied in detail [14]. The main prediction is that low-energy decay in B_{1g} channel below T_C will change from power law to linear by impurity, while the B_{2g} response keeps its linear form at low energies.

Recently there have been some new achievements, as shown in Fig. 12.9. The effect of magnetic impurity is totally different from that of non-magnetic impurity. At low concentrations, magnetic impurity has little influence on pair-breaking peaks while the peaks will be completely eliminated by non-magnetic impurity. The $A_{1g} + B_{1g}$ ERS is shown in Fig. 12.9(a), indicating no matter what impurity is doped, the pair-breaking peaks have little change. For pure B_{1g} symmetry, whose ERS is shown in Fig. 12.9(b), the pair-breaking peak is strongly suppressed by Zn doping, but survives with Ni-doping. These results lead to a theoretical conjecture

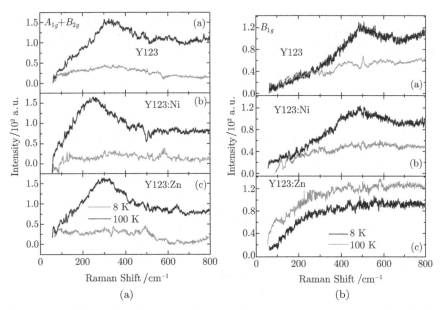

Fig. 12.9. ERS of YBCO system with magnetic and non-magnetic impurities. (a) is for $A_{1g} + B_{2g}$ channel and (b) B_{1g} channel. There is no obvious influence on the pair-breaking peak in $A_{1g} + B_{2g}$ channel by doping Zn and Ni, but Zn doping strongly suppresses pair-breaking peak in B_{1g} channel (b). Taken from Ref. [14].

based on $t - J$ model that A_{1g} electronic Raman spectrum directly reflects superconducting order parameter and B_{1g} represents d-wave CDW order parameter [15].

Actually, there are some different opinions on what contributes to A_{1g} channel. It was argued that there is a correlation between A_{1g} pair-breaking peak and the resonant mode measured by inelastic neutron scattering rather than B_{1g}. This demonstrates that anti-ferromagnetic (AF) spin fluctuations are involved in A_{1g} Raman scattering process in some way or even play a key role.

12.2.3. Normal State

12.2.3.1. Pseudogap

The pseudogap problem has been discussed a lot in copper oxide superconductors in both experimental and theoretical aspects. In the last two decades, more and more experimental evidence were accumulated and a consensus was reached that there definitely exists a pseudogap in the normal state. However, there are still many debates on its origin, not only theoretically, but also experimentally. Magnetic measurements such as nuclear magnetic resonance, neutron scattering and other experiments demonstrated that the pseudogap in normal state has a spin-relative origin, while transport measurements, tunneling spectrum, ARPES, etc., displayed that it is a charge gap. Electric Raman scattering can measure charge density–density correlation function. In principle, if pseudogap is related to charges, there should be some indications in ERS.

Figure 12.10 shows the ERS in underdoped Bi2212 system [16]. A broad peak located at about 600 cm^{-1} was observed in the normal state. It is not contributed by phonons. It disappears in both normal state and superconducting state under overdoping. Figure 12.10(a) displays its temperature dependence more clearly. The peak is enhanced gradually and passes through T_C smoothly. The features are considered as the evidence for a normal-state gap and they are consistent with other experimental techniques. We need to note that so far there are only a small amount of reports on pseudogap in Raman scattering. It remains unclear how pseudogap affects Raman scattering in cuprates.

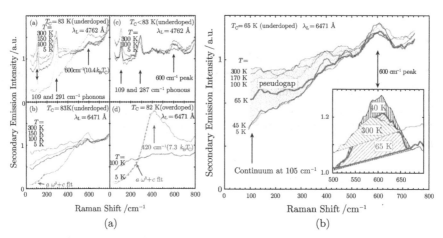

Fig. 12.10. ERS in underdoped Bi2212 system. (a) The ERS with different doping levels. (b) shows spectra from underdoped samples. A peak can be seen around 600 cm^{-1} even above T_C, which is regarded as a pseudogap feature. Taken from Ref. [16].

12.2.3.2. *Quantum critical point*

Figure 12.11 displays the electronic Raman spectra in normal state in B_{1g} and B_{2g} channels in Bi2212 system [17]. As a response function of

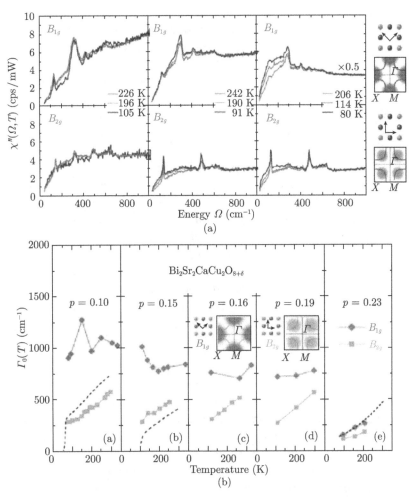

Fig. 12.11. (a) shows ERS of Bi2212 system in normal state at different temperatures. (b) shows the evolution of extrapolated static relaxation rate with doping. Using memory function approach [5], dynamical charge carrier relaxation rate can be extracted from the normal-state ERS and the static relaxation rate can be obtained by extrapolating to zero frequency. The data are compared with transport data (dashed line). Taken from Ref. [17].

charge density–density correlation, ERS can present dynamic information on charge carriers in principle. Using a so-called memory-function approach [5], one can extract a dynamic charge carrier relaxation rate hidden in normal-state Raman spectra. Then, static charge carrier relaxation rate can be obtained by extrapolating dynamic charge carrier relaxation rate to zero frequency, as shown by the red and green data points in Fig. 12.11. In principle, the static charge carrier relaxation rate, extrapolated from B_{1g} and B_{2g} spectra with the above method, could be compared with DC resistivity measurements. But it should be mentioned here that DC resistivity for a common metal is just an average in density of states (DOS) around whole Fermi surface, while the static relaxation rate extracted from B_{1g} and B_{2g} spectra corresponds to a DOS integral around the antinodes and nodes on the Fermi surface, respectively. So it gives the anisotropy of carrier dynamics, which demonstrates an interesting doping dependence in B_{1g} and B_{2g} symmetries: for B_{1g}, it behaves as an insulator (red data points in Fig. 12.11) in underdoped region and a metal in the overdoping region; for B_{2g}, it keeps a metallic behavior in the whole doping region. This behavior clearly indicates the anisotropy of charge carriers at nodes and antinodes. Furthermore, it can be seen that a metal–insulator transition occurs at a critical doping concentration in the antinodal region (B_{1g}), which is consistent with the critical point observed by the transport measurement and other experiments. So ERS experiments explain that the critical point is dominated by the charge carrier behavior at antinodes, which is of great importance to the understanding of quantum critical point.

12.2.3.3. *Electronic stripe phase*

As shown in Fig. 12.12, a Raman peak appears at very low energies in underdoped LSCO, which gets even sharper with decreasing temperature [18]. This peak remains very clear even far above T_C. So it may not be considered as a pair-breaking process. There is a very interesting characteristic that the peak appears in B_{2g} channel in heavily underdoped samples ($x = 0.02$) while it turns into B_{1g} channel in lightly underdoped region ($x = 0.10$) near optimal doping.

Combined with neutron scattering and other measurements, this low-energy peak can be understood in terms of electronic stripe phases. The sketches are shown in Fig. 12.12(b), in which the peak is connected with a collective excitation mode of electronic stripes driven by electromagnetic waves. In a heavy underdoping, neutron scattering, μSR and some other

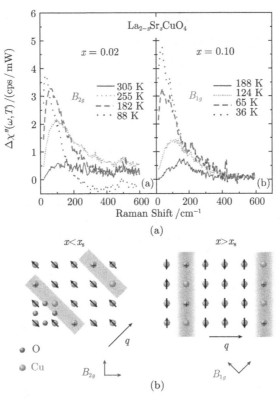

Fig. 12.12. ERS of underdoped LSCO system. One can see low-energy peaks in two underdoped samples. The peak appears in B_{2g} channel in heavily underdoped samples ($x = 0.02$) and turns into B_{1g} channel in lightly underdoped region ($x = 0.10$). Taken from Ref. [18].

experiments observed electronic stripes in the lower left sketch of Fig. 12.13, whose direction is diagonal, 45° relative to Cu–O bond direction. For B_{1g} symmetry, the polarization of incident light is either parallel or perpendicular to the direction of stripes, so the polarization of radiated scattering light is simply prohibited. Thus, one cannot observe the collective excitation mode of electronic stripes. On the other hand, for B_{2g} symmetry the polarization has a component along both incident and scattering directions. In this case, the collective excitation of the electronic stripes is permitted. In the lightly underdoped samples, some experiments reveal that the direction of electronic stripes is rotated by 45°, along Cu–O bond direction. Similarly, the collective excitation of electronic stripes is allowed in B_{1g} channel.

12.3. Raman Scattering of Two-Magnon Process

In a typical 2D square AF lattice like K_2NiF_4, in principle a magnon (or single-spin flip) can be excited after it absorbs incident photons with appropriate energy. However, a classical AF spin wave is acoustic-like, which means magnon energy also approaches zero when its momentum goes down to zero. For this reason, a single magnon can hardly be observed in light scattering. However, if two neighboring spins in a K_2NiF_4 spin lattice reverse simultaneously in a simple picture and meet the requirements of energy and symmetry, so-called two-magnon process is permitted in Raman scattering. In a microscopic picture, it looks like a double-exchange procedure between two spins on Cu ions bridged by intermediate oxygen ion.

In cuprate materials, two-magnon process can be observed in B_{1g} channel, which has a very high intensity AF parent compound, even much higher than most phonons, as shown in Fig. 12.13. As the simultaneous reverse of adjacent two spins breaks the nearest six antiferromagnetic couplings, two-magnon peak will appear at about $3J$ in energy. The ratio may slightly

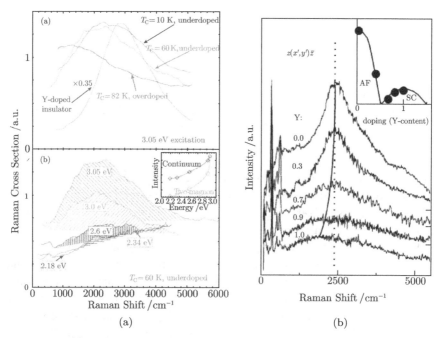

Fig. 12.13. (a) shows AF two-magnon peaks in Bi2212 samples, and (b) YBCO systems. Both of them show an evolution of two-magnon peak with doping.

deviate from 3, which depends on a detailed microscopic theoretical model. One can see this provides an opportunity to determine AF superexchange energy J. Actually, it is the most accurate method to measure J in cuprates so far. For insulating parent compounds of cuprates, two-magnon peak is around 3000 cm^{-1}, corresponding to a superexchange energy of $\sim$1000 cm^{-1} (about 125 meV). With increasing doping level, two-magnon peak becomes broader and moves to lower energies, which denotes the evolution of AF fluctuations with doping [16, 19].

A quite low-carrier doping in parent compounds will completely destroy long-range AF order. The remaining strong AF spin fluctuations will be further suppressed with approaching optimal doping. Consequently, two-magnon signal becomes much weaker, even disappears in some cases. However, weak two-magnon feature in underdoped HTSC demonstrates a novel and unexpected behavior under an applied magnetic field. Figure 12.14 shows two-magnon Raman peaks in LSCO ($x = 0.12$) with and without an applied field [20]. One can see that at low temperatures (below T_C) the weak two-magnon peak is strongly enhanced by a high magnetic field of 14

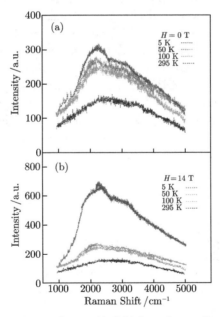

Fig. 12.14. The temperature and magnetic field dependence of two-magnon peak intensity in the underdoped LSCO. The two-magnon peak is dramatically boosted up under an applied field of 14 T in the superconducting state. Taken from Ref. [20].

T, which implies that AF correlation is strengthened largely. It is a really surprising result. Coincidentally, from neutron scattering experiments one can find that AF resonance mode at (π, π) is also largely enhanced when a high magnetic field of 14.5 T is applied [21], which is well consistent with two-magnon Raman scattering. Both experiments clearly reveal that AF spin fluctuations can be strongly enhanced by applied field in superconducting state. This is considered to be related to magnetic vortices driven by external magnetic field in superconducting state. Some theoretical models state that in a vortex core normal state is recovered, as a consequence it contributes to strong AF signal in two-magnon Raman and neutron scattering experiments.

12.4. Overview

Raman scattering is the response to the interaction between light and elementary excitations in material, so it can detect and provide plenty of information on quasi-particles such as electrons, optical phonons, magnons, excitons, plasmons, etc. Moreover, one can use the unique advantage of symmetry analysis to study anisotropic problems, which is particularly useful for d-wave superconductors. And we can obtain precise superexchange interaction by two-magnon process. Raman scattering has become one of the most fundamental experimental techniques in studying HTSC.

So far in the study of HTSC, Raman scattering has made important progress and consensuses have been reached in some aspects, such as d-wave pairing symmetry, superconducting gap and its doping dependence, AF correlation in two-magnon process, anisotropic properties of nodes and antinodes and so on. But there are still some problems that have not been well understood:

(1) The finite background in Raman spectra up to 1 eV which remains unknown till now.
(2) Two-gap observations that put forward a question on the origin of A_{1g} and B_{1g} contributions again.
(3) Pseudogap problem in the normal state in Raman scattering.
(4) The evolution of the pair-breaking peak from overdoped to underdoped state.
(5) The evolution of two-magnon peak with doping and its relation to AF spin fluctuations.

Raman scattering plays an important role in studying HTSC, meanwhile, the study of HTSC in return significantly contributes to the rapid developments of Raman scattering in both technique and theory, particularly in electronic Raman scattering. In the future, Raman scattering with extreme sample environments such as high pressure and magnetic field will provide us with more opportunities to reveal new phenomena, which will help us establish deeper insight into HTSC mechanism.

This work was supported by the 973 Program under Grant No. 2011CBA 00112, by the Natural Science Foundation of China under Grant Nos. 11034012 and 11004243, by the Fundamental Research Funds for Central Universities, and by the Research Funds of Renmin University of China.

References

[1] X. Tao, *d-Wave Supercondutor* (Sicence Press, Beijing, 2007).
[2] T. P. Devereaux, R. Hack, *Rev. Mod. Phys.* **79**, 175 (2007).
[3] F. Venturini, M. Opel, T. P. Devereaux et al., *Phys. Rev. Lett.* **89**, 107003 (2002).
[4] H. L. Liu, et al., *Phys. Rev. Lett.* **82**, 3524 (1999).
[5] M. Opel et al., *Phys. Rev. B* **61**, 9752 (2000).
[6] J. G. Naeini et al., *Phys. Rev. B* **59**, 9642 (1999).
[7] F. Venturini et al., *Phys. Rev. B* **66**, 060502(R) (2002).
[8] M. Kang et al., *Phys. Rev. B* **56**, R11427 (1997).
[9] A. Sacuto et al., *Phys. Rev. B* **58**, 11721 (1998).
[10] B. Stadlober et al., *Phys. Rev. Lett.* **74**, 4911 (1995).
[11] M. M. Quazilbash et al., *Phys. Rev. B* **72**, 214510 (2005).
[12] G. Blumberg et al., *Phys. Rev. Lett.* **88**, 10702 (2002).
[13] M. Le Tacon et al., *Nat. Phys.* **2**, 537 (2006).
[14] H. Martinho et al., *Phys. Rev. B* **69**, 180501(R) (2004).
[15] R. Zeyher and A. Greco, *Phys. Rev. Lett.* **89**, 17004 (2002).
[16] G. Blumberg et al., *Science* **278**, 1427 (1997).
[17] F. Venturini et al., *Phys. Rev. Lett.* **89**, 107003 (2005).
[18] L. Tassini et al., *Phys. Rev. Lett.* **95**, 117002 (2005).
[19] M. Rubhausen et al., *Phys. Rev. B* **56**, 14797 (1997).
[20] L. H. Machtouh, B. Keimer and K. Yamada, *Phys. Rev. Lett.* **94**, 107009 (2005).
[21] B. Lake et al., *Nature* **415**, 299 (2002).

13
Quasi-particle Excitations in High-T_C Cuprate Superconductors Probed by Specific Heat: Implications on the Superconducting Condensation

Yue Wang* and Hai-Hu Wen[†]

*Applied Superconductivity Center, State Key Laboratory
for Mesoscopic Physics, and School of Physics, Peking University,
Beijing 100871, China
[†]Center for Superconducting Physics and Materials,
National Laboratory of Solid State Microstructures,
and Department of Physics, Nanjing University, Jiangsu 210093, China

Among various experimental tools employed in the research of high-T_C superconductivity, low-temperature specific heat has the merit of probing the low-energy quasi-particle excitations and reflecting the bulk property of the sample. In this chapter, we first provide a brief description of the related background of specific heat studies in high-T_C cuprates and then summarize the recent experimental results in $La_{2-x}Sr_xCuO_4$ and $Bi_2Sr_{2-y}La_yCuO_{6+\delta}$ single crystals. Calorimetric evidences for d-wave symmetry of the order parameter throughout the phase diagram, an anomalous doping dependence of the superconducting gap, an intimate relationship between the pseudogap and superconductivity, and a "Fermi arc" ground state in the underdoped regime are discussed. A picture of the superconducting condensation is also proposed and applied to account for the doping evolution of some basic superconducting properties shown in the phase diagram.

13.1. Introduction

For more than twenty-five years, there has been no consensus on the origin of the high-temperature superconductivity in cuprate superconductors [1]. Compared with conventional low-T_C superconductors, high-T_C cuprate superconductors exhibit a series of new features. First, a large number of different experiments have consistently shown that high-T_C cuprate superconductors possess d-wave symmetry of the superconducting gap [2], not the s-wave symmetry as found in low-T_C superconductors. In momentum space, the d-wave superconducting gap is zero along the (0, 0)–(π, π) direction in the Brillouin zone, while it is maximum along the direction of (0, 0)–(π, 0). Another striking finding in high-T_C cuprates is the appearance of a pseudogap in the normal state, i.e. above the superconducting critical temperature T_C [3]. This phenomenon has been widely observed for underdoped and optimally doped cuprates by various measurements. The temperature at which the pseudogap sets in is usually labeled as T^*. In angle-resolved photoemission spectroscopy (ARPES), this pseudogap manifests itself as a partial loss of the density of states around the antinodes (π, 0), which leads to the Fermi surface of the sample forming arcs near the nodes [4]. The nature of the pseudogap and its relationship with the high-T_C superconductivity have been under intense debate, and the identification is argued to be crucial in determining the origin of the high-T_C superconductivity [5]. Generally speaking, currently there are two distinct opinions on the nature of the pseudogap [6]. Some suggest that the pseudogap may have an intimate relationship with the superconductivity, that is, the pseudogap state may be a precursor to the superconducting state. It is known that, for conventional BCS superconductors, Cooper pairing and coherent condensation occur simultaneously when the temperature chills to T_C, and after that the sample enters into the coherent superconducting state. For high-T_C cuprates, however, there is a possibility that the electron pairing may happen at a higher temperature than T_C while until the temperature decreases to T_C phase coherence establishes, leading to the appearance of a pseudogap above T_C. In contrast, others argue that the pseudogap may have a non-superconducting origin and coexist or compete with the superconductivity.

To help clarify the above issue, we note that, according to the results of ARPES, for underdoped high-T_C cuprates there would be a new gap

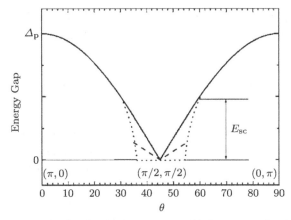

Fig. 13.1. Schematic plot for the pseudogap energy (Δ_p), superconducting energy scale (E_{sc}), and nodal gap slope [7]. The solid line represents the standard d-wave gap $\Delta = \Delta_p \cos(2\theta)$ with Δ_p the maximum gap value at (π, 0). The dotted line shows the pseudogap phase with the Fermi arc appearing around the nodes. The dashed line shows the possibility of a superconducting gap opening below T_C, which is characterized by the nodal gap slope v_Δ. As illustrated, if the superconductivity has nothing to do with the pseudogap, there may not be a correlation between the v_Δ and the Δ_p.

opening on the Fermi arc when the temperature decreases to below T_C. This new gap could be viewed as a superconducting gap or superconducting energy scale, Δ_{sc}, as shown in Fig. 13.1 (denoted as E_{sc} in the figure), since its opening accompanies the appearance of the coherent superconducting state [7]. Therefore, it would be useful for identifying the relationship between the pseudogap and superconductivity to investigate both the gap slope v_Δ of the Δ_{sc} near the nodes and the pseudogap Δ_p near the antinodes. If the pseudogap has nothing to do with the superconductivity, there may not be any correlation between v_Δ and Δ_p. On the other hand, if there is indeed some correlation between v_Δ and Δ_p, one could say there may be an intimate relationship between the pseudogap and high-T_C superconductivity.

In this chapter, we review some recent low-temperature specific heat experiments in high-T_C cuprates, which are shown to be helpful in clarifying the above issue. Specific heat, as a bulk measurement, is a sensitive probe of the low-energy excitations inside a sample. In what follows, before the experimental results are reviewed, some related theoretical background is outlined first.

13.2. Low-Temperature Specific Heat of a D-Wave Superconductor: Theoretical Background

13.2.1. Nodal Excitation Spectrum of a D-Wave Superconductor

Experiments have established that the superconducting state of high-T_C cuprates is still a coherent condensation of Cooper pairs, whose low-energy excitations could be well described in the framework of Fermi liquid theory. Figure 13.2(a) shows the polar plot of the standard $d_{x^2-y^2}$ superconducting gap, $\Delta(k) = \Delta_0 \cos(2\theta)$ with Δ_0 the maximum value along the antinodal direction ($k_x = 0$ or $k_y = 0$) and θ the azimuth angle ($\theta = 0$ along the k_x axis). Along the nodal direction ($k_x = \pm k_y$), the gap is zero. Low-temperature specific heat experiments mainly probe the low-energy quasi-particle excitations in the vicinity of the nodes. In this region, the quasi-particle energy spectrum is given by $E(k) = \hbar\sqrt{v_F^2 k_1^2 + v_\Delta^2 k_2^2}$, where $\hbar$ is Planck's constant, v_F and v_Δ are the energy dispersions, or quasi-particle velocities, along directions normal ($//\boldsymbol{k}_1$) and tangential ($//\boldsymbol{k}_2$) to the Fermi surface at the node, respectively. The v_Δ is related to the slope

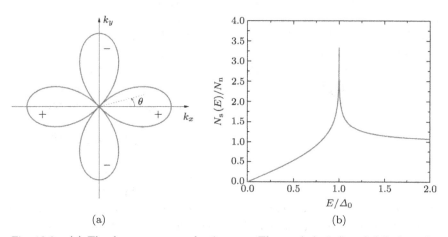

Fig. 13.2. (a) The d-wave superconducting gap. The symbols "+" and "−" show the phases of the different lobes. (b) Numerical simulations of the normalized density of states of a d-wave superconductor, $N_s(E)/N_n$, where N_n is the density of states in the normal state. Note the linear energy dependence of the $N_s(E)$ at low energy.

of the gap at the node as:

$$v_\Delta = \frac{1}{\hbar}\frac{\partial \Delta}{\partial k} = \frac{1}{\hbar k_F}\frac{\partial \Delta}{\partial \theta}\bigg|_{\text{node}},$$

where k_F is the Fermi wave vector. In this chapter, we simply call v_Δ as the nodal gap slope. For a standard $d_{x^2-y^2}$ gap, it is evident that:

$$\Delta_0 = \frac{1}{2}\hbar k_F v_\Delta. \tag{13.1}$$

The above energy spectrum leads to a quasi-particle density of states with linear dependence on energy:

$$N_s(E) = \frac{2}{\pi \hbar^2}\frac{1}{v_F v_\Delta}E, \tag{13.2}$$

as shown in Fig. 13.2(b). This in turn gives rise to a T^2 dependence of the electronic specific heat in zero magnetic fields [8,9]:

$$C_{el}(T) = \alpha T^2 = \frac{18\zeta(3)}{\pi}\frac{k_B^3}{\hbar^2}\frac{nV_{\text{mol}}}{d}\left(\frac{1}{v_F v_\Delta}\right)T^2, \tag{13.3}$$

where $\zeta(3) \approx 1.20$, k_B is Boltzmann's constant, n is the number of CuO_2 planes in a unit cell, d is the lattice constant along c-axis and V_{mol} is the volume of the unit cell per mol.

13.2.2. The Effect of a Magnetic Field: The Volovik Effect and Scaling Relations

High-T_C cuprates are strong type-II superconductors. In the mixed state, the magnetic field penetrates the sample as vortices. Around vortex normal cores, there are circulating supercurrents. Accordingly, two types of low-energy excitations could exist in the sample, that is, the quasi-particle excitations inside the vortex core and the excitations outside the vortex core. For d-wave superconductors, the semi-classical calculation of Volovik shows that the extended quasi-particle states outside the vortex core dominate the low-energy excitations [10]. This is correlated with the node structure of the d-wave superconducting gap as in momentum space, the density of states of these extended quasi-particles resides in the vicinity of the gap nodes. In the mixed state, due to the circulating supercurrents around vortex cores, the excitation spectrum of quasi-particles experiences a Doppler energy shift, $\delta E(\boldsymbol{k}) = \boldsymbol{v}_s \cdot \hbar \boldsymbol{k}$, where $v_s = \hbar/2mr$ is the velocity of the supercurrent with m the electron mass and r the distance away from the vortex core center. For isotropic s-wave superconductors, this energy shift is small compared with the superconducting gap Δ_0 and does not induce additional density of states at the Fermi level. For d-wave superconductors, however, this energy shift is comparable to the gap near the nodal region and induces new density of states at the Fermi level. For single vortex unit cell, the induced density of states is ($N_{\text{single}} \propto \int_\xi^R v_s 2\pi r dr$), where ξ is the coherence length and R is the distance between two neighboring vortices. As ($R \propto 1/\sqrt{H}$ for $H_{C1} \ll H \ll H_{C2}$) (H_{C1} and H_{C2} are the lower and upper critical fields, respectively), this gives $N_{\text{single}} \propto 1/\sqrt{H}$. On the other hand, the number of vortices is proportional to H. Therefore, it is obtained that the total density of states $N(0) \propto \sqrt{H}$. This is just the Volovik's relation for d-wave superconductors [10]. Specifically, in the mixed state the electronic specific heat $C_{\text{el}}(T,H)$ of a d-wave superconductor can be expressed as [10, 11]:

$$C_{\text{el}}(T,H) = \gamma(H)T = \gamma_N \sqrt{\frac{8}{\pi}} a \sqrt{\frac{H}{H_{C2}}} T, \qquad (13.4)$$

where γ_N is the coefficient of the electronic specific heat in the normal state, a is a numerical constant depending on the geometry of the vortex lattice. For square (triangle) vortex lattice $a = 0.5(0.465)$, on the other hand, the $C_{\text{el}}(T,H)$ can also be expressed in terms of the v_Δ as following [8, 9]:

$$\frac{C_{\text{el}}(T,H)}{T} = \gamma(H) = A\sqrt{H} = \frac{4k_B^2}{3\hbar}\sqrt{\frac{\pi}{\Phi_0}} \frac{nV_{\text{mol}}}{d} \frac{a}{v_\Delta} \sqrt{H}, \qquad (13.5)$$

where Φ_0 is the flux quantum. The prefactor A quantifies the strength of the field-induced electronic specific heat and, as we can see, is inversely proportional to the v_Δ.

Strictly speaking, the aforementioned Volovik's relation, that is, the $\sqrt{H}T$ dependence of the field-induced specific heat, holds only when $T/\sqrt{H} \ll T_C/\sqrt{H_{C2}}$. When the temperature is relatively high and the magnetic field is relatively low, that is, $T/\sqrt{H} > T_C/\sqrt{H_{C2}}$, $C_{\text{el}}(T,H)$ of the sample in fact consists of a T^2 term and a linear-in-H term [12]. To cover the above two limits, Simon and Lee pointed out that in the whole temperature and magnetic field range, $C_{\text{el}}(T,H)$ of a d-wave superconductor obeys the following scaling relation [13]:

$$\frac{C_{\text{el}}(T,H)}{T\sqrt{H}} = F\left(\frac{T}{\sqrt{H}}\right), \qquad (13.6)$$

where $F(x)$ represents the scaling function. At present, there is no exact analytical form for $F(x)$, but an empirical scaling function (neglecting the T^2 term) has been proposed [14]:

$$\frac{C_{\mathrm{el}}(T,H)}{T\sqrt{H}} = Af(y), \quad f(y) = \frac{1}{1+y}, \quad y = \eta\sqrt{\frac{H_{\mathrm{C2}}}{H}}\frac{T}{T_{\mathrm{C}}}, \qquad (13.7)$$

where η is a parameter of order unity. When $y \ll 1$ (low-T and high-H), $f(y) \to 1$, and the above scaling relation reduces to Volovik's relation $C_{\mathrm{el}}(T,H)/T = A\sqrt{H}$. Whereas for $y \gg 1$ (high-T and low-H), $f(y) \to 1/y$, and the above scaling relation describes the region where the specific heat shows a linear H dependence.

13.2.3. *Effect of Impurity Scattering*

In the above description of the low-energy quasi-particle excitations, we have assumed that the d-wave superconductors are in the "clean limit". In real samples, there is usually impurity scattering owing to the presence of chemical impurities or other defects. When the strength of such impurity scattering is not very weak and the sample is in the so-called "dirty limit", some argue that the superconducting gap near the nodal region would be suppressed, which then leads to the low-energy quasi-particle excitations deviating from the theoretical predictions in the clean limit [15]. The calculation of Kübert and Hirschfeld even showed that, when accounting for the impurity scattering, the field-induced specific heat $C_{\mathrm{el}}(T,H) = \gamma(H)T$ of a dirty d-wave superconductor will not follow Volovik's relation $\gamma(H) = A\sqrt{H}$, instead it will follow an $H \lg H$ dependence [11]:

$$\frac{\gamma(H)}{\gamma_{\mathrm{N}}} \cong \frac{\Delta_0}{8\gamma_0}a^2\left(\frac{H}{H_{\mathrm{C2}}}\right)\lg\left[\frac{\pi}{2a^2}\left(\frac{H_{\mathrm{C2}}}{H}\right)\right], \qquad (13.8)$$

where γ_0 is the so-called pair-breaking parameter.

Recently, however, new theoretical work suggested that the impurity scattering may not be able to affect the low-energy quasi-particle excitations in high-T_{C} cuprates, owing to the strong correlation effect [16]. In the work of Garg et al. [16], dependence of the density of states of a d-wave superconductor on the impurity concentration n_{imp} has been studied by including strong electron correlations. It has been found that, when ignoring the correlation effect, the low-energy density of states for nodal quasi-particle excitations varies with the strength of disorder and increases as

$n_{\rm imp}$ increases, consistent with conventional expectations. In contrast, when taking the correlation effect into account, it is shown that the low-energy density of states becomes insensitive to the impurity scattering and the nodal quasi-particles are actually protected against disorder.

13.2.4. Specific Heat Anomaly at $T_{\rm C}$

The specific heat anomaly of a d-wave superconductor can also be evaluated within the BCS scheme. Figure 13.3 shows the temperature dependence of the electronic specific heat $C_{\rm el}(T)$ for a d-wave superconductor, including the superconducting anomaly at $T_{\rm C}$, in comparison with that of an s-wave superconductor. The curves are obtained by assuming the weak coupling limit, namely, $\Delta_0/k_{\rm B}T_{\rm C} = 2.14$ for d-wave superconductor [17] and $\Delta_0/k_{\rm B}T_{\rm C} = 1.76$ for s-wave superconductor. It is seen that at low temperatures, as pointed out in Section 13.2.1, the $C_{\rm el}(T)$ of the d-wave superconductor follows a T^2 dependence, different from the exponential dependence for the s-wave superconductor. At $T_{\rm C}$, the specific heat jump of the d-wave superconductor is $\frac{\Delta C}{\gamma T_{\rm C}}|_{T_{\rm C}} = 0.95$, which is lesser than the value $\frac{\Delta C}{\gamma T_{\rm C}}|_{T_{\rm C}} = 1.43$ of the s-wave superconductor.

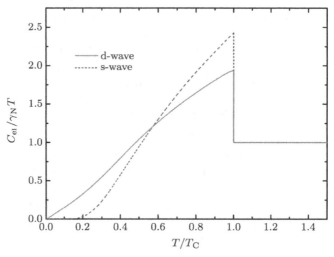

Fig. 13.3. Temperature dependence of the electronic specific heat, $C_{\rm el}(T)$, for a d-wave superconductor (solid line) or an s-wave superconductor (dashed line) in the weak coupling limit. $\gamma_{\rm N}$ denotes the coefficient of the specific heat in the normal state.

13.3. Experimental Background

13.3.1. *Experimental Techniques: Principle of the Relaxation Method*

There are different experimental techniques to measure the specific heat, which may be roughly categorized into two classes, adiabatic methods or non-adiabatic methods. Adiabatic methods include heat pulses and continuous heating methods. Relaxation method or ac calorimetry belongs to the non-adiabatic class. Different techniques have their own advantages and disadvantages [18, 19]. Relaxation method is suitable for measuring small samples at low temperatures and has been widely used in most specific heat measurements. Below, we briefly introduce the principle of it.

A typical relaxation calorimeter is illustrated schematically in Fig. 13.4. In the measurement, the sample and the platform are well connected by thermal conductive grease. They are suspended in vacuum by several thin wires, which provide the necessary electrical connections and a weak thermal link between the platform and the surroundings. With thermal power P being applied to the platform, the temperature of the platform (sample), T, evolves with time, t, and satisfies the following equation:

$$C_{\text{total}} \frac{dT}{dt} = -K_{\text{w}}(T - T_{\text{both}}) + P(t), \qquad (13.9)$$

where C_{total} is the total heat capacity of the sample and the platform, K_{w} is the thermal conductivity of the thin wire and T_{both} is the temperature of the thermal bath. The solution of Eq. (13.9) shows that T is exponentially dependent on t. Specifically, with ΔP the change in P, the change in the temperature of the platform, ΔT, is given by:

$$\Delta T = \Delta T_0 (1 - e^{-t/\tau}), \qquad (13.10)$$

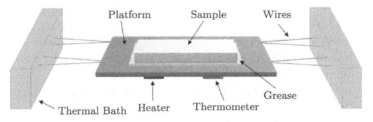

Fig. 13.4. Schematic of a relaxation calorimeter.

where ΔT_0 is final temperature change after equilibrium and $\tau = C_{\text{total}}/K_{\text{w}}$ is a time constant called the relaxation time constant. By fitting the experimental data to Eq. (13.10), one can obtain the above two parameters, ΔT_0 and τ. Then, the heat capacity can be calculated as $C_{\text{total}} = K_{\text{w}}\tau$ with the relation $K_{\text{w}} = \Delta P/\Delta T_0$. Finally, by subtracting the heat capacity of the platform and the grease from C_{total}, one can obtain the heat capacity of the sample.

13.3.2. Data Analysis: Separation of Different Contributions

In most experiments, the measured specific heat C comes from all elemental excitations of the sample. This means that to investigate the specific heat associated with one particular type of excitations, one needs to separate it from other contributions to C. In general, the total specific heat of a metallic system at low temperatures may contain the following terms [19, 20]:

$$C = C_{\text{ph}} + C_{\text{el}} + C_{\text{mag}} + C_{\text{hyp}}, \tag{13.11}$$

where C_{ph} represents the lattice contribution (phonon specific heat), C_{el} is the electronic specific heat, C_{mag} represents the contribution associated with paramagnetic centers, and C_{hyp} is a hyperfine contribution arising from an interaction of nuclear magnetic moments with a magnetic field.

Some detailed examinations of the above terms on their temperature or magnetic field dependence in typical metallic or superconductor systems have already been given in previous review articles [18, 19, 20]. As for C_{hyp}, it is known that its presence is usually confined to temperatures below about 1 K [20], a temperature range not covered by most of the specific heat measurements reviewed in this chapter (performed at T above 1 K or 2 K). Hence, in this section we choose to briefly describe the modeling of the other two non-electronic terms, that is, C_{ph} and C_{mag}, which is important in correctly determining the interested C_{el} term in specific heat studies of high-T_{C} cuprates.

C_{ph} makes a major contribution to C in high-T_{C} cuprates, and the reliable separation of C_{ph} and C_{el} is usually the first concern when dealing with the specific heat data. It is known that in most cases C_{ph} can be well described by the Debye-T^3 law at low temperatures, namely:

$$C_{\text{ph}} = \frac{12\pi^4}{5} N k_{\text{B}} \left(\frac{T}{\Theta_D}\right)^3 \equiv \beta_3 T^3, \tag{13.12}$$

where N is Avogadro constant and Θ_D is the Debye temperature. As the temperature rises, C_{ph} will show deviations from this simple cubic temperature dependence. One way to account for such deviations is to include higher-order terms such as T^5, T^7, etc., by considering the phonon dispersion. Another way is to consider excitations of the optical phonon mode, whose contribution to the specific heat can be expressed as:

$$C_E = mNk_B \left(\frac{T_E}{T}\right)^2 \frac{e^{-T_E/T}}{(1-e^{-T_E/T})^2}, \tag{13.13}$$

where m is the number of oscillators per unit cell and T_E is the Einstein temperature. At even higher temperatures, the modeling of C_{ph} would become more difficult and the uncertainty in its determination would become correspondingly larger.

One important feature of C_{ph} is that it shows no magnetic field dependence. Taking advantage of this can sometimes facilitate the data analysis, especially when studying the variation of C_{el} in different magnetic fields, as one can subtract the zero-field specific heat from the in-field one to remove C_{ph} or fit all the specific heat data in different fields simultaneously with C_{ph} constrained to be the same (the so-called "global fit").

For many high-T_C cuprates, the accurate determination of C_{el} could be further complicated by the appearance of C_{mag}. This term is usually seen at low temperatures as an upturn in zero-field or a broad peak in magnetic fields when plotting the specific heat data as C/T vs. T. The paramagnetic centers associated with impurities or defects inside the sample are responsible for this contribution. When the concentration of these paramagnetic centers is low and the interaction among them is not significant, C_{mag} can be well described by a Schottky anomaly. The Schottky anomaly of free spin-1/2 particles is given by:

$$C_{sch}(T,H) = n_{sch} \left(\frac{g\mu_B H}{k_B T}\right)^2 \frac{e^{g\mu_B H/k_B T}}{(1+e^{g\mu_B/k_B T})^2}, \tag{13.14}$$

where n_{sch} is the effective concentration of the particles, μ_B is the Bohr magneton and g is the Landé g factor.

In the early years of research on high-T_C superconductivity, it has already been noted that most cuprate materials such as $YBa_2Cu_3O_{7-\delta}$ (YBCO) and $La_{2-x}Sr_xCuO_4$ (LSCO) show a specific heat term in zero-field which is linear in temperature [18, 20]. This is the so-called residual linear specific heat, denoted as $\gamma(0)T$ here. At present, there are different opinions on its origin, and people in the community have not reached a consensus [15, 19]. One school of thought is that $\gamma(0)T$ may be associated with the impurity scattering effect. As mentioned in Section 13.2.3, according to traditional theory, impurities or defects in the sample would serve as additional scattering centers for quasi-particles and this would cause suppression of the superconducting gap near the nodes and then the presence of a finite zero-energy density of states at the Fermi level, that is, a non-zero $\gamma(0)$. Another line of thought suggests that as the residual specific heat shows linear-T dependence, just like the electronic specific heat in a normal metal, the observation of this term in high-T_C cuprates may indicate the existence of normal, non-superconducting regions in the sample.

13.3.3. Overview of Previous Specific Heat Studies in High-T_C Cuprates

13.3.3.1. Low-temperature specific heat: Evidence for d-wave symmetry of the superconducting gap

Now it is generally perceived that the superconducting gap in hole-doped high-T_C cuprates has d-wave symmetry. This agreement is reached based upon a large number of experiments conducted with diverse experimental techniques, such as ARPES [21], phase-sensitive tricrystal experiment [2], microwave [22], and so on. Among them, specific heat is a bulk probe which is not sensitive to surface imperfections or abnormalities of the sample. Although it cannot give information on the phase of the order parameter, the detecting of low-lying quasi-particle excitations makes it a sensitive probe of the superconducting gap near the nodes for a d-wave superconductor. Owing to these features, the evidence of d-wave pairing from specific heat has attracted considerable interest and been regarded as indispensable for a complete view of the superconducting properties of high-T_C cuprates [15].

As shown in Section 13.2, theoretical work suggested that the electronic specific heat of a d-wave superconductor at low temperatures follows the relation $C_{\text{el}}(T) = \alpha T^2$ in zero-field and $C_{\text{el}}(T, H) = A\sqrt{H}T$ in fields when $H \ll H_{C2}$. The experimental identification of these two effects has centered the low-temperature specific heat studies in high-T_C cuprates to identifying the d-wave symmetry of the superconducting gap. In 1994, Moler et al. first reported the identification of both αT^2 and $A\sqrt{H}T$ terms (Fig. 13.5) in

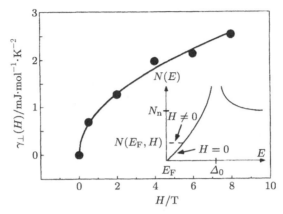

Fig. 13.5. Magnetic field dependence of the $\gamma(H)$ in a nearly optimally doped YBCO single crystal. The solid line represents the Volovik relation $\gamma(H) = A\sqrt{H}$ with $A = 0.91$ mJ·mol^{-1}·K^{-2}·T$^{-0.5}$. Inset shows a schematic plot of the density of states. Taken from Ref. [23].

the low-temperature specific heat of a nearly optimally doped YBCO single crystal [23], through a global fit to the data in different magnetic fields. And the value of the coefficient α or A was found to be in reasonable agreement with theoretical descriptions of a d-wave pairing state in this material. In the following 10 years, more specific heat measurements had been performed by different research groups on YBCO [24–27] and LSCO [28–30]. While in some studies the αT^2 term in zero-field had not been resolved from the data, presumably owing to its small magnitude ($\alpha \sim \gamma_N/T_C$ in theory) or difficulties in the data analysis, the $\sqrt{H}$ dependence of the C_{el} in fields had been confirmed in most cases, therefore giving more evidence for line nodes in the superconducting gap in these prototypical high-T_C compounds.

The proposed scaling relation of the C_{el} for a d-wave superconductor, Eq. (13.6), had also been tested by several authors in YBCO [25, 26] and LSCO [28–30]. Figure 13.6 shows an example for YBCO by Wright et al. [25]. Overall, although there are still some inconsistencies among the reports in certain aspects such as the form of the scaling function, the specific heat data for nearly optimally doped samples were found to satisfy the predicated scaling relation Eq. (13.6), which constituted further evidence for the d-wave pairing state in high-T_C cuprates.

For more detailed reviews of the experimental progress outlined above, the readers are encouraged to refer to the review articles that have appeared in recent years [15, 19].

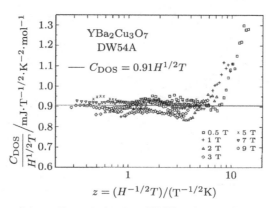

Fig. 13.6. A test of the scaling relation in a YBCO polycrystalline sample close to the optimal doping. The electronic specific heat C_{el} is denoted as C_{DOS} in the figure. Taken from Ref. [25].

13.3.3.2. *Specific heat anomaly at T_C*

There have been a lot of efforts to determine the specific heat anomaly (or jump) to investigate the nature of the superconducting transition in high-T_C cuprates. For early measurements, the reader is referred to some thorough reviews, for example, by Junod [18] and Phillips *et al.* [20]. For more recent measurements, a detailed review by Fisher *et al.* [19] can be recommended. Here, we only outline some typical results obtained in recent years. As mentioned before, the C_{ph} is usually difficult to be accurately modeled at elevated temperatures. For cuprates, the high T_C further implies that $C_{ph} \gg C_{el}$ in the vicinity of the T_C. All these make it challenging to reliably separate the C_{el} from the large C_{ph} background to study the superconducting transition at T_C in traditional specific heat experiments where the raw data contain both phonon and electronic contributions.

Loram and co-workers have used a differential calorimeter to remove the C_{ph} background and determine the C_{el} in several high-T_C compounds [31, 32]. In their technique, the difference in the specific heat between a superconducting sample studied and a non-superconducting reference sample can be measured. By choosing the reference sample to show similar phonon specific heat as that of the superconducting sample, the C_{el} of the superconducting sample can be extracted. It is noted that by this technique the C_{el} can be determined in a wide temperature range, not restricted to near T_C. Figure 13.7 shows the temperature dependence of the electronic specific heat coefficient γ in YBCO determined by Loram *et al.* [31]. Specific

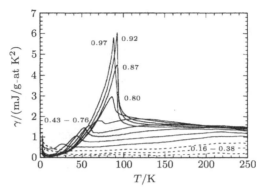

Fig. 13.7. Electronic specific heat coefficient γ vs. T for $YBa_2Cu_3O_{6+x}$. The x values of the samples are shown in the figure. Taken from Ref. [31].

heat anomalies are observed for all superconducting samples but found to show different characteristics with the variation of hole concentration. In the overdoped regime ($x \geq 0.92$), the superconducting anomaly is found to resemble that of a strong coupling BCS superconductor. With decreasing doping, however, the height of the anomaly falls quickly, and for underdoped samples ($0.43 \leq x \leq 0.76$), the anomalies are quite small and found to be much less than the weak coupling BCS predication. It is also evident that the specific heat anomaly becomes broader toward underdoping.

The magnetic field has also been used to separate the superconducting anomaly from the C_{ph} as the latter is field independent. Figure 13.8 shows a compilation of the specific heat anomaly for various superconductors by Junod et al. [33]. The data measured in the largest available magnetic field have served as a reference and have been subtracted, which effectively removes the C_{ph} contribution near T_C and makes the field dependence of the superconducting anomaly more transparent. It shows that for low-T_C superconductor $Nb_{77}Zr_{23}$ (Fig. 13.8(a)), a moderate magnetic field of 10 T can fully suppress the superconductivity and that the superconducting anomaly is BCS-like. For high-T_C cuprates YBCO, $Bi_2Sr_2CaCu_2O_8$ (Bi2212) and $Bi_2Sr_2Ca_2Cu_3O_{10}$ (Bi2223) (Figs. 13.8(b)–13.8(f)), as the upper critical field H_{C2} is quite large, a magnetic field of similar magnitude cannot fully suppress the superconductivity to help expose the C_{el} in zero field from T_C down to low temperatures. Nevertheless, the superconducting anomaly and its field dependence in the vicinity of the T_C can be clearly observed. It is shown from Figs. 13.8(b)–13.8(d) that the specific heat anomaly in YBCO

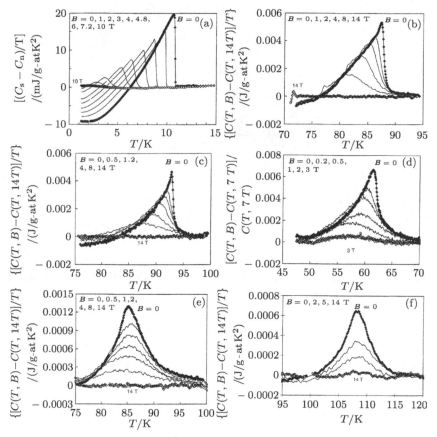

Fig. 13.8. Specific heat anomaly in different magnetic fields for various superconductors. The curves measured in the largest available magnetic field are used as a baseline and subtracted. (a) $Nb_{77}Zr_{23}$, (b) $YBa_2Cu_3O_7$, (c) $YBa_2Cu_3O_{6.92}$, (d) $YBa_2Cu_3O_{6.6}$, (e) $Bi_{2.12}Sr_{1.9}Ca_{1.06}Cu_{1.96}O_8$ (Bi2212) and (f) $Bi_{1.84}Pb_{0.34}Sr_{1.91}Ca_{2.03}Cu_{3.06}O_{10.1}$ (Bi2223). Taken from Ref. [33].

changes from being BCS-like to being more symmetric about T_C as the doping level decreases from the overdoped to the underdoped regime. Continuing this evolution, the specific heat anomaly in Bi2212 and Bi2223 (Figs. 13.8(e) and 13.8(f)) is nearly symmetric about T_C and the peak position is nearly field independent. This systematic variation in the shape of the specific heat anomaly has been discussed by some authors in the context of a BCS to BEC cross-over as the doping deceases in high-T_C cuprates [33, 34].

13.4. Low-Temperature Specific Heat Results in $La_{2-x}Sr_xCuO_4$ (LSCO) and $Bi_2Sr_{2-y}La_yCuO_{6+\delta}$ (La-Bi2201)

As pointed out in Section 13.3.3.1, in the first 10 years since the work of Moler *et al.* in 1994, the low-temperature specific heat experiments to investigate the symmetry of the superconducting gap had been primarily made on YBCO and LSCO. For YBCO, the measurements were performed either on single crystals [23, 24, 26, 27] or polycrystals [25], with the doping level of the sample confined in a range from nearly optimally doped to slightly overdoped. For LSCO, while the measurements had been carried out on samples covering a wider range of the doping [28–30], there was only one study reported on single crystal samples and in this study, the field-dependent specific heat for the underdoped sample was found to deviate from the $\sqrt{H}$ behavior and instead follow the relation Eq. (13.8) as predicted for "dirty-limit" superconductors [30].

In the phase diagram of high-T_C cuprates, the underdoped regime has continuously attracted a lot of interest from the community, as many highly unusual properties of high-T_C cuprates are observed in this regime in both the normal and superconducting states. Therefore, more exploration of the low-temperature specific heat in this part of the phase diagram would be helpful not only in clarifying the issues raised in previous specific heat studies, but also in providing new thermodynamic information on the nature of the high-T_C superconductivity. On the other hand, it is known that high-T_C cuprates are anisotropic superconductors with the H_{C2} along the c-axis much smaller than the one along the basal ab-plane. According to Eq. (13.4), this implies that, in magnetic fields, the C_{el} probed in a polycrystalline sample reflects mostly the contribution from the powders with c-axis roughly parallel to the fields, which are only a portion of the sample. Therefore, more quantitative analysis of the C_{el} data, such as to determine the v_Δ via Eq. (13.5), cannot be reliably made based on measurements of polycrystalline samples. It is necessary for one to resort to measurements on single crystals to fulfill such a purpose.

The above forms the background and motivation for our recent low-temperature specific heat work on LSCO and $Bi_2Sr_{2-y}La_yCuO_{6+\delta}$(La-Bi2201) [7, 35–41]. For LSCO, the measurements were made on high-quality single crystals with the doping ranging from the under doped, through the optimally doped, to the overdoped regimes of the phase diagram [35, 36], and for La-Bi2201, the measurement was performed on a nearly optimally doped single crystal [40]. In what follows, we give a brief review and discussion on these experimental results and their implications on the superconducting state of high-T_C cuprates.

13.4.1. Evidence for a d-wave Pairing State Throughout the Phase Diagram

Among the families of high-T_C cuprates, LSCO is a material in which one can vary the hole-doping concentration p across essentially the whole superconducting phase diagram by varying the Sr content x of the sample ($p = x$), which makes it an attractive candidate for studying systematic trends with doping. Furthermore, in terms of specific heat experiment, LSCO has an additional advantage over other materials such as YBCO in that for this system, the high-quality samples usually show weak or absence of C_{mag} contribution to the specific heat at low temperatures, which can greatly facilitate the data analysis in determining the interested C_{el} term.

Figure 13.9 summarizes the coefficient of the field-induced specific heat, $\gamma(H) = C_{\text{el}}(T,H)/T$, in LSCO across the whole superconducting phase diagram [35, 36]. For crystals where there is no discernible contribution of C_{mag} to the specific heat, the $C_{\text{el}}(T,H)$ has been determined in a straightforward manner by subtracting the zero-field data $C(T,0)$ from the in-field data $C(T,H)$ and then extrapolating the subtraction to zero-temperature limit as in this case where we have $C(T,H) - C(T,0) = C_{\text{el}}(T,H) - \alpha T^2$ [7, 35]. For crystals where a small contribution of C_{mag} is observed in the

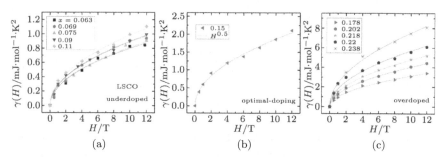

Fig. 13.9. The coefficient of the field-dependent specific heat, $\gamma(H)$, in LSCO across the whole superconducting phase diagram [38]. Solid lines are fits to $\gamma(H) = A\sqrt{H}$ (as indicated in (b)) according to the d-wave pairing theory, with the parameter A depending on the doping level of the sample.

data, the $C_{el}(T, H)$ was determined in another way by fitting the data in different fields to Eq. (13.11) with the C_{hyp} component omitted as the experiment was conducted down to about 2 K [36]. It is noted that in Ref. [36] the fit was performed to the data in each field individually and there was no αT^2 term resolved in zero-field data, similar to previous studies on polycrystalline samples [28]. Recently it has been found, however, that, if a global fit to all the data in different fields was performed, a small αT^2 term in zero field could indeed be identified [42]. This just demonstrates the difficulty in separating the αT^2 term from the total specific heat when the C_{mag} contribution is present in the data. On the other hand, it is noted that, as the αT^2 term extracted in the global fit is rather small, the $C_{el}(T, H)$ determined in the global fit are nearly the same as that in the individual-field fit, which shows the reliability of its identification.

In Fig. 13.9, the solid lines represent the fit of the data to $\gamma(H) = A\sqrt{H}$ according to the d-wave pairing theory. It is clear that the agreement between the data and the fit is rather good for all the samples investigated (doping $p \sim 0.06 - 0.24$), which constitutes a compelling evidence for a predominant d-wave superconducting gap throughout the phase diagram. We note that recent phase-sensitive tricrystal experiments have also showed the robust nature of the d-wave pairing in high-T_C cuprates over a wide doping range ($p \sim 0.07 - 0.21$) [43], which is consistent with our specific heat result. In Fig. 13.10(a), we have further highlighted the

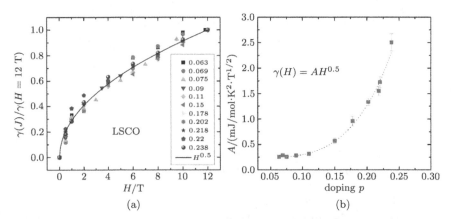

Fig. 13.10. (a) Field dependence of $\gamma(H)$ in LSCO normalized by the value at 12 T. (b) Doping dependence of the prefactor A in LSCO. The dotted line is a guide to the eye. Ref. [39].

universal $\sqrt{H}$ dependence of the $\gamma(H)$ in LSCO by plotting the normalized $\gamma(H)/\gamma(H = 12T)$ for all the samples. As mentioned above, in a previous specific heat study by Nohara et al. on LSCO single crystals [30], the $\gamma(H)$ of the underdoped or optimally doped sample was found to follow the $H \lg H$ dependence as predicted for a d-wave superconductor in the "dirty limit", not the $\sqrt{H}$ dependence as shown in Fig. 13.9. This discrepancy, we believe, may arise from the differences in both the sample quality and the procedure adopted in the data analysis. Unlike our work, in the study of Nohara et al., the zero-field specific heat of the crystals contains a sizeable contribution of C_{mag}, and to extract the $C_{\text{el}}(T, H)$ subtraction $C(T, H//c) - C(T, H//ab)$ rather than $C(T, H//c) - C(T, 0)$ has been performed [30]. In fact, in the specific heat study of YBCO it was even shown that using the subtraction $C(T, H//c) - C(T, H//ab)$ to separate $C_{\text{el}}(T, H)$ may lead to erroneous conclusions on the field dependence of the $\gamma(H)$ [19, 26, 44].

More recently, we have measured the low-temperature specific heat on a nearly optimally doped La-Bi2201 single crystal ($y \sim 0.4$) [40]. It is known that for this monolayer high-T_C cuprate, the phonon specific heat C_{ph} is considerably larger than for YBCO or LSCO, and for many samples the specific heat shows the contribution of C_{mag} at low temperatures, which makes it a challenging task to resolve the C_{el} for this material. Thanks to the high quality of our crystal which shows no discernible C_{mag} term, we have been able to carry out a global fit to the data in different fields to successfully identify the C_{el} contribution. It was found that in zero field, the αT^2 term has the coefficient $\alpha = (0.10 \pm 0.04)$ mJ $\cdot$ mol^{-1} $\cdot$ K^{-3} and in magnetic fields, as shown in Fig. 13.11, the $\gamma(H)$ obeys an $A\sqrt{H}$ dependence with $A = (0.90 \pm 0.06)$ mJ $\cdot$ mol^{-1} $\cdot$ K^{-2} $\cdot$ T$^{-0.5}$, both of which are in conformity with the d-wave pairing theory [40]. This therefore provides the bulk evidence for d-wave symmetry of the superconducting gap in optimally doped La-Bi2201.

13.4.2. *Bulk Measure of the Superconducting Gap and its Anomalous Doping Dependence*

In Fig. 13.9, it can be seen that the magnitude of the $\gamma(H)$ in LSCO shows a systematic variation with the doping; as the doping increases, the $\gamma(H)$ becomes progressively larger at a given field. To quantify its systematic change, the prefactor A in the fit of $\gamma(H) = A\sqrt{H}$ has been determined and

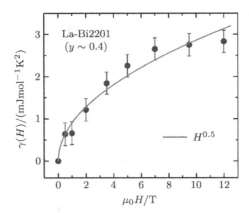

Fig. 13.11. Field dependence of $\gamma(H)$ in a nearly optimal-doped La-Bi2201 single crystal ($y \sim 0.4$). The solid line is the fit to $\gamma(H) = A\sqrt{H}$ for d-wave pairing with $A = 0.90$mJ $\cdot$ mol^{-1} $\cdot$ K^{-2} $\cdot$ T$^{-0.5}$. Ref. [40].

plotted in Fig. 13.10(b). It is clear that A goes up continuously as the doping is increased from the underdoped to the overdoped regimes. Within the d-wave framework, Eq. (13.5) tells us that the magnitude of A is inversely proportional to the slope of the superconducting gap at the node v_Δ, which, as given by Eq. (13.1), is in turn proportional to the magnitude of the superconducting gap at the antinode Δ_0. Therefore, the continuous increase of A shown in Fig. 13.10(b) implies a persistent decline of Δ_0 with the increase of hole concentration for LSCO.

Owing to the single-crystal nature of the sample, a quantitative determination of the Δ_0 from the above experimental values of A can be made by using Eqs. (13.1) and (13.5). Figure 13.12 summarizes the Δ_0 determined from the specific heat as a function of doping for both LSCO and La-Bi2201 [41]. For LSCO, the lattice parameters used in Eq. (13.5) are $n = 2$, $d = 13.28$ Å, and $V_{\text{mol}} = 58$ cm^3 $\cdot$ mol^{-1}, while for La-Bi2201, $n = 2$, $d = 24.4$ Å, and $V_{\text{mol}} = 106.8$ cm^3 $\cdot$ mol^{-1}. In further converting v_Δ into Δ_0 through Eq. (13.1), the Fermi wave vector k_F along the nodal direction $(0, 0)$-(π, π) has been quoted from ARPES experiments. It has been shown by recent ARPES studies that k_F has a value of 0.7 Å^{-1} and is nearly doping-independent for LSCO [45] and $k_F \approx 0.74$ Å^{-1} for optimally doped La-Bi2201 [46]. Therefore, this suggests that there are no adjustable parameters when deriving the v_Δ or Δ_0 from A by the use of Eqs. (13.1) and (13.5).

Figure 13.12 explicitly shows that, starting from the overdoped regime, the Δ_0 keeps increasing as the doping level of the sample decreases. Such

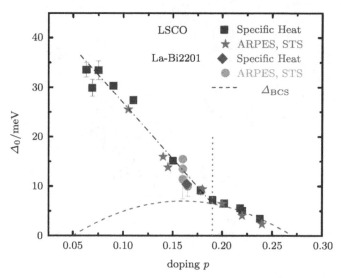

Fig. 13.12. Δ_0 vs. doping p plot of the superconducting gap for LSCO and La-Bi2201, as determined from the field dependence of the specific heat. For $p = 0.19$, the Δ_0 is determined for LSCO based on the value of $A=1.2$ mJ · mol^{-1} · K^{-2} · T$^{-0.5}$ as reported by Nohara et al. [30]. The Δ_0 obtained from some ARPES and scanning tunneling spectroscopy (STS) studies are also shown for comparison (stars for LSCO [47, 48] and circles for La-Bi2201 [49–51], respectively). The dashed line is the weak-coupling d-wave BCS gap form $\Delta_{BCS} = 2.14 k_B T_C$ with $T_C/T_C = 1 - 82.6(p - 0.16)^2$ and $T_C^{max} = 38$ K. The vertical dotted line marks the doping level $p = 0.19$. The dashed dot line is a guide to the eye [41].

a trend holds in almost the whole superconducting phase diagram, including in the underdoped regime. This means that while in the overdoped regime the Δ_0 exhibits a similar doping dependence to the T_C, both increasing with decreasing p, in the underdoped regime, the Δ_0 shows a doping dependence opposite to the T_C, as the latter is known to decrease with diminishing p. This anti-correlation between Δ_0 and T_C suggests an unconventional superconducting state in the underdoped regime since, as we know, for conventional superconductors the T_C scales with the pairing strength, namely the Δ_0, in a BCS fashion. In Fig. 13.12, a BCS d-wave gap $\Delta_{BCS} = 2.14 k_B T_C$ in the weak-coupling approximation [17] has also been plotted for comparison. When $p \geq 0.19$, it is seen that the superconducting gap Δ_0 follows the BCS predication rather well, indicating a weak-coupling superconducting state in this overdoped regime. As the p decreases from 0.19, however, deviation of the Δ_0 from the weak-coupling BCS behavior becomes progressively large. The doping $p = 0.19$ may be a critical doping point signifying a marked change in the doping dependence of the Δ_0.

Traditionally, the Δ_0 can be measured by ARPES or scanning tunneling spectroscopy (STS) experiment. Thus, it is desirable to compare the Δ_0 determined from these techniques with the one from specific heat. Figure 13.12 illustrates such a comparison by plotting together the Δ_0 from several recent ARPES and STS studies for both LSCO [47, 48] and La-Bi2201 [49–51]. It is evident that these determinations of the Δ_0 are in excellent quantitative agreement with the result of specific heat, which further confirms the reliable measurement of the Δ_0 from specific heat and, in particular, a growing pairing strength toward the underdoping. Moreover, as in STS and some ARPES studies, the Δ_0 is determined directly at the antinode, while in specific heat it is derived from the nodal gap slope v_Δ with the assumption $\Delta = \Delta_0 \cos(2\theta)$; this agreement also suggests that the superconducting gap essentially follows the standard d-wave form in high-T_C cuprates, which has been questioned recently by some investigators [52].

13.4.3. Evidence for an Intimate Relation Between Pseudogap and High-T_C Superconductivity

As discussed at the beginning of this chapter, the nature of the pseudogap and its relationship with the superconductivity in high-T_C cuprates are still under intense debate and the identification of them is widely believed to be crucial in identifying the pairing mechanism of superconductivity [5, 6]. With the knowledge of the nodal gap slope v_Δ and its doping dependence, one may be able to take a step further in investigating these issues, as the v_Δ is unambiguously a property associated with superconductivity and comparison of it with the established pseudogap behavior may give important clues on the possible connections between the two. For LSCO, the pseudogap energy scale, or the pseudogap temperature T^* has been systematically studied by a series of experiments. Figure 13.13 shows the T^* in LSCO determined from resistivity and magnetic susceptibility measurements (quoted from [3]) and its comparison with the v_Δ. It is seen that the T^* and v_Δ show a very similar doping dependence and can be simply scaled through a relation of $k_B T^* \approx 2.18 \Delta_0$ as shown in Fig. 13.13 by the solid line, which is remarkable as the T^* and v_Δ are determined from totally different experiments [7]. This strongly suggests that the pseudogap has an intimate relationship with the superconductivity. Note that this conclusion is in line

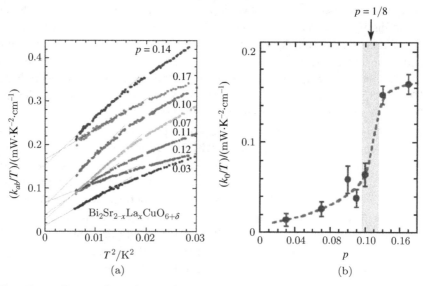

Fig. 13.13. Doping dependence of the pseudogap temperature T^* (open symbols) quoted from [3] and the nodal gap slope v_Δ (solid diamonds) derived from specific heat in LSCO [7]. $T^* - \rho$ and $T^* - \chi$ represent the pseudogap onset temperature determined from resistivity and magnetic susceptibility experiments, respectively. For more details, please refer to [3]. It is found that the T^* and the v_Δ can be correlated through a simple relation of $k_B T^* \approx 2.18\Delta_0$ (solid line), suggesting a close relationship between the pseudogap and the superconductivity.

with the results of Sutherland et al. who extracted the nodal gap slope v_Δ from thermal conductivity measurements on YBCO and found that it grows with underdoping and scales closely with the pseudogap measured by ARPES [53].

The similar trend of T^* and v_Δ with variation of the doping, particularly in the underdoped regime, favors the notion that the pseudogap may have a superconducting origin. It indicates that what is responsible for the formation of the pseudogap may also be the source for electron pairing in the superconducting state. In this regard, it is worth pointing out that the above experimental observations seem to be consistent with the predications of one particular theoretical model, namely the SU(2) slave-boson theory [54]. According to this theory which is proposed based on the general resonating-valence-bond (RVB) picture, the spin-singlet pairing in the RVB state causes the formation of the pseudogap and also lends its pairing strength to the mobile electrons, making them naturally pair and then condense at T_C.

13.4.4. Residual Linear-T Specific Heat in Zero-Field: Implications on Phase Separation

The specific heat experiments in LSCO also provide an opportunity to identify the origin of the residual linear-T term in zero magnetic fields, $\gamma(0)T$, from a doping-dependent perspective. As pointed out in Section 13.3.2, at present there is no consensus on where this anomalous term comes from for high-T_C cuprates. Either impurity scattering effect or phase separation has been invoked to account for its appearance. Fig. 13.14(a) summarizes the doping variation of the $\gamma(0)$ determined in our measurements [35, 36] and in previous studies [28–30] for LSCO. The coefficient of the normal-state specific heat γ_N reported in Ref. [55, 56] is also plotted for comparison. It is shown that around the optimal-doping point the samples seem to have the lowest $\gamma(0)$; with either underdoping or overdoping, the $\gamma(0)$ increases and becomes comparable to the γ_N when approaching the edges of the superconducting dome. According to the impurity-scattering scenario [11], this rapid increase of the $\gamma(0)$ from the optimal doping point would indicate a stronger scattering effect from impurity or disorder on d-wave nodal quasi-particles toward either underdoping or overdoping, which, in turn, would indicate a breakdown of the $\sqrt{H}$ dependence of the field-induced specific heat for samples located near both ends of the superconducting dome. Obviously, this expectation is not consistent with the experiments: as shown by Fig. 13.9, the $\sqrt{H}$ dependence of the C_{el} holds rather well for LSCO throughout the whole superconducting phase diagram, even for the very underdoped ($p = 0.063$) and highly overdoped ($p = 0.238$) samples. Furthermore, it was found that detailed numerical estimations of the density of states induced by the impurity scattering would give values of $\gamma(0)$ much lower than the experimental ones [36], again arguing against impurity scattering as the main source of $\gamma(0)$.

The appearance of small normal-state regions in the sample, i.e. a phase separation into superconducting and non-superconducting regions seems to better explain the experimental observations. According to this scenario, the increase of the $\gamma(0)$ with doping moving away from the optimal-doping point suggests an increase of non-superconducting regions in the sample and accordingly a decrease of the superconducting volume fraction. In the

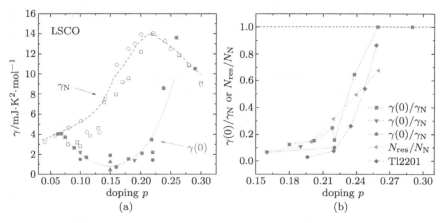

Fig. 13.14. (a) Coefficient of the residual specific heat, $\gamma(0)$ (solid symbols) and the coefficient of the normal-state specific heat, γ_N (open symbols) in LSCO as a function of doping. The $\gamma(0)$ reported in [35, 36] (squares), [28] (circles), [29] (up triangles), and [30] (down triangles) are shown. For γ_N, the values determined by Matsuzaki et al. [55] (squares) and Momono et al. [56] (circles) are shown. (b) $\gamma(0)$ in the overdoped regime normalized by γ_N [36]. The normalized residual spin Knight shift N_{res}/N_N, another probe of the residual density of states, in LSCO from [60] is shown together. For comparison, the $\gamma(0)/\gamma_N$ for overdoped Tl2201 [62] is also plotted.

vicinity of the edges of the superconducting dome, the superconducting volume fraction becomes so small that most of the sample is occupied by normal-state regions, leading to the value of the $\gamma(0)$ being very close to the γ_N. On the other hand, as the increase of the specific heat in fields is a property exclusive to the superconducting regions, its $\sqrt{H}$ dependence can be well preserved in spite of the appearance of the normal-state regions, explaining the coexistence of a well-defined $\sqrt{H}$ dependence in fields and a large $\gamma(0)$ in zero field.

In fact, a phase separation occurring in the superconducting state has been suggested by a number of experiments for high-T_C cuprates, particularly in the overdoped regime. Early muon spin relaxation (μSR) experiments in overdoped $Tl_2Ba_2CuO_{6+\delta}$ (Tl2201) showed that the superconducting carrier density divided by the effective mass decreased with increasing doping, suggesting the appearance of unpaired carriers at $T = 0\,\text{K}$ [57]. Subsequently, other experiments such as optical conductivity [58, 59], Knight shift [60] (as shown in Fig. 13.14(b)) and magnetization [61] also indicated the existence of non-superconducting regions in overdoped YBCO or LSCO. As illustrated in Fig. 13.14(b), the consistency among the results from specific heat and the above-mentioned experiments in different families of materials suggests that phase separation may be a generic property of the overdoped high-T_C cuprates.

13.4.5. *Implications on the Superconducting Condensation*

13.4.5.1. *Evidence for a "Fermi arc" ground state in the underdoped regime*

Below the pseudogap temperature T^* and above T_C, ARPES experiments showed that the Fermi surface of the underdoped cuprates are disconnected segments known as Fermi arcs residing near the nodal points owing to the appearance of the pseudogap around the antinodes [4]. When the doping of the sample increases, these Fermi arcs grow toward the antinodes and eventually connect with each other to form a large Fermi surface for the overdoped samples [45]. This anomalous property of the Fermi surface in underdoped regime has attracted a lot of interest from the community. When the temperature is lowered to below T_C, these Fermi arcs are gapped out as the superconductivity sets in and a d-wave superconducting gap is fully developed, precluding one from directly observing their evolution toward the zero-temperature limit. Some suggested that the four Fermi arcs observed in the pseudogap state would shrink to four nodal points at $T = 0$ K if the superconductivity could be completely suppressed, producing a novel, nodal liquid (metal) phase of the pseudogap ground state [63]. Whether this really happens in high-T_C cuprates or instead the length of the Fermi arc still remains finite at $T = 0$ K is an important issue in investigating the nature of the pseudogap and the occurrence of superconductivity. In what follows we show that specific heat is in fact very helpful in addressing this issue, owing to its bulk sensitivity to the low-energy density of states. Importantly, it is found that the specific heat results in high-T_C cuprates point to an arc metal, rather than a nodal metal, picture of the pseudogap ground state, that is, there should be finite density of states in the $T = 0$ pseudogap phase [7, 37].

The first piece of evidence for an arc metal state is given by the specific heat measurements in zero fields. In the study of Loram *et al.* [62] and Matsuzaki *et al.* [55], the coefficient of the electronic specific heat, $\gamma = C_{el}/T$, has been determined in LSCO for temperatures below and above T_C by using the differential calorimetry. Figure 13.15 shows the result of

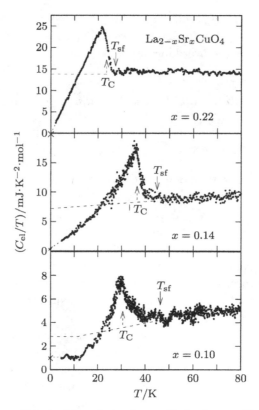

Fig. 13.15. The coefficient of the electronic specific heat, $\gamma = C_{el}/T$, in LSCO for temperatures below and above T_C. The dashed lines represent the γ in the hypothetical normal state, $\gamma_N(T)$ at $T < T_C$ determined based on the entropy conservation. The finite values of γ_N at $T = 0$ K suggest an arc metal ground state of the pseudogap phase. Taken from Ref. [55].

Matsuzaki *et al.* [55]. To infer the normal-state coefficient γ_N below T_C, a simple linear extrapolation of the γ_N above T_C down to $T = 0$ K has been made, as shown by the dashed lines in Fig. 13.15. The validity of this extrapolation has been checked with the law of entropy conservation [55]. It is known that the superconducting transition in zero field is a second-order phase transition and thus obeys the constraint of $S_S(T_C) = S_N(T_C)$ where the entropy:

$$S_S(T_C) = \int_0^{T_C} \gamma_S dT \quad \text{and} \quad S_N(T_C) = \int_0^{T_C} \gamma_N dT.$$

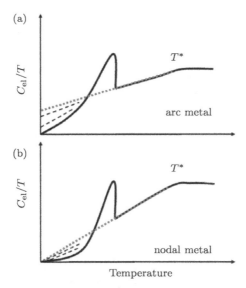

Fig. 13.16. Schematic plot for temperature dependence of the electronic specific heat coefficient C_{el}/T for arc metal (a) and nodal metal (b) in the pseudogap ground state (red dotted lines) [37]. The two short dashed lines in the zero-temperature limit illustrate how C_{el}/T changes with increasing magnetic field (from bottom to up) if the pseudogap ground state is an arc metal (a) or a nodal metal (b).

It is found that the γ_N determined by the above method indeed satisfies the entropy conservation, which is manifested by the equal of the excess and deficit areas of the γ_S vs. T curve compared with the extrapolated γ_N curve. In Fig. 13.15, it is clear that at $T = 0$ K the γ_N inferred above has a finite value, suggesting an arc metal ground state of the pseudogap phase.

Another piece of evidence for an arc metal state comes from the field dependence of the specific heat. In Fig. 13.16, the different response in specific heat to magnetic fields (dashed lines) has been shown schematically by assuming the pseudogap ground state of the sample to be an arc metal or a nodal metal [37]. In the scenario of an arc metal state, an increase in the specific heat coefficient γ with increasing fields would be expected, while in the scenario of a nodal metal state, such change in the γ at $T = 0$ K would not be expected. As shown in Section 13.4.1, a field-induced electronic specific heat $\gamma(H)T$, with the coefficient $\gamma(H) \propto \sqrt{H}$ in the zero-temperature limit, has been universally observed in superconducting LSCO single crystals, including the very underdoped samples. This, on the one hand, gives the bulk evidence for the d-wave symmetry of the superconducting gap

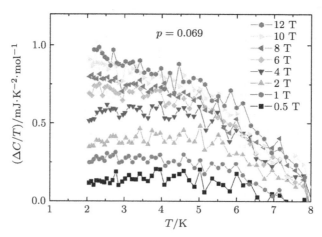

Fig. 13.17. Change of the low-temperature specific heat for the very underdoped LSCO ($p = 0.069$) in different magnetic fields [39]. Extra quasi-particle density of states induced by the field is clearly indicated, contradicting the expectation of a nodal metal state.

as discussed previously, and on the other hand implies that, according to Eq. (13.4), the exposed normal state would have a finite value of γ_N when the magnetic field increases to the H_{C2} to fully suppress the superconductivity, consistent with the expectation of an arc metal ground state. Figure 13.17 shows in particular the change of the low-temperature specific heat for the very underdoped LSCO ($p = 0.069$) in different magnetic fields [7, 35, 37]. It is clearly seen that the coefficient of the field-induced specific heat increases with the increasing of magnetic field, indicating the increasing generation of quasi-particle density of states. This contradicts the expectation of a nodal metal state but agrees with that of an arc metal state in the pseudogap phase.

It is noted that the conclusion of a Fermi arc state at $T = 0$ K has also been supported by other types of experiments such as the nuclear magnetic resonance (NMR). In the NMR measurements on La-Bi2201, Zheng et al. have found that for the nearly optimally doped sample ($T^* \sim 150$ K), the $1/T_1T$ is non-zero at $T \sim 1$ K when the superconductivity is fully suppressed by a high magnetic field of 43 T, showing that there is still substantial density of states at the Fermi level in the normal ground state [64]. Only when the field is removed and the superconductivity is recovered again, this density of states disappears. It is seen that this result is well consistent with the above specific heat experiments, indicating an arc metal normal state.

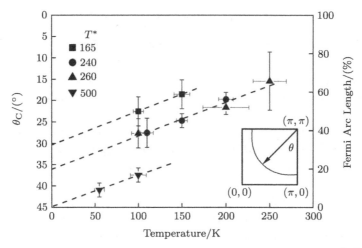

Fig. 13.18. Fermi-arc length extracted from the raw data of Knigel et al. [65] plotted vs. temperature (symbols) [66]. With the exception of the most underdoped sample, the Fermi-arc lengths extrapolate to finite values at $T = 0$ K. Inset: Fermi surface showing the angle θ.

Through ARPES experiment, Kanigel et al. even measured the length of the Fermi arc in Bi2212 above T_C and determined its evolution with doping and temperature [65]. When plotting it as a function of reduced temperature $t = T/T^*$, they found that the Fermi arc length seems to extrapolate linearly to zero at $t = 0$. Based on this, they suggested that the pseudogap ground state at $T = 0$ K is a nodal metal (liquid) [65], which is apparently in contrast with the above conclusion of specific heat and NMR. Recently, Storey et al. re-examined the photoemission data of Kanigel et al. and found that, against the claim of Kanigel et al., their data in fact suggest a non-zero length of the Fermi arc in the normal ground state [66]. By demonstrating this, Storey et al. plotted the Fermi arc length data of Kanigel et al. as a function of the absolute temperature T, not the reduced temperature t, as shown in Fig. 13.18 [66]. It is seen that, except for the most underdoped sample ($T^* = 500$ K), the Fermi arc length extrapolates to a non-zero value at $T = 0$ K for all other samples and, moreover, it increases as the doping level of the sample increases, precisely consistent with the observation of specific heat. The scaling analysis of Kanigel et al. has actually concealed all these important experimental details [66].

13.4.5.2. *Indication of a second superconducting energy scale*

From field-dependent specific heat on LSCO, it has been shown in Sections 13.4.2 and 13.4.3 that the nodal gap slope v_Δ, or equivalently the superconducting gap maximum Δ_0 at the antinode, keeps rising with decreasing doping and tracks the pseudogap Δ_p in the underdoped regime. This shows that the magnitude of the superconducting pairing strength becomes larger toward underdoping and according to the conventional theory, this would lead to even higher superconducting transition temperatures T_C for underdoped samples than for optimal- or overdoped samples, which apparently contradicts the experimental observation that with decreasing doping from overdoped side, the T_C reaches its maximum value at the optimal doping point and then falls down with further decreasing doping in the underdoped regime. This contrast between the expectation and the experimental observation in the underdoped regime suggests that, apart from the energy scale $\Delta_0(\Delta_p)$ representing the pairing strength, there may be another lower energy scale in underdoped high-T_C cuprates determining the occurrence of the superconductivity. Based on the ARPES findings and the above discussion on the pseudogap ground state, we suggest that this second energy scale is related to the arc feature of the Fermi surface in the underdoped regime [7]. As pointed out before, ARPES shows that above T_C, the Fermi surface of the underdoped cuprate consists of disconnected arcs centered at the nodal region owing to the pseudogap formation around the antinodes [4]. When decreasing temperature to below T_C, these Fermi arcs are further gapped as the sample enters into the superconducting state with a d-wave gap fully opened from node to antinode. We propose that associated with this gapping of the Fermi arc, there is an effective superconducting energy scale (gap) Δ_{sc} present for underdoped cuprates, where the Δ_{sc} is defined as the gap opened at the tip of the Fermi arc [7]. It is the Δ_{sc}, not the Δ_0 that determines most of the superconducting properties such as the T_C of the underdoped cuprates. In other words, it is the pairing of the carriers on nodal Fermi arcs that triggers the coherent superconductivity and defines an effective superconducting gap Δ_{sc} for underdoped cuprates. Figure 13.19 shows a schematic plot of the Δ_{sc}, Δ_0 (Δ_p) and their evolution with the doping [37]. When the temperature is above T_C, the pseudogap Δ_p opens around the antinodes as shown by blue dashed curves, leaving the Fermi surface to be arcs at the nodal region (red dashed lines). Below T_C, the carriers on the nodal Fermi arc are paired to open a new gap as shown by

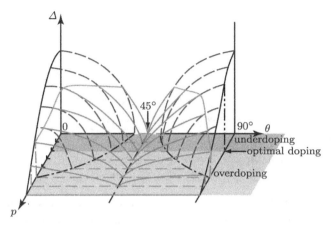

Fig. 13.19. Schematic plot for the doping p dependence of different energy scales in high-T_C cuprates [39]. The angle θ is defined associated with the Fermi surface in momentum space with $\theta = 0°$ and $90°$ corresponding to the antinodes and $\theta = 45°$ to the nodal point. Above T_C, the Fermi surface is truncated forming arcs near the nodal region (red dashed lines) due to the opening of the pseudogap Δ_p around the antinodes (blue dashed curves). Below T_C, the carriers on the Fermi arc are paired to open a new gap (red solid curves) with its slope v_Δ tracking the pseudogap and its value at the end point of the Fermi arc signifying an additional superconducting energy scale Δ_{sc}. According to this, while the gap slope v_Δ or the superconducting gap maximum Δ_0 keeps increasing with the underdoping as shown by the solid black curve, the effective superconducting energy scale Δ_{sc} as shown by the green solid curve decreases in the underdoped regime owing to the shrink of the Fermi arc, explaining the decrease of the T_C with decreasing doping in this part of the phase diagram.

red solid curves, with Δ_{sc} the value at the end point of the arc (green solid curve). Because the nodal gap slope v_Δ tracks the pseudogap Δ_p, the red solid curve smoothly joins the dashed blue curve at Δ_{sc}.

It is important to see that, unlike the Δ_0 which is only determined by the v_Δ, the Δ_{sc} is determined by both the nodal gap slope v_Δ and the Fermi arc length [7]. If we use k_{arc} to represent the length of each Fermi arc segment at zero-temperature in the Brillouin zone, we can have $\Delta_{sc} \propto v_\Delta k_{arc}$, or more precisely $\Delta_{sc} \approx (\hbar/\pi) v_\Delta k_{arc}$. On the other hand, as discussed before, the length of the Fermi arc is a measure of the residual density of states in the arc metal ground state of the pseudogap phase. By assuming a uniform distribution of the density of states along the Fermi arc, we can readily obtain that the density of states $N(E_F)$ in the pseudogap phase can be expressed as $N(E_F) = 2n k_{arc}/\pi^2 \hbar v_F$. Furthermore, as the coefficient of the normal-state electronic specific heat is given by $\gamma_N =$

$(\pi^2/3)k_B^2 N(E_F)V_{mol}$, one can have:

$$\gamma_N = \frac{2k_B^2}{3\hbar}\frac{n}{d}\frac{V_{mol}}{v_F}k_{arc} \qquad (13.15)$$

for the pseudogap phase at $T = 0$ K. With increasing doping in the underdoped regime, the γ_N increases as the k_{arc} increases approximately linearly, as found by experiments shown in Fig. 13.14(a). With Eq. (13.1), the Δ_{sc} can further be expressed as:

$$\Delta_{sc} \approx \frac{3\hbar^2}{2\pi k_B^2}\frac{d}{n}\frac{v_F}{V_{mol}}v_\Delta \gamma_N. \qquad (13.16)$$

It is clearly seen that $\Delta_{sc} \propto v_\Delta \gamma_N$, which shows that in the underdoped regime, the establishment of the superconductivity is governed by not only the pairing strength (measured by v_Δ) but also the density of states available for superconducting condensation (measured by γ_N). In the following Section 13.4.5.3, we shall show that the doping evolution of basic superconducting properties such as the T_C and the superconducting condensation energy E_C for underdoped cuprates can be successfully accounted for by invoking the presence of this effective superconducting energy scale Δ_{sc}.

13.4.5.3. *Doping dependence of T_C, the superconducting condensation energy E_C, and the upper critical field H_{C2}*

In the underdoped regime, the coherent superconductivity disappears with the appearance of the Fermi arc at the nodal region as the temperature is increased up to T_C. As the effective superconducting energy scale Δ_{sc} is defined associated with the gapping of the Fermi arc, it also closes at $T = T_C$ with the recovery of the Fermi arc. It is noted that this concomitance of the disappearance of the coherent superconductivity with the close of the Δ_{sc} resembles the conventional BCS picture that the vanishing of the superconductivity is accompanied by the closure of the superconducting energy gap and vice versa. Indeed, recent ARPES experiments on underdoped Bi2212 showed that the energy gap opened on the Fermi arc region follows a BCS-like temperature dependence and closes at $T = T_C$ [67]. These suggest that when considering the superconducting properties associated with the pairing of the carriers on the nodal Fermi arcs, a general BCS framework may be a reasonable approximation. Note that, however, this by no means states that the superconducting pairing on the Fermi arc has a phonon origin or that the superconducting transition over the entire underlying Fermi surface, i.e. from node to antinode, can be described by the conventional BCS theory.

We first discuss the variation of the T_C with doping in terms of the Δ_{sc}. In this regard, it is noted that, as shown in Fig. 13.12, for overdoped LSCO ($p \geq 0.19$) the T_C scales with the superconducting gap Δ_0 in agreement with the weak-coupling BCS prediction $\Delta_0 = 2.14 k_B T_C$. By assuming this relation to hold in the underdoped regime as well, that is, $\Delta_{sc}/k_B T_C \approx 2.14$, with Eq. (13.16) one can have:

$$T_C \approx \frac{3\hbar^2}{4.28\pi k_B^3} \frac{d}{n} \frac{v_F}{V_{mol}} v_\Delta \gamma_N = \beta v_\Delta \gamma_N, \qquad (13.17)$$

which shows that in the underdoped regime the T_C, like the Δ_{sc}, depends on both the v_Δ and the γ_N. Figure 13.20 shows a comparison of the experimental T_C with the one calculated based on Eq. (13.17) ($\beta = 0.7445$ K$^3 \cdot$ mol/Jm) for LSCO [7, 37]. It is seen that the agreement between the two is rather good, particularly for the underdoped regime, which further validates the defining of an effective superconducting energy scale associated with the nodal Fermi arc. In the underdoped regime, although the v_Δ (Δ_0) increases with decreasing doping, the length of the nodal Fermi arc or the γ_N decreases, which overwhelms the rise of the v_Δ and results in the decline of the Δ_{sc} and T_C toward underdoping. In the overdoped regime, the doping variation of the γ_N becomes weak and the Δ_{sc} or T_C is mainly governed by the v_Δ.

The superconducting condensation energy E_C, which measures the energy lowered in the superconducting state compared with the normal state, can also be evaluated in the picture of Fermi arc condensation [39]. In the BCS framework, the E_C is approximately given by $E_C = (\varepsilon/2) N(E_F) \Delta_0^2$ with $\varepsilon \cong 0.4$ for a d-wave superconductor. It is noted that this expression can also be expected to hold in general models for d-wave superconductivity [68]. In the underdoped regime, with the effective superconducting gap Δ_{sc}, we have:

$$E_C = \frac{\varepsilon}{2} N(E_F) \Delta_{sc}^2 = \varepsilon \frac{27}{8\pi^4} \frac{\hbar^4}{k_B^6} \frac{d^2 v_F^2}{n^2 V_{mol}^3} v_\Delta^2 \gamma_N^3 = \eta v_\Delta^2 \gamma_N^3, \qquad (13.18)$$

which shows that $E_C \propto v_\Delta^2 \gamma_N^3$. In experiments, the E_C can be determined in specific heat through integrating the entropy difference $S_N - S_S$ between

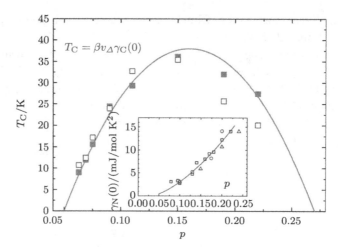

Fig. 13.20. Comparison of the measured T_C (solid squares) with the calculated one from Eq. (13.17) ($\beta = 0.7445 \text{K}^3 \text{mols/Jm}$) (open squares) in LSCO [37]. The solid line represents the empirical relation $T_C/T_C^{\max} = 1 - 82.6(p-0.16)^2$ with $T_C^{\max} = 38$ K. Inset: the γ_N in the pseudogap ground state used in the calculation (open symbols), quoted from [55]. The solid line is a fit to $\gamma_N = m(p-p_C)^n$ with $m = 182.6$, $p_C = 0.03$, and $n = 1.54$.

$T = 0$ and $T = T_C$ such that:

$$E_C = \int_0^{T_C} (S_N(T) - S_S(T)) \mathrm{d}T.$$

Using this method Loram et al. [32] and Matsuzaki et al. [55] have even evaluated the E_C in LSCO across a wide doping range. In Fig. 13.21, a comparison of the calculated E_C from Eq. (13.18) with their measured data in LSCO has been made [39]. To assist the comparison, the doping p has been normalized by the concentration p^m where the E_C shows a maximum value E_C^m because the doping level p^m reported in the two groups are slightly different. It is shown that the experimental data can be well reproduced by the calculation based on the picture of Fermi arc condensation.

The upper critical field H_{C2} is another important parameter of the superconducting state. It signifies the minimum field scale to make the sample transit from the superconducting (mixed) state to the normal state. In thermodynamics, it indicates that with increasing H the electronic density of states of the sample increases and when $H = H_{C2}$ the normal-state density of states is recovered. For underdoped cuprates, as pointed out, when increasing temperature up to T_C, it is the density of states on nodal Fermi arcs that is restored as the coherent superconductivity vanishes. Therefore,

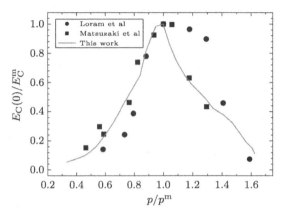

Fig. 13.21. Comparison of the measured E_C (solid symbols) with the calculated one from Eq. (13.18) (solid curve) in LSCO [41]. The experimental values are from Refs. [32, 55] and normalized by its maximum E_C^m at $p = p^m$.

it is natural to define the H_{C2} of the underdoped cuprates as the field scale to recover the same density of states at $T = 0$ K which, as discussed above, is characterized by the γ_N determined for the arc metal ground state. For a d-wave superconductor, with Eq. (13.4) and $\gamma(H) = A\sqrt{H}$, one can have:

$$H_{C2} = \frac{8a^2}{\pi}\left(\frac{\gamma_N}{A}\right)^2, \qquad (13.19)$$

which shows that $H_{C2} \propto (v_\Delta \gamma_N)^2$ as $A \propto 1/v_\Delta$. By using the γ_N as shown in Fig. 13.14(a) (also shown in the inset of Fig. 13.20) and the prefactor A determined in field-dependent specific heat as shown in Fig. 13.10(b), the doping evolution of the H_{C2} can be mapped for LSCO, as shown in Fig. 13.22(a) [38]. It is seen that the H_{C2} thus obtained forms a dome-like shape with the doping variation, much similar to the T_C. In Fig. 13.22(a), the H_{C2} of LSCO determined from high-field resistivity [69], Nernst effect [70], and diamagnetic magnetization [71] measurements are also plotted. A rough consistency among different experimental determinations is shown to hold for most part of the phase diagram. The exception is for the very underdoped region where the H_{C2} obtained from Nernst or torque measurements is higher than that from specific heat or resistivity and notably varies continuously with p down to $p = 0.03$, i.e. below the threshold $p \sim 0.055$ for coherent superconductivity. It may indicate that in this very region, the H_{C2} probed by Nernst or torque measurements corresponds to a higher field scale to destroy the phase-incoherent superconducting condensate [38].

The consistency of the H_{C2} from various experiments suggests the properness of its determination in specific heat and in turn the validity of the Fermi arc condensation picture. In the underdoped regime, the H_{C2} decreases with decreasing doping, reflecting the depletion of the density of states participating in the superconducting condensation. It is noted that a decrease of the H_{C2} with underdoping has also been shown in other families of cuprates. Gao et al. even investigated the reversible magnetization in YBCO by systematically changing the doping concentration of the sample [72]. Based on a scaling analysis of the data, the value of the H_{C2} was obtained and it was found that the H_{C2} decreases roughly linearly with decreasing doping in the underdoped regime [72], which is similar to the finding by Kim et al. in Bi2212 [73]. These suggest that it is a generic property of high-T_C cuprates that the H_{C2} declines with underdoping.

Figure 13.22(b) shows the coherence length ξ calculated from the H_{C2} shown in Fig. 13.22(a) with $H_{C2} = \Phi_0/(2\pi\xi^2)$ [38]. It is clear that from optimal doping point, the ξ grows toward either underdoping or overdoping, as expected from the doping dependence of the H_{C2}. Previously,

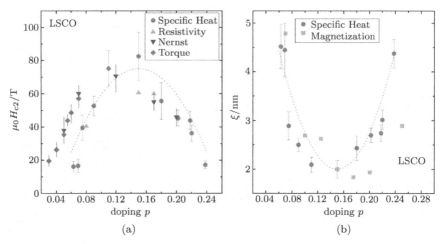

Fig. 13.22. (a) The upper critical field H_{C2} for LSCO obtained from specific heat (circles), in comparison with that from high-field resistivity [69] (up triangles), Nernst effect [70] (down triangles) and torque magnetometry [71] (diamonds). The dotted curve is a guide to the eye. It is seen that, like the T_C, the H_{C2} forms a dome-like shape. (b) The coherence length ξ for LSCO derived from the specific heat measurement of the H_{C2} with $H_{C2} = \Phi_0/(2\pi\xi^2)$ (circles), in comparison with that from the magnetization measurement by Wen et al. [74] (squares). The dotted curve is a guide to the eye. From optimal doping point, the ξ grows toward either underdoping or overdoping.

Wen et al. even determined the ξ in LSCO thin films over a wide doping range by using detailed low-temperature magnetization measurements [74], which are also plotted in Fig. 13.22(b). The reasonable consistency between the two suggests again the reliability of the H_{C2} determination in specific heat. Furthermore, the growing of the ξ with underdoping is actually another manifestation of the existence of an effective superconducting energy scale Δ_{sc} [38]. It is known from the textbook that the ξ can also be expressed as $\xi = \hbar v_F/\beta_0 \Delta$ in terms of the superconducting energy gap Δ with β_0 a numerical constant [75]. For high-T_C cuprates, ARPES showed that the v_F is nearly doping independent [76], which means that $\xi \propto 1/\Delta$. For underdoped cuprates, it is seen that if the superconducting gap Δ is simply the Δ_0, one would expect a decrease of the ξ with decreasing doping, which apparently contradicts the experimental results shown in Fig. 13.22(b), while with Δ_{sc} as the effective superconducting gap, one would naturally obtain a growing of the ξ, consistent with the experiment, toward underdoping as the Δ_{sc} decreases with decreasing p.

13.4.5.4. *Additional remarks*

In the above, we have suggested that, in view of the specific heat results, in the absence of superconductivity, the ground state of the pseudogap phase should have finite density of states for underdoped cuprates. Linked with the well-established ARPES' results on various cuprate materials, it is proposed that this finite density of states resides on the nodal Fermi arcs and its depletion during the superconducting transition is associated with the gapping of the Fermi arcs which produces an effective superconducting energy scale Δ_{sc} governing the basic superconducting properties such as the T_C and E_C in the underdoped regime. Recently, quantum oscillations [77] have been identified in underdoped YBCO, which, according to the conventional paradigm, suggests the Fermi surface to be closed pockets. This indication of a closed Fermi surface is contrary to the notion of Fermi arcs from the ARPES as cited above. Several proposals have been put forward to reconcile the conflicting result of quantum oscillation and ARPES studies. Some suggested that the Fermi pocket inferred from the quantum oscillations might indeed exist for underdoped cuprates, but the ARPES could not see it for certain reasons such as owing to a vanishing spectral weight of the outer section of the pocket [78, 79]. On the other hand, it is noted that recently a different point of view has also emerged. It has been argued that only finite segments of a Fermi surface, that is, Fermi arcs, can also give rise to quantum oscillations [80]. By assuming the Fermi arcs terminated by a pairing gap, it has been shown that the oscillations exhibit features mimicking those observed in YBCO cuprates [80]. We note that the assumption made in this study is just similar to what has been suggested in discussing the low-temperature specific heat, that is, the appearance of a Fermi arc ground state with the antinodal pseudogap having a superconducting origin.

In considering the Cooper pairing of the carriers on nodal Fermi arcs, we have adopted a traditional BCS framework to evaluate its effect in determining the basic superconducting properties of the underdoped cuprates. The justification of this approach has been discussed before and provided by the good agreement between the experiments and calculations of the T_C and E_C. It should be emphasized that, as pointed out earlier, this does not imply that the conventional BCS theory would also hold in describing the superconducting transition over the whole underlying Fermi surface, i.e. from node to antinode, as there may be a dichotomy between the nodal and antinodal electronic excitations in the underdoped cuprates because of the presence of pseudogap near the antinodes. In fact, from a large number of different experiments performed around the T_C, it has been known that the superconducting transition exhibits many unusual features in the underdoped cuprates which are difficult to be explained in terms of the BCS theory. For example, recent Nernst [70, 81, 82] or torque magnetization [83, 84] experiments have indicated that the regime of the superconducting fluctuations may occupy an anomalously large temperature interval above T_C in underdoped cuprate compounds. Such apparent deviation from the BCS expectations has also been probed by specific heat measurements [31, 33] performed in the vicinity of the T_C, as briefly reviewed in Section 13.3.3.2. In recent experiments on La-Bi2201, it was revealed that the specific heat anomaly at T_C also showed a hump-like structure with a long tail persisting far into the normal state for underdoped samples, and the entropy associated with the superconducting transition was found to be roughly conserved only when the long tail part had been taken into account [85]. These findings also indicate an extended regime of fluctuating superconductivity in the normal state and prohibit us from using the conventional BCS picture to understand the superconducting transition in underdoped cuprates. It seems likely that a final, full description of the superconducting transition in high-T_C cuprates would need the unveiling of origins of the Fermi arc and pseudogap phenomena which are still mysterious.

13.5. Summary

In this article, we have briefly reviewed some recent experimental progress on low-temperature specific heat studies in high-T_C cuprates. It has been shown that the low-temperature specific heat experiments can not only probe the symmetry of the superconducting gap, but also provide quantitative information on the superconducting state. In particular, by investigating the systematic doping evolution of the low-temperature specific heat in LSCO, an intimate relation between the pseudogap and high-T_C superconductivity has been suggested. The presence of an effective superconducting energy scale Δ_{sc} is also indicated, which supports the nodal Fermi arc condensation picture in the underdoped regime. Further investigation along this direction is desired in future studies.

References

[1] M. R. Norman, *Science* **332**, 196 (2011).
[2] C. C. Tsuei and J. R. Kirtley, *Rev. Mod. Phys.* **72**, 969 (2000).
[3] T. Timusk and B. Statt, *Rep. Prog. Phys.* **62**, 61 (1999).
[4] M. R. Norman et al., *Nature* **392**, 157 (1998).
[5] J. Orenstein and A. J. Millis, *Science* **288**, 468 (2000).
[6] M. R. Norman et al., *Adv. Phys.* **54**, 715 (2005).
[7] H. H. Wen et al., *Phys. Rev. B* **72**, 134507 (2005).
[8] M. Chiao et al., *Phys. Rev. B* **62**, 3554 (2000).
[9] I. Vekhter et al., *Phys. Rev. B* **64**, 064513 (2001).
[10] G. E. Volovik, *JETP Lett.* **58**, 469 (1993).
[11] C. Kbert and P. J. Hirschfeld, *Solid State Commun.* **105**, 459 (1998).
[12] N. B. Kopnin and G. E. Volovik, *JETP Lett.* **64**, 690 (1996).
[13] S. H. Simon and P. A. Lee, *Phys. Rev. Lett.* **78**, 1548 (1997).
[14] G. E. Volovik, *JETP Lett.* **65**, 491 (1997).
[15] N. E. Hussey, *Adv. Phys.* **51**, 1685 (2002).
[16] A. Garg et al., *Nat. Phys.* **4**, 762 (2008).
[17] H. Won and K. Maki, *Phys. Rev. B* **49**, 1397 (1994).
[18] A. Junod, in *Physical Properties of HTSC II*, edited by D. Ginsberg (World Scientific, Singapore, 1990).
[19] R. A. Fisher et al., in *Handbook of High-Temperature Superconductivity: Theory and Experiment*, edited by J. R. Schrieffer and J. S. Brooks (Springer-Verlag, New York, 2007).
[20] N. E. Phillips et al., in *Progress in Low Temperature Physics*, edited by D. F. Brewer (Elsevier Science Publishers B. V., Amsterdam, 1992).
[21] A. Damascelli et al., *Rev. Mod. Phys.* **75**, 473 (2003).
[22] W. N. Hardy et al., *Phys. Rev. Lett.* **70**, 3999 (1993).

[23] K. A. Moler et al., *Phys. Rev. Lett.* **73**, 2744 (1994).
[24] K. A. Moler et al., *Phys. Rev. B* **55**, 3954 (1997).
[25] D. A. Wright et al., *Phys. Rev. Lett.* **82**, 1550 (1999).
[26] Y. Wang et al., *Phys. Rev. B* **63**, 094508 (2001).
[27] H. Gao et al., *Physica C* **432**, 293 (2005).
[28] S. J. Chen et al., *Phys. Rev. B* **58**, R14753 (1998).
[29] R. A. Fisher et al., *Phys. Rev. B* **61**, 1473 (2000).
[30] M. Nohara et al., *J. Phys. Soc. Jpn.* **69**, 1602 (2000).
[31] J. W. Loram et al., *Phys. Rev. Lett.* **71**, 1740 (1993).
[32] J. W. Loram et al., *J. Phys. Chem. Solids* **62**, 59 (2001).
[33] A. Junod et al., *Physica C* **317–318**, 333 (1999).
[34] Q. Chen et al., *Phys. Rep.* **412**, 1 (2005).
[35] H. H. Wen et al., *Phys. Rev. B* **70**, 214505 (2004).
[36] Y. Wang et al., *Phys. Rev. B* **76**, 064512 (2007).
[37] H. H. Wen and X. G. Wen, *Physica C* **460-462**, 28 (2007).
[38] Y. Wang and H. H. Wen, *Europhys. Lett.* **81**, 57007 (2008).
[39] H. H. Wen, *J. Phys. Chem. Solids* **69**, 3236 (2008).
[40] Y. Wang et al., *Phys. Rev. B* **83**, 054509 (2011).
[41] Y. Wang et al., *J. Phys.: Conf. Series* **400**, 022133 (2012).
[42] Y. Wang et al., unpublished.
[43] C. C. Tsuei et al., *Phys. Rev. Lett.* **93**, 187004 (2004).
[44] B. Revaz et al., *Phys. Rev. Lett.* **80**, 3364 (1998).
[45] T. Yoshida et al., *Phys. Rev. B* **74**, 224510 (2006).
[46] M. Hashimoto et al., *Phys. Rev. B* **77**, 094516 (2008).
[47] T. Nakano et al., *J. Phys. Soc. Jpn.* **67**, 2622 (1998).
[48] M. Shi et al., *Phys. Rev. Lett.* **101**, 047002 (2008).
[49] J. M. Harris et al., *Phys. Rev. Lett.* **79**, 143 (1997).
[50] J. H. Ma et al., *Phys. Rev. Lett.* **101**, 207002 (2008).
[51] J. Q. Meng et al., *Phys. Rev. B* **79**, 024514 (2009).
[52] T. Kondo et al., *Phys. Rev. Lett.* **98**, 267004 (2007).
[53] M. Sutherland et al., *Phys. Rev. B* **67**, 174520 (2003).
[54] P. A. Lee et al., *Rev. Mod. Phys.* 78(2006), 17.
[55] T. Matsuzaki et al., *J. Phys. Soc. Jpn.* **73**, 2232 (2004).
[56] N. Momono et al., *Physica C* **233**, 395 (1994).
[57] Y. J. Uemura et al., *Nature* **364**, 605 (1993).
[58] J. Schtzmann et al., *Phys. Rev. Lett.* **73**, 174 (1994).
[59] S. Uchida et al., *Phys. Rev. B* **53**, 14558 (1996).
[60] S. Ohsugi et al., *Physica C* **282–287**, 1373 (1997).
[61] H. H. Wen et al., *Phys. Rev. Lett.* **85**, 2805 (2000).
[62] J. W. Loram et al., *Physica C* **235–240**, 134 (1994).
[63] L. Balents et al., *Int. J. Mod. Phys. B* **12**, 1033 (1998).
[64] G.-q. Zheng et al., *Phys. Rev. Lett* **94**, 047006 (2005).
[65] A. Kanigel et al., *Nat. Phys.* **2**, 447 (2006).
[66] J. G. Storey et al., *Phys. Rev. B* **78**, R140506 (2008).
[67] W. S. Lee et al., *Nature* **450**, 81 (2007).

[68] P. W. Anderson et al., Science **268**, 1154 (1995).
[69] Y. Ando et al., Phys. Rev. B **60**, 12475 (1999).
[70] Y. Wang et al., Phys. Rev. B **73**, 024510 (2006).
[71] L. Li et al., Nat. Phys. **3**, 311 (2007).
[72] H. Gao et al., Phys. Rev. B **74**, R020505 (2006).
[73] G. C. Kim et al., Phys. Rev. B **72**, 064525 (2005).
[74] H. H. Wen et al., Europhys. Lett. **64**, 790 (2003).
[75] M. Tinkham, *Introduction to Superconductivity* (McGraw-Hill, New York, 1996).
[76] X. J. Zhou et al., Nature **423**, 398 (2003).
[77] N. Doiron-Leyraud et al., Nature **447**, 565 (2007).
[78] S. Chakravarty et al., Phys. Rev. B **68**, R100504 (2003).
[79] T. D. Stanescu et al., Phys. Rev. Lett. **101**, 066405 (2008).
[80] T. Pereg-Barnea et al., Nat. Phys. **6**, 44 (2010).
[81] Z. A. Xu et al., Nature **406**, 486 (2000).
[82] H. H. Wen et al. Europhys. Lett. **63**, 583 (2003).
[83] Y. Wang et al., Phys. Rev. Lett. **95**, 247002 (2005).
[84] L. Li et al., Phys. Rev. B **81**, 054510 (2010).
[85] H. H. Wen et al., Phys. Rev. Lett. **103**, 067002 (2009).

14
Recent Results on the 2D Hubbard, $t-J$ and Gossamer Models and Relevance to High-Temperature Superconductivity

Gang Su

College of Physical Sciences, Graduate University of Chinese Academy of Sciences, Beijing 100049, China

In this chapter, we mainly address a number of numerical and analytical advances on 2D Hubbard and $t-J$ models as well as the relevant physical findings related to high-temperature superconductivity in the past 20 years. The gossamer superconductor model and its progress are also mentioned. The possible connections with the recently discovered iron-based superconductors are discussed as well. Finally, my personal understanding and reflection on the high-temperature superconducting theory are briefly described.

14.1. Introduction

It is now a good opportunity to recall some advances of high-temperature superconducting theories that have been proposed in more than 20 years since the discovery of cuprate superconductors. I was requested to give an overview of the progresses of the $t-J$ model and gossamer superconductivity. As the $t-J$ model has a deep and historic link with the Hubbard model, I will first describe the advances on the studies of the Hubbard model, and then on the $t-J$ model.

I have to remark here that there are countless papers on the Hubbard and $t-J$ models published in internationally recognized journals and conferences over the past 20 years, and if one makes a web search on Google with key words "Hubbard model" and "$t-J$ model", millions of results will be found, illustrating on one side that the research works on these two models are quite rich and prosperous. Therefore, an overall review of the advances on the Hubbard and $t-J$ models is an impossible mission. As a choice, I will concentrate on some recent numerical and analytical results. Even so, for the numerical works, there are still numerous results that are contained in thousands of articles. In this situation, I am going to introduce some representative works that I personally believe are interesting and inspiring, for which it is unavoidable to bring in my personal biases. In particular, I should say that those works are not mentioned in the following not because of their unimportance, but really owing to the limitation of size and taste.

Some people may be newcomers to the field of high-temperature superconductivity, who are probably unclear why high-temperature superconductivity has a relationship with the Hubbard model and $t-J$ model. In fact, there is a historic reason, as will be seen below. This chapter is organized as follows. First, I will give a brief overview of some numerical and analytical results of 2D Hubbard model and $t-J$ model. One will see that some results are quite controversial. Then, I will discuss gossamer superconductor proposed by Laughlin and its advances. Finally, I shall give some arguments and outline prospects.

High-temperature cuprate-oxide superconductors have been disclosed by two Swiss scientists Bednorz and Müller [1] in 1986. The title of their seminar paper is "Possible High-T_C superconductivity in the Ba–La–Cu–O system", showing that they were, at that time, not quite sure if the high superconducting transition temperature observed to be around 30 K really appeared in such oxide ceramic materials. They believed that the phenomenon they observed was reminiscent of the percolative superconductivity, and could be attributed to 2D superconducting fluctuations of double perovskite layers. Two or three months later after the appearance of their paper, three research groups globally, S. Uchida's group from University of Tokyo, Japan, Paul Chu's group from Houston University, USA, and Zhong-Xian Zhao's group from Institute of Physics, Chinese Academy of Sciences, confirmed independently, at almost the same time, the phenomenon that Bednorz and Müller observed. It was shown that superconductivity above the transition temperature 40 K indeed exists in the cuprate-oxide ceramic materials. The importance of these three groups' works can also be reflected from the citations [1] and [2].

Upon confirmation of Bednorz and Müller's result, Anderson [3] proposed an interesting model (now known as RVB model) for high-temperature oxide superconductors, which was published in *Science* in February 1987. There, he suggested that the undoped cuprate oxide La_2CuO_4 has the ground state of resonating valence bond (RVB) state with peculiar magnetic properties, which is also the long-sought state that was hypothesized by himself for the ground state of the spin-1/2 Heisenberg anti-ferromagnet on the triangular lattice, and is an insulating quantum spin liquid. He pointed out that the system with low spin, low dimensionality and magnetic frustrations tends to favor such a magnetic insulating state. When this insulating state is heavily doped with holes, the preexisting singlet pairing electrons become the charged superconducting electron pairs, enabling the system to exhibit superconductivity. He believed that, although weak electron–phonon interactions may favor this state, superconductivity may primarily arise from electronic and magnetic interactions.

As in the large U limit, the Hubbard model at half-filling can be mapped onto an effective spin-1/2 Heisenberg anti-ferromagnetic model, Anderson thought that the reference ground state of these two equivalent models should be the RVB state. When holes are doped into the RVB state, the Mott insulating state becomes superconducting. Based on this assumption, Anderson and his coworkers [2, 4] proposed the RVB mean-field theory for high-temperature superconductivity based on the Hubbard model in the large U limit. Since this theory had a very influential impact on the community of high-temperature superconductors at that time, the Hubbard model jumped on the central stage of investigations on high-temperature superconducting mechanism. It is now known that the parent compound of undoped cuprate oxide superconductors has an anti-ferromagnetic long-range order (LRO), while RVB state is a quantum spin liquid with short-range order and has no LRO. Therefore, Anderson had made modifications on his RVB theory later.

It is of interest that, on May 1, 1987, PRL published two independent articles on high-temperature superconductivity, one on the RVB high-temperature superconducting mean-field theory coauthored by Anderson *et al.*, [2] and another written by Emery [5]. The latter proposed, for the first time, a high-temperature superconducting theory based on the three-band Hubbard model. Emery thought that, based on this three-band model, the charge carriers come from the holes on O(2p) states, which couple strongly with the local spins on Cu sites to form Cooper pairs. Within the framework of Bardeen–Cooper–Schrieffer (BCS) mean-field theory, Emery found a d-wave superconducting state. He believed that his three-band theory quite differs from Anderson's RVB theory, because his theory can recover the anti-ferromagnetic LRO at half-filling in the insulating limit, while Anderson's RVB state is dimerized and has no LRO.

After that, the $t-J$ model came out. In 1988, Zhang and Rice [6] presented a celebrated work. They started from the three-band Hubbard model, and introduced a singlet mechanism that could map the three-band Hubbard model onto an effective single-band $t-J$ model. They suggested that the Cu–O hybridization could bind a hole strongly on each square of O atoms to the central Cu^{2+} ion to form a local singlet, which is called the Zhang-Rice singlet nowadays. The $t-J$ model consists of two parts, the first part being the kinetic energy of holes, and the second part describing the anti-ferromagnetic exchange interactions. As cuprate oxide high-temperature superconductors have layered perovskite structures, a common view is that the motion of holes in CuO_2 plane plays a key role in forming superconductivity. In the Hubbard model, there are four states for electrons: empty ($|0\rangle$), double occupied ($|\uparrow\downarrow\rangle$), spin up ($|\uparrow\rangle$), and spin down ($|\downarrow\rangle$) states. In the large U limit, the Hubbard model can be equivalently mapped onto the $t-J$ model, in which there are only three states, say empty ($|0\rangle$), spin up ($|\uparrow\rangle$), and spin down ($|\downarrow\rangle$) states, and the double occupancy is forbidden.

In 2006, Scalapino [7] summarized the physical properties of the 2D Hubbard model based on the results of numerical simulations, and discussed the relationship between the 2D Hubbard model and high-T_C cuprate oxide superconductivity. In the following, let us retrospect some numerical and analytical results on the 2D Hubbard model in the past 20 years, and discuss whether the 2D Hubbard model can accommodate superconducting LRO and phase separation.

14.2. Two-Dimensional Hubbard Model

In order to describe the itinerant electron magnetism in transition metals, and to attempt to solve the issue of why the 3D itinerant electrons in transition metals can form localized magnetic moments, in 1963 J. Hubbard [8] proposed a model given by:

$$H = -t \sum_{\langle i,j \rangle \sigma} (c_{i\sigma}^\dagger c_{j\sigma} + \text{h.c.}) + U \sum_i n_{i\uparrow} n_{i\downarrow}$$

where $c_{i\sigma}^\dagger$ is the creation operator of an electron with spin $\sigma(=\uparrow,\downarrow)$ on the ith site. This Hamiltonian comprises of two parts: the first is the kinetic energy, describing the hopping process of electrons with t the hopping matrix element, and the second is the potential term, describing the on-site Coulomb correlations between electrons, represented by U. The number density of electrons with spin σ is defined by $n_{i\sigma} = c_{i\sigma}^\dagger c_{i\sigma}$. Owing to the competition between the kinetic energy and on-site Coulomb interaction, the correlated electrons could have various physical properties. It can be seen that when U is small, i.e. $U < U_C \sim W$ (where W is the band width of electrons), the kinetic energy predominates, where the system exhibits metallic property; and when U becomes larger, $U \geq U_C$, the Hubbard model can exhibit a metal–insulator transition, and the system enters into an insulating state, because in this case the on-site Coulomb repulsive interaction U is so strong that the electronic band splits into two sub-bands, the upper Hubbard band and the lower Hubbard band, and a gap opens in the middle of the bands. At the half-filling, the Fermi level is located just in the mid of the gap, leading to the system displaying an insulating property, where electrons are localized on the sites, and the indirect exchange interactions between these localized electrons enable the system in the anti-ferromagnetic insulating state (i.e. the so-called Mott state). When t and U are comparable, the system will demonstrate complicated behaviors. It was shown that anti-ferromagnetism and superconductivity could be induced from this system under proper circumstances. Therefore, although it looks apparently quite simple, the Hubbard model is really complicated, and it contains much enriched physics.

In 1985, Hirsch [9] performed a Monte Carlo (MC) simulation on the Hubbard model in the large U limit, and found that in the strong repulsive interactions the Hubbard model can generate the effective, attractive interactions between two neighboring electrons with opposite spins, where it was observed that the anisotropic singlet pairing correlation function tends to enhance, while the triplet pairing correlation is suppressed. By means of the second-order perturbation theory, he derived the $t-J$ model directly from the large U Hubbard model. In 1989, Hirsch and Tang [10] studied the magnetic properties of the 2D Hubbard model by using the MC method. They uncovered that at the half-filling, the 2D Hubbard model can exhibit an anti-ferromagnetic LRO for any repulsive Coulomb interactions when temperature T is down to $T = t/20$; and when away from the half-filling by doping, the anti-ferromagnetic correlations immediately disappear. It should be noticed that this is the result obtained

26 years ago. Now we know that this numerical result appears to be problematic.

In 1989, Scalapino et al. [11] adopted the conserving approximations and the Bethe–Salpeter equation to self-consistently explore the particle–particle correlation function, and obtained the low-temperature phase diagram of the 2D Hubbard model. Based on the numerical results on a square lattice with 16 × 16 sites, they found that the d-wave singlet pairing correlation is enhanced when temperature is decreased, and there are commensurate and non-commensurate magnetic correlations in the low-doping region. In the same year, they [12] utilized the quantum Monte Carlo (QMC) technique to study the particle–hole magnetic and particle–particle susceptibilities in terms of the dressed Green functions and interaction vertices, and found that near half-filling ($\langle n \rangle = 0.875$) the d-wave singlet pairing interaction is the most attractive, and the extended nearest neighbor s-wave pairing interaction is weakly attractive, while the singlet s_{xy}, d_{xy} and triplet p_x interactions are repulsive.

In 1995, Scalapino [13] presented a review article discussing the d-wave pairing in cuprate oxide superconductors, where he attempted to answer two questions: (i) why could the d-wave pairing occur in these materials? (ii) how can we know it is d-wave pairing? What is the mechanism leading to the d-wave pairing? Through diagrammatic expansions as well as QMC simulations on the 2D Hubbard model, it was observed that near half-filling, the spin-fluctuation exchanges via the singlet $d_{x^2-y^2}$ channel could induce an effective electron–electron attractive interaction; owing to the strong short-range anti-ferromagnetic correlations, the 2D Hubbard model can display $d_{x^2-y^2}$-wave pairing fluctuations. He further pointed out that, although it is not known whether the 2D Hubbard model, when doped with holes, can have a low-temperature superconducting phase, this model shows $d_{x^2-y^2}$-wave pairing fluctuations as evidenced by the existing numerical results.

In 1997, Zhang and coworkers [14] used the constrained-path QMC method to study the 2D Hubbard model. They disclosed that when the lattice size is small and the interaction is weak, the system has a signature of d-wave superconductivity, where the pairing fluctuations are strong; when the size and interaction increase, the long-range correlation vanishes. Therefore, they concluded that in the 2D Hubbard model, the d-wave pairing LRO does not exist.

In 2005, Tremblay et al. [15] utilized the variational perturbation theory to investigate the competition between anti-ferromagnetism and superconductivity in the $t - t' - t'' - U$ Hubbard model with next-nearest neighbor and next-next-nearest neighbour hoppings by means of large scale computer simulations. They found that the primary features of the overall ground state phase diagram and the single-particle excitation spectra of the hole- and electron-doped high-temperature superconductors can be reproduced from this model. When the electron density is greater than the half-filling, the d-wave superconducting order parameter increases while the anti-ferromagnetic order decreases with increasing electron density; when the density is smaller than the half-filling, the superconducting order parameter decreases while the anti-ferromagnetic order increases with increasing electron density. In both cases, when the superconducting order becomes the maximum, the anti-ferromagnetic order becomes the minimum. These results were obtained for the lattice size of 2×3 and 2×4 and $U = 6t$, $8t$ and $12t$. For different U, the obtained results are the same. It should be noted that the lattice size is still very small for this calculation.

It is known that there is not yet a consensus on the electron pairing mechanism of high-temperature superconductors in the community of superconductivity till now. Nevertheless, Scalapino [7] pointed out that, some numerical results show that the 2D Hubbard model already exhibits many features seen in these materials, but several interesting issues that people always try to answer still remain: Is T_C really high? Can we detect the mutual influence of stripe phase and superconducting phase? Why does the pseudogap appear? What is the relationship between the pseudogap and superconductivity? What are the roles of ions between layers and single chains? Do phonons contribute to superconductivity? Surely, these questions are still open now.

In 2007, Anderson [16] gave a brief review on the progress of high-temperature superconducting theory during the past 20 years since the discovery of cuprate oxide superconductors. He proposed a complicated phase diagram in the temperature (T)-doping concentration (x) plane for high-temperature superconductors, which differs slightly from the previous one. Here, two quantum critical points (QCPs) are indicated. In this phase diagram, above the superconducting region, there are vortex liquid state and RVB state, which are separated by a characteristic temperature T^*_{Ong} line. On the left side of the QCP II is the anti-ferromagnetic insulating state, and on the right side of the QCP I is the normal metal (Landau

Fermi liquid); a strange metal state appears in the optimized and overdoped regions. Anderson said, "I am going to have the temerity to urge on you that the central problems of high-T_C theory have been solved". In another paper, Anderson [17] stated that a lot of high-T_C superconducting pairing theories go on the wrong track. It manifests in the high-T_C pairing mechanism are still being actively debated.

In the preceding discussions, we are only associated with the repulsive Hubbard model. Certainly, the Hubbard model with attractive interactions ($U < 0$) is also of great interest and also has wide applications. Scalettar et al. [18] gave a phase diagram of the 2D attractive Hubbard model in terms of numerical calculations. They found that, at half-filling, the ground state of the system has both superconducting and charge density wave (CDW) LROs, but the transition temperature is zero; and away from the half-filling, the system can exhibit a Kosterlitz–Thouless (KT) phase transition into a superconducting state at finite temperatures, where the superconducting pairing correlation functions decay spatially in a power law. By using the QMC method, Scalapino et al. [19] studied the 2D attractive Hubbard model and found that, away from the half-filling through doping, the s-wave pairing correlations are enhanced, while the CDW correlations are suppressed. No phase separation was observed in this model. Away from the half-filling, the pair-field correlations are consistent with the KT scaling behavior, and there exists a transition temperature T_C which has a peak near the half-filling. At exactly half-filling, the transition temperature vanishes.

Concerning the phase separation in the 2D Hubbard model, I substantiated in 1996 that [20], by using a rigorous method, the 2D Hubbard model on a square lattice cannot have a phase separation at any finite temperature and at arbitrary filling fraction in spite of repulsive and attractive Coulomb interactions. This statement is consistent with the QMC numerical result.

In 1998, we rigorously proved a theorem [21]: At any finite temperature, the 2D Hubbard model with narrow bands (including next nearest neighbor, next next nearest neighbor hopping, etc.) does not exhibit the d-wave pairing LRO for both repulsive and attractive Coulomb interactions and for arbitrary electron fillings. This theorem suggests that it is inappropriate to use the 2D Hubbard model to directly describe high-temperature superconductivity.

Two years after the appearance of our theorem, a group [22] invoked a dynamical cluster approximation to numerically study the possibility of d-wave superconductivity in the 2D Hubbard model. They found that at sufficiently low temperature, the 2D Hubbard model has a stable d-wave superconducting solution with off-diagonal LRO (ODLRO). They estimated the maximal superconducting transition temperature T_C to be about 150 K with a doping concentration $\delta \sim 0.2$, and observed that the doping dependence of the transition temperature is well in agreement with the generic phase diagram of high-T_C superconductors. It is obvious that this numerical result completely contradicts the above-mentioned mathematically rigorous theorem. To eliminate this contradiction, in 2001 I wrote a comment [23] on this paper, in which their numerical results were re-analyzed and compared with the existing QMC results. It was found that all existing QMC and other numerical results reveal that the 2D Hubbard model can display either superconducting fluctuations or the enhancement of superconducting long-range correlations, but no definite sign of d-wave pairing LRO has been found in those numerical results. By looking carefully at their calculations, I noted that the size effect may be the key factor. Their calculations were made on a small lattice size, but the conclusion appears to be made for a large lattice size, which leads to a contradiction. I assumed that their finding may be a KT phase transition, not a genuine superconducting transition. In 2005, this group [24] recalculated this system, and found that the d-wave pairing LRO indeed does not occur in the 2D Hubbard model at finite temperature. In addition, Laughlin [25] also pointed out that there is no persuasive evidence that supports superconductivity in the 2D Hubbard model, and Müller [26] shared the same view.

On the other hand, our theorem can be taken as a criterion to judge if the numerical calculations on the 2D Hubbard model are reliable. The previous QMC results were thus re-examined and some inconsistencies were found [27]. In 2007, a group from Japan [28] published an article entitled "Does simple 2D Hubbard model account for high-T_C superconductivity in copper oxides?" They found that in the ground state of the 2D repulsive Hubbard model, the d-wave superconducting correlation function decays spatially in a power law ($\sim r^{-3}$), and the upper boundary for the long-range correlation function is around 10^{-3}. This suggests that the pairing correlation is so weak that it cannot account for cuprate high-T_C superconductivity. Thus, if someone starts directly from the 2D Hubbard model to study high-T_C superconductivity, the resulting conclusion will be strongly questioned.

14.3. The 2D $t-J$ Model

Now let us look at the 2D $t-J$ model. This model describes the motion of doped holes on the anti-ferromagnetic background in the CuO_2 plane, and the Hamiltonian is given by:

$$H = -t \sum_{\langle i,j \rangle \sigma} [(1 - n_{i\bar{\sigma}})c_{i\sigma}^{\dagger} c_{j\sigma}(1 - n_{j\bar{\sigma}}) + \text{h.c.}] + J \sum_{\langle i,j \rangle} \left(\boldsymbol{S}_i \cdot \boldsymbol{S}_j - \frac{1}{4} n_i n_j \right).$$

This model consists of two parts. The first part is the hopping term (the kinetic energy term), and the second part is the anti-ferromagnetic exchange interaction (the potential term). In this model, as mentioned above, the double occupancy is forbidden. The competition of these two terms governs the dynamic motion of holes in doped Mott insulators, thereby resulting in very rich physics. At half-filling, the $t-J$ model recovers the spin-1/2 Heisenberg anti-ferromagnetic model. Moreover, the $t-J$ model can be derived from the Hubbard model in the large U limit. This is also the reason that Anderson proposed his celebrated RVB theory based on the Hubbard model. Interested readers can refer to my previous paper for a derivation of the $t-J$ model [29].

Tracing back to the history of the $t-J$ model, one may find that at the early stage it was introduced to study the properties of the magnetic metals strongly correlated with the non-Fermi liquid ground state [30]. Later, it was also utilized to explore the heavy fermion superconductivity [7] and anti-ferromagnetic correlations of nearly localized Fermi liquids [31]. Anderson and coworkers [4] started from the $t-J$ model and developed a RVB mean-field theory for high-temperature superconductivity. This mean-field theory was quite influential at that time, based on which, many results can be obtained, but most of them differ from what the experiments in the copper oxides. Another mean-field theory was also proposed [32]. In 1988, Zhang and coworkers [33] proposed a renormalized RVB mean-field theory where, by considering the Gutzwiller approximation and introducing the doping dependent renormalization factor, they have renormalized the kinetic energy of holes and the spin–spin coupling term, and derived an effective renormalized Hamiltonian. According to this Hamiltonian, the d-wave superconducting order parameter as a function of the doping fraction is found to be similar to that observed experimentally in high-temperature superconductors. Besides, the doping dependent variational parameter they calculated looks similar to the behavior of the crossover temperature T^* at which the pseudogap appears in the normal state of the underdoped high-T_C superconductors. In the same year, another group [34] also developed a d-wave superconducting mean-field theory by utilizing the auxiliary boson mean-field method. There are many others in literature.

Apart from the analytical studies on the $t-J$ model, there are a great number of numerical studies on lattices with small size. In 1993, by making QMC calculations, Dagotto et al. [35] found that, when the electron density is 1/2, the ground state of the $t-J$ model reveals a strong signature of d-wave superconductivity. This statement was made based on the calculated results on the static pairing correlation function, Meissner effect and magnetic flux quantization. The zero-temperature phase diagram in the $J/t - x$ plane of the $t-J$ model was given, where the Fermi liquid, binding, d-wave superconducting and phase separation regions were identified. Near the half-filling, the calculations became difficult, and ferromagnetic and anti-ferromagnetic regions were found. They also gave the spatial dependence of the d-wave pairing correlation function, and observed a signature of superconductivity.

However, a group [36] utilized a Power-Lanzos method they developed to recalculate the d-wave pairing correlation function in the 2D $t-J$ model, and found that, in the parameter regions under interest J/t ≤0.5, the 2D $t-J$ model does not exhibit a d-wave LRO. Meanwhile, they adopted the same method and calculated the Hubbard model, giving the same result as that of Zhang et al. [14], which is that no d-wave LRO exists in this model. By applying the variational MC method to the projected d-wave superconducting state, Paramakanti et al. [37] studied the large U Hubbard model, and uncovered that when the doping x is in the range of $0 < x \leq 0.35$, the so-obtained superconducting order parameter as a function of x looks very similar to the non-monotonic behavior of $T_C(x)$ that was experimentally observed for cuprate oxide superconductors. The variational parameter as a function of x was seen to resemble the hump scale at $(\pi, 0)$ and T^* in ARPES, which is very similar to that obtained by Zhang et al. with aid of renormalized mean-field theory. Besides, it was found that the projection leads to the incoherent part of spectral function, and the quasi-particle weight Z along the nodal direction was disclosed to be $Z \sim x$ from the singular behavior of the moment. When $x \rightarrow 0$, the Fermi velocity remains finite; the Drude weight and superfluid density were found to be consistent with experiments, and $D_{\text{low}} \sim Z$. This and Sorella et al.'s [38] works reminded Anderson to re-propose his famous renormalized RVB theory.

By using various methods including QMC, density matrix renormalization group (DMRG), Lanczos, etc., Sorella *et al.* [38] studied the possible superconductivity in the 2D $t-J$ model. They found that, in 2D, quantum fluctuations are not sufficient to suppress the superconductivity. Starting from a d-wave BCS variational wave function that was based on the RVB wave function with a Jastrow factor, they found numerically that the 2D $t-J$ model has the d-wave superconductivity at zero temperature. This result, upon publication, gained a critical comment by Lee *et al.* [39]. The comment pointed out that the data that Sorella *et al.* obtained were consistent with their results obtained in 1998. When the successive Lanczos steps were used to improve the variational wave function, the corrections from each additional Lanczos step should be carefully treated, while Sorella *et al.* did not examine the corrections from the steps by applying the extrapolation method to the case with slow convergence. They believed that Sorella *et al.*'s conclusion was overstated. When replying to this comment, Sorella *et al.* [40] stated that their results were not consistent with Lee *et al.*'s data. To verify the reliability of their results, they studied four holes on a lattice with 26 sites, and the systematic checks for the largest diagonalization results were made. They found that their data were consistent with those of Lanczos diagonalization and DMRG, thereby eliminating the suspicions on their 2002 results. From these studies, one may feel that, owing to the complexity of strongly correlated electrons, most numerical calculations are limited to small lattice sites, and that different numerical methods usually give various errors. As a result, when people use the small-size results to extrapolate the property of a large-size system, controversial conclusions are often obtained. It appears that whether the 2D $t-J$ model can have the d-wave superconductivity in the ground state or not is still under active debate.

Kivelson *et al.* [41] utilized the high-temperature series expansion method to explore the d-wave superconducting susceptibility of the 2D $t-J$ model, and disclosed that the $t-J$ model does not support high-temperature superconductivity. Up to the 10th-order of expansion, and for temperature down to the order of J, they found that in the unphysical parameter regime ($t < J$), a larger superconducting susceptibility can be obtained, which increases with decreasing temperature; while in the physically interesting parameter regime ($t > J$), the superconducting susceptibility is small and decreases with decreasing temperature. Their conclusion is that the $t-J$ model does not support high-temperature superconductivity.

In 2007, they published an erratum that pointed out an error in the series expansion. When the error was corrected, their original conclusion that the $t-J$ model does not support high-temperature superconductivity still remained, though not as strong as before. In addition, Lee et al. [42] adopted the variational Monte Carlo, mean-field theory and exact diagonalization methods, respectively, studied the effect of the next nearest-neighbor and next next nearest-neighbor hopping terms (i.e. t' and t'' terms) on the superconducting correlations in the 2D $t-J$ model, and uncovered from all calculated results that the d-wave superconducting correlations are slightly suppressed by t' and t'' terms in the underdoped region, but are strongly enhanced in the overdoped and optimally doped regions. The optimized T_C is the consequence of the balance of these two opposite trends. In this work, the authors emphasized that the t' term plays an essential role in enhancing the superconducting correlations.

In 2006, Putikka and Luchini [43] utilized the high-temperature series expansion method up to the 12th-order to investigate the hole doping dependence of the d-wave superconducting correlation length, and observed that upon dropping out the unconnected diagrams that were usually kept in previous calculations when calculating the four-point correlation function, the d-wave pairing correlation length is rapidly increased with the behavior $\sim \exp[\text{const.}/\sqrt{T/J}]$ with decreasing temperature; it was found that the d-wave superconducting correlation length develops a broad peak as a function of doping in the optimal region ($x = 0.25, T/J = 0.25, J/t = 0.4$), but the s- and d_{xy}-wave pairing correlation lengths are small and have no such broad peaks. In low doping, the anti-ferromagnetic spin correlation tends to suppress the d-wave superconducting correlation length. They believed that the superconducting correlations are robust in the 2D $t-J$ model.

From these results, one may see that whether the 2D $t-J$ model superconducts or not is still under controversy. In spite of the complexity of strongly correlated electronic systems, an important factor is that the kinetic energy (t) term is commensurate with the anti-ferromagnetic exchange J term in the $t-J$ model, where the perturbative theory does not apply. Such a confused situation is conceivable because no reliable, non-perturbative analytical methods are available for the 2D interacting quantum many-particle systems at the moment.

In 2005, I published an article [44] entitled "Is the Two-Dimensional $t-J$ Model Adequate for High-Temperature Superconductivity?" When it was published, the title was changed by the editor to "Investigation of the

Adequacy of the Two-Dimensional $t-J$ Model for High Temperature Superconductivity". In this paper, I pointed out that it is a common consensus that the 2D $t-J$ model indeed captures some basic features of cuprate superconductors, which were also supported by some numerical and meanfield results. However, the recent results for superconducting correlations, whenever numerical or analytical, are more or less controversial, showing evidently that whether the 2D $t-J$ model is appropriate for cuprate oxide superconductors is still under active debate.

Two reasons led to our discussions on this issue. The first is that although the $t-J$ model can be derived from the Hubbard model in the large U limit, a number of numerical results in 1990s show that these two models do not coincide in some parameter regimes, so to speak. The $t-J$ model is quite different from the Hubbard model to some extent. The second is that whether or not the $t-J$ model could superconduct at zero and finite temperature is very controversial, and such controversies are still under way, as reflected from my preceding discussions. To answer this question, I utilized Bogoliubov's inequality, and proved in a rigorous way a theorem that the 2D $t-J$ model cannot exhibit the d-wave superconducting LRO at any finite temperature. If there is an excited energy gap in this model, this above conclusion is still valid at zero temperature. It can be seen in literature that several variational wave functions such as Anderson's RVB wave function, d-wave RVB wave function, d-wave BCS wave function, and projected d-wave BCS function, etc., were proposed, which were applied to the $t-J$ model. But the question of whether the 2D $t-J$ model superconducts in the ground state remains, and no consensus has been reached up to now. In fact, the key to the question is that, even if the superconducting LRO exists in the ground state of the 2D $t-J$ model, when any finite temperature is tuned onto the system, such a superconducting LRO should be destroyed due to the thermal fluctuations. Moreover, the high-temperature superconductivity in cuprate oxides is actually a finite-temperature phenomenon. In this sense, the 2D $t-J$ model cannot be a proper model for an overall description of the cuprate oxide superconductivity. Also in this paper [44], I made a footnote, stating that it is proven that the 2D $t-J$ model does not exhibit an anti-ferromagnetic LRO at any finite temperature. Since the anti-ferromagnetic ordering phase exists in the region with very low doping ($x < 0.05$) for cuprate oxide superconductors, the 2D $t-J$ model also cannot be applied to describe the anti-ferromagnetic LRO in the cuprate oxides. However, the above results do not necessarily mean that the 2D $t-J$ model cannot be used to describe some features of cuprate oxides. In fact, one may see that regardless of zero or finite temperature cases, a number of analytical and numerical calculations show that the 2D $t-J$ model indeed grasps some essential characteristics of cuprate oxides. For an overall description of high-temperature superconductivity, some other factors should be considered.

14.4. Gossamer Superconducting Model

Now let us look at Laughlin's gossamer superconducting model [25]. The word "gossamer" means "light, thin, delicate". Therefore, a gossamer superconductor can be viewed as a "thin superconductor". As discussed in preceding sections, a common strategy to formulate superconductivity is that one may first start from a given Hamiltonian, regardless of the single- or three-band Hubbard model or the $t-J$ model, then solve the ground-state properties of this Hamiltonian by using various analytical and numerical approaches, or assume a variational ground-state wave function, and finally get the theoretical result that is afterward compared with the existing experiments, to justify the model Hamiltonian until an agreement between theoretical and experimental results is reached. In 2002, Laughlin disseminated a preprint in which he proposed the concept of gossamer superconductivity. In contrast to the traditional route, the Hamiltonian of gossamer superconductivity was induced from a presumed state. The basic idea is that a partially projected RVB state with a particular projection operator is first assumed, and then this state is required to satisfy the basic properties of BCS superconducting state, which is also imposed to be the ground state of a proposed BCS-like Hamiltonian with quasi-particle operators. In this way, a new Hamiltonian for superconductivity is introduced, whose ground state is exactly the partially projected RVB state. This Hamiltonian differs from the Hubbard and $t-J$ models, which contain strong on-site Coulomb repulsion between electrons (the Hubbard U term). Such a superconducting state has the following characters [25]: a full and intact d-wave tunneling gap, quasi-particle photoemission intensities which are strongly suppressed, a suppressed superfluid density and an incipient Mott–Hubbard gap. In underdoped cuprate superconductors, the superconductivity can persist deep into the insulating state in low-doping region, and coexists with the anti-ferromagnetism. Owing to the low superfluid density in this region, the superconducting LRO is disrupted, and it failed to conduct.

Laughlin believed that the RVB theory was incorrect because it encouraged superconductivity as a universal aspect of anti-ferromagnetism, which was confusing. He thought that the conventional spin density wave (SDW) ground state was a perfect prototype of a quantum anti-ferromagnet while the SDW state contains no superconductivity. A gossamer superconductor contains on-site Coulomb repulsive interactions, and is physically equivalent to a dilute gas of bosons, which is very unstable and with a thin superfluid density.

In gossamer superconductivity, although the possibility of spin-up and spin-down electrons occupying the same site is largely suppressed, it still has non-zero finite value. Because of the incomplete projection, even at the half-filling, the partially projected RVB state is still superconducting, which is the essential distinction between the gossamer superconducting state and the RVB state [45]. For the latter, at half-filling the double occupancy of electrons is completed projected out, and the half-filled RVB state is a Mott-insulating state, and superconductivity can appear only away from the half-filling.

In 2003, Laughlin *et al.* [46] published a paper in which they detailed the derivation of the Hamiltonian, and found that the gossamer superconducting Hamiltonian consists of three terms: the first term contains the chemical potential, kinetic energy and Coulomb correlation term (the Hubbard U term); the second is the superconducting term, including the pairing interactions between electrons with a complicated form; and the third contains the on-site electron pairing operator that breaks the $U(1)$ symmetry, which cancels because of the symmetry of coefficients with respect to momentum. By separating the on-site and off-site terms in the first term of the Hamiltonian, they obtained the expression for the on-site Coulomb correlation energy U that is observed to be dependent on the projection parameter.

In 2003, Zhang [45] started directly from the $t-J-U$ model to study the physical properties of gossamer superconductivity. He constructed a gossamer superconducting variational ground-state wave function that is the partially projected BCS wave function in which the double occupied states are partially allowed. Then, by utilizing the renormalized mean-field theory and variational Monte Carlo technique, he found that at half-filling, the ground state is a gossamer superconductor for small repulsive Coulomb potential U, and is a Mott insulator for large repulsive U. He also observed that the gossamer superconducting state is essentially similar to the RVB superconducting state, except that away from the half-filling, the chemical potential is pinned in the middle of lower and upper Hubbard bands. This can be viewed as an explanation of the gossamer superconductivity.

Gan et al. [47] gave a further comparative study of gossamer and RVB superconductivity, and presented an interesting phase diagram. They uncovered that, at half-filling, there is a critical U_C, and when $U > U_C$, the system is in a Mott insulating state; when $U < U_C$, the system is in a gossamer superconducting state, i.e. in the gossamer superconductor the double occupied states are allowable; in the critical situation of $U = U_C$, there is a first-order phase transition from the gossamer superconducting state to the Mott-insulating state, which results from the competition between the kinetic energy and spin exchange interactions. Away from the half-filling, they disclosed that the system enters the superconducting state; when $U > U_C$, it is in the RVB superconducting state; while $U < U_C$, it is in the gossamer superconducting state. It was found that the gossamer superconducting state is smoothly connected with the RVB superconducting state.

In 2003, Coleman [48] lifted the veil of gossamer superconductivity, and pointed out that the RVB state is a Mott-insulating state where the double occupancy is prohibited, and when the Mott-state is doped with holes, it becomes the RVB superconducting state where the unoccupied states exist. While the gossamer superconducting state connects the insulating and superconducting states, where both double occupied and unoccupied states are possible, suggesting that although electrons do not move as freely as those in the superconducting state, they can move in the gossamer state. He also mentioned that, in Laughlin's gossamer theory, phase fluctuations, which are very important for understanding the enhancement of the superconducting transition temperature with increasing superfluid density, are not considered. But in any case, this uses a new perspective to understand an old issue: how an insulator can become a high-temperature superconductor, there is now a possibility that the Mott-insulating ground-state of copper oxides can be a more delicate version of the superconductor than it becomes upon doping [48].

Recently, there has been a breakthrough on iron-based superconductors (e.g. 1111 materials: $LnO_{1-x}F_xFeAs$, Ln=La, Ce, Pr, Nd, Sm, Gd; 122 materials: $Ba_{1-x}K_xFe_2As_2$, etc.) with superconducting transition temperatures larger than $50\,K$. It was shown that in these superconductors there are strong anti-ferromagnetic fluctuations, and the doping of F appears to suppress the instability of anti-ferromagnetic SDW, leading to the appearance of superconductivity. In conventional superconductors, the introduction of magnetic elements tends to destroy superconductivity. The first-principles calculations [49] showed that for the undoped parent compound LaOFeAs, the ground state is an ordered anti-ferromagnet, with the magnetic momentum of Fe around $2.3\,\mu_B$, and these materials have strong Coulomb correlations, whose electronic structures depend closely on the magnetic states. Upon doping with F, it was found that the anti-ferromagnetism coexists with superconductivity. Therefore, in iron-based superconductors, there exist both anti-ferromagnetic interactions and Coulomb correlations, and the Hamiltonian of these materials could be phenomenologically described by some versions of the $t-J-U$ model. In this way, the gossamer superconductivity appears to have some connections with Fe-based high-temperature superconductors.

14.5. Concluding Remarks

In this chapter, we have reviewed briefly the advances on the 2D Hubbard and $t-J$ models, as well as the gossamer superconductivity in recent years. Based on these results, one may ask a question: As it has been proven rigorously that the 2D Hubbard model and 2D $t-J$ model cannot have d-wave superconducting LRO at any finite temperature, can superconductivity appear in these models through the Kosterlitz–Thouless (KT) phase transition? The answer to this question is NO. Actually, by using numerical and analytical methods, some authors obtained superconductivity from the 2D Hubbard model or $t-J$ model. Since these results are in conflict with our rigorous theorems, they conceived that such superconductivity could be realized via the mechanism of KT phase transition, as mentioned before. Nonetheless, the KT phase transition cannot be taken as the mechanism of superconductivity. This is because the superconducting state is a condensed state of Cooper pairs, possessing the true pairing LRO, namely the off-diagonal LRO (ODLRO), indicating that the superconducting correlation function tends to a non-zero finite value when the two electron pairs are infinitely separated in spatial space, characterized by the ODLRO. While in the KT phase transition, the correlation function in spatial space decays in a power law, and tends to zero when the electron pairing operators are infinitely separated, indicating a character of quasi-LRO, not a true superconducting LRO. In other words, the KT phase transition describes an important topological phase transition, not a superconducting transition. In their seminar paper [50], Kosterlitz and

Thouless also pointed out clearly that the KT transition cannot occur in a superconductor. By acknowledging this fact, one may conclude that the 2D $t-J$ model and Hubbard model cannot be used to describe the high-temperature superconductivity at finite temperature. However, this does not exclude the possibility that the Hubbard and $t-J$ models can be applied to describe some features of high-temperature superconductors.

Then, what might be the microscopic Hamiltonian for high-temperature superconductors? This is a highly fundamental and challenging but still unsolved issue. A widely accepted Hamiltonian must satisfy a few minimum conditions [44]: (i) it should reproduce the main features of the SC and anomalous normal-state properties of the cuprates, such as the sophisticated phase diagram in the $T-x$ plane, as shown in Fig. 14.1; (ii) it should interpret the pairing mechanism and the origin of the d-wave symmetry of the SC gap; (iii) it should explain well the origin of the pseudogap phenomenon in the normal states; (iv) it should be concise and simple in form.

Since the discovery of cuprate superconductors in 1986, a great progress on understanding the physical properties of these materials has been gained, and many interesting theoretical proposals have been made. These advances obtained during the past 20 years have led to a leap in development of condensed matter physics. However, a consensus on a key issue in understanding high-temperature superconductivity, say the pairing mechanism of electrons, has not yet been reached. And the microscopic Hamiltonian for high-temperature superconductors is still not well accepted. Therefore,

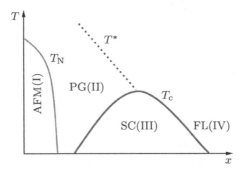

Fig. 14.1. Schematic phase diagram of high-temperature superconductors in the temperature (T)-doping concentration (x) plane. (I) anti-ferromagnetic (AFM) phase; (II) pseudogap (PG) phase; (III) superconducting (SC) phase; (IV) Fermi liquid (FL) phase; T_N the Neel temperature for AFM order; T^* the crossover temperature from PG phase to a strange metal.

to completely solve the pairing mechanism of high-temperature superconductivity, we still have a long way to go.

I am grateful for the partial support from the NSFC and the CAS.

References

[1] J. G. Bednorz and K. A. Müller, *Z. Phys. B* **64**, 189 (1986).
[2] P. W. Anderson, G. Baskaran, Z. Zou and T. Hsu, *Phys. Rev. Lett.* **58**, 2790 (1987).
[3] P. W. Anderson, *Science* **235**, 1196 (1987).
[4] G. Baskaran, Z. Zou and P. W. Anderson, *Solid State Commun.* **63**, 973 (1987).
[5] V. J. Emery, *Phys. Rev. Lett.* **58**, 2794 (1987).
[6] F. C. Zhang and T. M. Rice, *Phys. Rev. B* **37**, 3759 (1988).
[7] D. J. Scalapino, *J. Supercon. Novel Magn.* **19**, 195 (2006).
[8] J. Hubbard, *Proc. Roy. Soc. (London) A* **276**, 238 (1963).
[9] J. E. Hirsch, *Phys. Rev. Lett.* **54**, 1317 (1985).
[10] J. E. Hirsch and S. Tang, *Phys. Rev. Lett.* **62**, 591 (1989).
[11] N. E. Bickers, D. J. Scalapino et al., *Phys. Rev. Lett.* **62**, 961 (1989).
[12] S. R. White, D. J. Scalapino et al., *Phys. Rev. B* **39**, 839 (1989).
[13] D. J. Scalapino, *Phys. Rep.* **250**, 329 (1995).
[14] S. Zhang, J. Carlson and J. E. Gubernatis, *Phys. Rev. Lett.* **78**, 4486 (1997).
[15] D. Senechal, P.-L. Lavertu, M.-A. Marois and A.-M. S. Tremblay, *Phys. Rev. Lett.* **94**, 156404 (2005).
[16] P. W. Anderson, *Physica C* **460–462**, 3 (2007).
[17] P. W. Anderson, *Science* **316**, 1705 (2007).
[18] S. T. Scalettar et al., *Phys. Rev. Lett.* **62**, 1407 (1989).
[19] A. Moreo and D. J. Scalapino, *Phys. Rev. Lett.* **66**, 946 (1991).
[20] G. Su, *Phys. Rev. B* **54**, R8281 (1996).
[21] G. Su and M. Suzuki, *Phys. Rev. B* **58**, 117 (1998).
[22] Th. Maier et al., *Phys. Rev. Lett.* **85**, 1524 (2000).
[23] G. Su, *Phys. Rev. Lett.* **86**, 3690 (2001).
[24] Th. Maier et al., *Phys. Rev. Lett.* **95**, 237001 (2005).
[25] R. B. Laughlin, *Philos. Magaz.* **86**, 1165 (2006).
[26] K. A. Müller, *J. Phys.: Cond. Mat.* **19**, 251002 (2007).
[27] H.-G. Matuttis and N. Ito, *Inter. J. Mod. Phys. C* **16**, 857 (2005).
[28] T. Aimi and M. Imada, *J. Phys. Soc. Japan* **76**, 113708 (2007).
[29] G. Su, *J. Phys. A: Math. Gen.* **26**, L139 (1993).
[30] K. A. Chao, J. Spalek, and A. M. Oles, *J. Phys. C* **10**, L271 (1977); *Phys. Rev. B* **18**, 3453 (1978).
[31] C. Gross, R. Joynt, and T. M. Rice, *Phys. Rev. B* **36**, 381 (1987).
[32] Y. Isawa, S. Maekawa, and H. Ebisawa, *Physica B* **148**, 391 (1987); M. Cyrot, *Solid State Commun.* **62**, 821 (1987).

[33] F. C. Zhang, C. Gross, T. M. Rice, and H. Shiba, *Supercond. Sci. Tech.* **1**, 36 (1988).
[34] G. Kotliar and J. Liu, *Phys. Rev. B* **38**, 5142 (1988).
[35] E. Dagotto and J. Riera, *Phys. Rev. Lett.* **70**, 682 (1993).
[36] C. T. Shih, Y. C. Chen, H. Q. Lin, and T. K. Lee, *Phys. Rev. Lett.* **81**, 1294 (1998).
[37] A. Paramekanti, M. Randeria and N. Trivedi, *Phys. Rev. Lett.* **87**, 217002 (2001).
[38] S. Sorella *et al.*, *Phys. Rev. Lett.* **88**, 117002 (2002).
[39] T. K. Lee, C. T. Shih, Y. C. Chen, and H. Q. Lin, *Phys. Rev. Lett.* **89**, 279702 (2002).
[40] S. Sorella *et al.*, *Phys. Rev. Lett.* **89**, 279703 (2002).
[41] L. P. Pryadko, S. A. Kivelson, O. Zachar, *Phys. Rev. Lett.* **92**, 067002 (2004); **98**, E 069901 (2007).
[42] C. T. Shih, T. K. Lee, R. Eder, C.-Y. Mou, and Y. C. Chen, *Phys. Rev. Lett.* **92**, 227002 (2004).
[43] W. O. Putikka and M. U. Luchini, *Phys. Rev. Lett.* **96**, 247001 (2006).
[44] G. Su, *Phys. Rev. B* **72**, 092510 (2005).
[45] F. C. Zhang, *Phys. Rev. Lett.* **90**, 207002 (2003).
[46] B. A. Bernevig, R. B. Laughlin and D. A. Santiago, *Phys. Rev. Lett.* **91**, 147003 (2003).
[47] J. Y. Gan, F. C. Zhang, and Z. B. Su, *Phys. Rev. B* **71**, 014508 (2005).
[48] P. Coleman, *Nature* **424**, 625 (2003).
[49] Chao Cao, P. J. Hirschfeld, and Hai-Ping Cheng, *Phys. Rev. B* **77**, 220506 (2008).
[50] J. M. Kosterlitz and D. J. Thouless, *J. Phys. C* **6**, 1181 (1973).

15
The High-Temperature Superconducting Cuprates Physics: The "Plain Vanilla" Version of RVB*

Yue Yu[†] and Ru-Shan Han[‡,§]

[†]*Institute of Theoretical Physics, Chinese Academy of Science, Beijing 100190, China*
[‡]*School of Physics, Peking University, Beijing 100871, China*
[§]*China Center of Advanced Science and Technology, Beijing 100080, China*

15.1. Introduction

In order to understand the cuprate superconductivity, P. W. Anderson put forward the resonating valence bond (RVB) superconducting theory [1], and thought a Gutzwiller projected BCS wave function as an approximate ground state [2]. Recent work by Paramekanti et al. [3] has shown that this variational approach gives a semi-quantitative understanding of the doping dependences of a variety of experimental observables in the superconducting state of the cuprates. Here, we will introduce a summary chapter on this issue, recently put forward by Anderson. In this chapter, the authors revisit these issues using the "Renormalized Mean Field Theory" of Zhang et al. [4] based on the Gutzwiller approximation in which the kinetic and superexchange energies are renormalized by different doping-dependent factors g_t and g_S, respectively. They point out a number of consequences of this early mean field theory for experimental measurements which were not available

---

*P. W. Anderson *et al.*, *J. Phys. Cond. Mat.* **16**, R755 (2004); arXiv: cond. Mat. /0311467.

when it was first explored, and observe that it is able to explain the existence of the pseudogap, properties of nodal quasi-particles and approximate spin–charge separation, the latter leading to large renormalizations of the Drude weight and superfluid density. They use the Lee-Wen theory [5] of the phase transition as caused by thermal excitation of nodal quasi-particles, and also obtain a number of further experimental confirmations. Finally, they remark that superexchange, and not phonon, is responsible for d-wave superconductivity in the cuprates.

The RVB liquid was suggested in 1973 by Anderson et al. [6, 7] as a possible quantum state for anti-ferromagnetically coupled $S = 1/2$ spins in low dimensions. Their ideas were based on numerical estimates of the ground state energy. Instead of orienting the separate atomic magnets, oppositely directed sublattices, in the liquid they were supposed to form singlet "valence bonds" in pairs, and regain some of the lost anti-ferromagnetic exchange energy. Such states form the basis of Pauling's early theories of aromatic molecules (such as benzene, as well as of his unsuccessful theories of metals), and are fair descriptions for Bethe's (1931) anti-ferromagnetic linear chain [8]. The $S = 1/2$ anti-ferromagnetic Heisenberg model arises naturally in Mott insulators. Unlike conventional band insulators, Mott insulators have an odd number of electrons per unit cell and are insulating by virtue of the strong Coulomb repulsion between two electrons on the same site. Virtual hopping favors anti-parallel spin alignment, leading to anti-ferromagnetic exchange coupling J between the spins [9]. In the RVB picture, $S = 1/2$ is important because strong quantum fluctuations favor singlet formation rather than the classically ordered Neel state.

In 1986 the high-T_C cuprates were discovered, and the crystal electronic structures were soon determined, i.e. the square planar CuO_2 lattice. In the "undoped" condition, where the Cu is stoichiometrically Cu^{2+}, the CuO_2 plane is just such an anti-ferromagnetic coupled Mott insulator. These planes are weakly coupled to each other. Anderson [1], in response to this discovery, showed that an RVB state could be formally generated as a Gutzwiller projection of a BCS pair superconducting state. This is a much more convenient and suggestive representation than those based on atomic spins, and it immediately makes a connection with superconductivity.

The method of Gutzwiller [2] was initially proposed as a theory of magnetic metals, in conjunction with the Hubbard model [10]. His proposal was to take into account the strong local Coulomb repulsion of the electrons by taking a simple band Fermi sea state and simply removing, by projection,

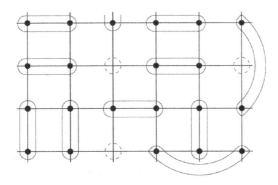

Fig. 15.1. Indicates RVB configuration, and displays singlet pairing and partial hole doping. Multi-body ground-state wave function is a linear superposition of these configurations [6, 7]. Taken from Anderson et al. (2004).

all (or, in the early version, a fraction) of the components in it which have two electrons on the same site. When one projects a half-filled band in this way, the result is to leave only singly occupied sites with spins. Anderson's new idea was to project a BCS paired superconducting state; then, the spins would be paired up in singlet pairs to make a liquid of pair "bonds" (see Fig. 15.1).

But of course, with exactly one spin at every site, this state is a Mott insulator, not a metal. Such an RVB liquid state is of rare occurrence in real Mott insulators, which usually exhibit either anti-ferromagnetic long-range order as in the cuprates, or possibly have ordered "frozen" arrays of bonds, i.e. valence bond crystals rather than liquids. However, the importance of the RVB liquid was the suggestion that, as this state is doped with added electrons or holes, the resulting metal would be a high T_C superconductor, retaining the singlet pairs but allowing them to carry charge and support supercurrents. The motivation for the pairing would be the anti-ferromagnetic superexchange of the original Mott insulator.

For over a decade and a half, a number of theorists have been trying to implement this suggestion of Anderson's along a bewildering variety of routes. Recently, the Gutzwiller–RVB wavefunction approach was revived by Paramekanti et al. [3] who used careful numerical methods to calculate many quantities of direct experimental relevance. Their results turned out to correspond remarkably well with the experimental phenomena observed in the cuprates across a very broad spectrum of types of datum, a spectrum which was simply not available in 1987–1988 when the original work was done.

All of these works rely on one basic assumption, which has gone unquestioned among a large fraction of theorists concerned with this problem from the beginning. This is the assumption that the physics of these materials is dominated by the strong repulsive interactions of a single non-degenerate band of electrons on the CuO_2 planes, and is specifically not at all similar to that of the conventional BCS superconductors. In the latter, the direct electron interactions are heavily screened, and the lattice vibrations play the dominant role. This summary chapter's authors thought it would be difficult to support that it is the demonstration of d-wave superconductivity in particular that makes phonons major players, even though there are some notable physicists, such as Mott et al. [11], who disagree. The phonon mechanisms are local in space, extended in time, making the dynamic screening mechanism emphasized by Schrieffer and Anderson [12] relevant and leading to s-wave pairing. The more electrons there are per unit cell, the better this mechanism and fails for monovalent metals. d-Wave pairing, on the other hand, is essentially non-local in space and deals with strong repulsions by conventional space avoidance. Furthermore, it is now known that the energy gap in high-T_C superconductors is much larger than that predicted by BCS theory, making it obvious that a phonon cannot be the key player. These considerations suggested the use of models where the strong repulsive correlations are emphasized, specifically the Hubbard model, which takes a strong on-site repulsion as the only interaction. The Hubbard model can be transformed by a perturbative canonical transformation into a block-diagonal form in which double occupancy is excluded and replaced by an exchange interaction between neighbouring sites. This procedure converges well for sufficiently strong on-site interaction U [13], but presumably fails at the critical U for the Mott transition; the singly occupied "undoped" case is unquestionably a Mott insulator in the cuprates and this transformation *ipso facto* works. The further simplified $t-J$ model is often used; for refined calculations it has been argued that this simplification may be too great, but for the semi-quantitative purposes of this chapter we will at least think in terms of that model.

The Mott-insulator-based theory for the cuprates has been expressed in a variety of forms other than straightforward Gutzwiller projection, and the author here focuses on properties of the ground state and of low-lying excitations, which by good fortune include the basic physics of T_C. The authors take some pages to introduce some main methods, including Hamiltonian, Gutzwiller approximation, renormalized mean field theory and the related variational Monte Carlo, and compare the main results with the experimental results. Following that, we will do a brief introduction for the main points, and interested readers can refer to the original.

15.2. The Method

Hamiltonian:

$$H = -\sum_{\langle ij\rangle,\sigma} t_{(ij)}(c_{i\sigma}^\dagger c_{j\sigma} + c_{j\sigma}^\dagger c_{i\sigma}) + U\sum_i n_{i\uparrow}n_{i\downarrow}.$$

The U is very small and can be eliminated as:

$$e^{iS}He^{-iS} = H_{t-J} = \text{PTP} + J\sum_{ij} S_i \cdot S_j.$$

In the precision half-full:

$$H = J\sum_{\langle ij\rangle} S_i \cdot S_j.$$

1D Hubbard model, this model is exactly solvable, and its solution is a class of RVB. 2D RVB is the variational ground state of $t-J$ model (see Fig. 15.2).

Like a Fermi liquid, half-filled RVB becomes unstable at low temperatures under the limit. In some systems, some may become possible: derivation from the Mott insulators, doped or self-doped near the critical point, RVB superconductor.

From the Hubbard model to $t-J$ model, there are:

$$H = -\sum_{\langle ij\rangle\sigma} t_{ij}c_{i\sigma}^\dagger c_{j\sigma} + U\sum_i n_{i\uparrow}n_{i\downarrow} \equiv T + V,$$

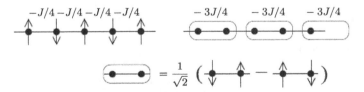

Fig. 15.2. 2D RVB: The variational ground state of $t-J$ model.

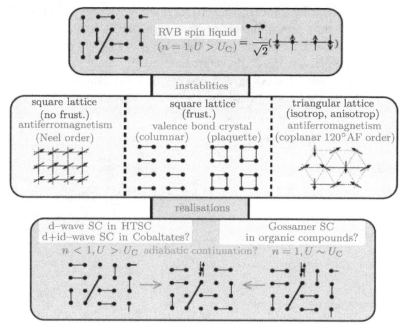

Fig. 15.3. Predictions for high-T_C superconductors based on the RVB. Taken from Ref. [14].

$$T = T_{\text{h}} + T_{\text{d}} + T_{\text{mix}},$$

$$T_{\text{h}} = -\sum_{\langle ij\rangle\sigma} t^{\text{h}}_{ij}(1-n_{i\bar{\sigma}})c^{\dagger}_{i\sigma}c_{j\sigma}(1-n_{j\bar{\sigma}}),$$

$$T_{\text{d}} = -\sum_{\langle ij\rangle\sigma} t^{\text{d}}_{ij} n_{i\bar{\sigma}} c^{\dagger}_{i\sigma} c_{j\sigma} n_{j\bar{\sigma}},$$

$$T_{\text{mix}} = -\sum_{\langle ij\rangle\sigma} t^{\text{d}}_{ij} n_{i\bar{\sigma}} c^{\dagger}_{i\sigma} c_{j\sigma}(1-n_{j\bar{\sigma}}) - \sum_{\langle ij\rangle\sigma} t^{\text{mix}}_{ij}(1-n_{i\bar{\sigma}}) c^{\dagger}_{i\sigma} c_{j\sigma} n_{j\bar{\sigma}}.$$

Partial Gutzwiller projection operator is:

$$\Pi(x) = \prod_i (1-(1-x)n_{i\uparrow}n_{i\downarrow}) = \sum_{l=0}^{M} x^l P_l = x^D,$$

where M is the number of lattice,

$$D = \sum_i n_{i\uparrow} n_{i\downarrow},$$

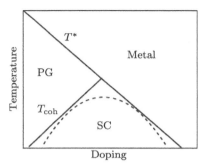

Fig. 15.4. Phase diagram. T_{coh} is coherent temperature, T_C is the minimum of T^* and T_{coh}, and T^* is from the formation of RVB singlet.

projection operator is

$$P_l = \sum_{\{i_1,\cdots,i_l\}} \left[v_{i_1}\cdots v_{i_l} \prod_j^l (1-v_j) \right],$$

projecting a state into Hilbert space, and l is a fixed double occupancy, here

$$v_i = n_{i\uparrow}n_{i\downarrow}.$$

Obviously,

$$P_0 = \Pi(0).$$

The above formula expresses a full Gutzwiller projection operator. For convenience, we introduce the following symbols:

$$P_l(x) = x^l P_l, \quad P_{\eta i} = \sum_{l \geq i} P_l(x),$$

$$H_{\text{eff}}^{(M)} = \Pi(x)[T_{\text{h}} + T_{\text{d}} + J + V]\Pi(x),$$

$$J = \sum_{\langle ij \rangle} J_{ij}\left(S_iS_j - \frac{1}{4}n_in_j - \frac{1}{2}n_{i\uparrow}n_{i\downarrow}n_j - \frac{1}{2}n_in_{j\uparrow}n_{j\downarrow} + n_{i\uparrow}n_{i\downarrow}n_{j\uparrow}n_{j\downarrow}\right),$$

$$J_{ij} = 4|t_{ij}^{\text{mix}}|^2/U.$$

Gutzwiller projection wave function:

$$|\Psi_{\text{RVB}}\rangle = P_N P_G |\text{BCS}\rangle,$$
$$|\text{BCS}\rangle = \prod_k (u_k + v_k c_{k\uparrow}^\dagger c_{-k\downarrow}^\dagger)|0\rangle.$$

In the real space,

$$\Pi(x) \left[\sum_{R_j, R_l'} a(R_j - R_l') c^\dagger_{R_j,\uparrow} c^\dagger_{-R_l',\downarrow} \right]^{N_e/2} |0\rangle,$$

$$a(r) = \sum_k a_k \cos(k \cdot r), \quad a_k = \frac{v_k}{u_k}.$$

Gutzwiller approximation is as follows:

$$\frac{\langle \Psi_0 | P_G \hat{O} P_G | \Psi_0 \rangle}{\langle \Psi_0 | P_G P_G | \Psi_0 \rangle} \approx g_0 \frac{\langle \Psi_0 | \hat{O} | \Psi_0 \rangle}{\langle \Psi_0 | \Psi_0 \rangle},$$

$$\langle c^\dagger_{i\uparrow} c_{j\uparrow} \rangle_\Psi = \langle (1-\hat{n}_{i\uparrow}) c^\dagger_{i\uparrow} (1-\hat{n}_{j\downarrow}) c_{j\uparrow} \rangle_\Psi$$

$$\approx \tilde{g}_t \langle (1-\hat{n}_{i\downarrow}) c^\dagger_{i\uparrow} (1-\hat{n}_{j\downarrow}) c_{j\uparrow} \rangle_{\Psi_0}$$

$$\approx \tilde{g}_t (1-n^0_{i\downarrow})(1-n^0_{j\downarrow}) \langle c^\dagger_{i\uparrow} c_{j\uparrow} \rangle_{\Psi_0},$$

$$g_t = \frac{[n_{i\uparrow}(1-n_j) n_{j\uparrow}(1-n_i)]^{1/2}}{[n^0_{j\uparrow}(1-n^0_{j\uparrow}) n^0_{i\uparrow}(1-n^0_{i\uparrow})]^{1/2}}.$$

When

$$n_{i\uparrow} = n_{i\downarrow} = n/2, \quad g_t = \frac{1-n}{1-n/2},$$

$$\langle S_i S_j \rangle_\Psi = -g_s \langle S_i S_j \rangle_{\Psi_0},$$

$$g_s^\pm = \frac{1}{(1 - 2n^0_\uparrow n^0_\downarrow/n)^2}, \quad g_s^z = \frac{1}{(1 - 2n^0_\uparrow n^0_\downarrow/n)^2}, \quad g_s = \frac{1}{(1-n/2)^2},$$

renormalized t–J model and renormalized Mean field theory are as

$$\tilde{H}_{t-J} = -g_t t \sum_{\langle i,j \rangle, \sigma} (c^\dagger_{i,\sigma} c_{j,\sigma} + c^\dagger_{j,\sigma} c_{i,\sigma}) + g_s J \sum_{\langle i,j \rangle} S_i S_j,$$

$$\tilde{\Delta}_r \equiv \langle c^\dagger_{i\uparrow} c^\dagger_{i+r\downarrow} - c^\dagger_{i\downarrow} c^\dagger_{i+r\uparrow} \rangle_{\Psi_0}, \quad \tilde{\xi}_r \equiv \sum_\sigma \langle c^\dagger_{i\sigma} c_{i+r\sigma} \rangle_{\Psi_0},$$

$$|\Psi_0\rangle = \prod_k (u_k + v_k c^\dagger_{k\uparrow} c^\dagger_{-k\downarrow}) |0\rangle,$$

$$v_k^2 = \frac{1}{2}\left(1 - \frac{\xi_k}{E_k}\right),$$

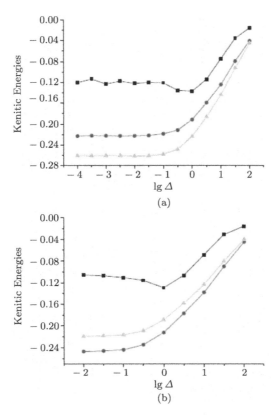

Fig. 15.5. Kinetic energy as a function of $\lg \Delta$ [24].

$$\Delta_k = \frac{3g_s J}{4}(\tilde{\Delta}_x \cos k_x + \tilde{\Delta}_y \cos k_y),$$

$$\xi_k = -\left(2g_t t + \frac{3g_s J}{4}\tilde{\xi}_x\right)\cos k_x - \left(2g_t t + \frac{3g_s J}{4}\tilde{\xi}_y\right)\cos k_y - \mu.$$

Variational Monte Carlo (VMC)

In order to get an accurate Gutzwiller projection wave function, we use VMC to improve them:

$$\langle \hat{O} \rangle_\Psi = \frac{\langle \Psi_0 | P_G P_N \hat{O} P_N P_G | \Psi_0 \rangle}{\langle \Psi_0 | P_G P_N P_G | \Psi_0 \rangle}$$

$$= \sum_{\alpha,\beta} \langle \alpha | \hat{O} | \beta \rangle \frac{\langle \Psi_0 | \alpha \rangle \langle \beta | \Psi_0 \rangle}{\langle \Psi_0 | P_N P_G | \Psi_0 \rangle}$$

$$= \sum_\alpha \left(\sum_\beta \frac{\langle\alpha|\hat{O}|\beta\rangle\langle\beta|\Psi_0\rangle}{\langle\Psi_0|\alpha\rangle} \right) \frac{|\langle\Psi_0|\alpha\rangle|^2}{\langle\Psi_0|P_N P_G|\Psi_0\rangle}$$

$$= \sum_\alpha f(\alpha)p(\alpha),$$

$$f(\alpha) = \sum_\beta \frac{\langle\alpha|\hat{O}|\beta\rangle\langle\beta|\Psi_0\rangle}{\langle\Psi_0|\alpha\rangle},$$

$$p(\alpha) = \frac{|\langle\Psi_0|\alpha\rangle|^2}{\langle\Psi_0|P_N P_G|\Psi_0\rangle},$$

$$|\alpha\rangle = c^\dagger_{R_1,\uparrow} \cdots c^\dagger_{R_N,\uparrow} c^\dagger_{R_1,\downarrow} \cdots c^\dagger_{R_N,\downarrow}|0\rangle.$$

(In order to renormalize $t-J$ model, taking the variational parameters: giving the bare energy gap parameter, and doping you can change the spin configuration, and you can get the minimum energy, and change the bare energy gap parameter to get best value.)

15.3. Overview

We use a graphical representation to summarize the VMC calculation process by Hubbard model and $t-J$ model [19] (see Fig. 15.6).

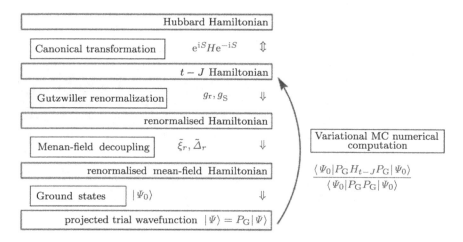

Fig. 15.6. A graphical representation to summarize the VMC calculation process.

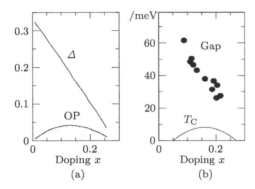

Fig. 15.7. (a) d-Wave superconducting energy gap amplitude and the superconducting order parameter (OP) as a function of doping concentration x in the $t-J$ model [15]. (b) The energy gap and T_C as a function of doping by ARPES measurements, the corresponding relationship of x and T_C according to the formula: $T_C/T_C^{max} = 1 - 82.6(x - 0.16)^2$, $T_C^{max} = 95$ K [16]. Taken from Anderson et al. (2004).

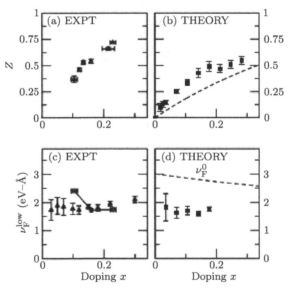

Fig. 15.8. (a) Nodal quasi-particle Z as a function of doping x for Bi2212 sample by ARPES, x is calculated by the empirical formula of T_C [17] ($T_C^{max} = 95$ K); (b) $Z(x)$ is the prediction of Paramekanti VMC calculations; dashed line is the Gutzwiller approximation: $Z(x) = 2x/(1+x)$; (c) low-temperature nodal Fermi velocity v_F^{low} is the Bi2212 ARPES data (empty square) [22] and LSCO (empty triangle) [20], which are independent of the doping; (d) the predicted renormalized v_F^{low}, as a function of x [3]; the dashed line is the Fermi velocity v_F^0 [21] of bare band. (b) taken from Anderson et al. (2004).

15.4. Conclusion

To sum up, our basic assumptions as to the physics of the cuprates, together with a mean-field theory that is a little less manageable than BCS theory, seem to give a remarkably complete picture of the unusual nature of the superconducting state. The RVB state is still a pairing state between electrons. It is made clear in the recent studies of a partially projected BCS state by Laughlin and Zhang. [15]. Furthermore, its low-lying excitations are well-defined quasi-particles which dominate low-energy physics. Thus, the RVB state is in some ways rather conventional. What is unusual is the reduction of the superfluid density and the quasi-particle spectral weight. With increasing degrees of projection, the state evolves from pairing of quasi-particles to one which is better understood as a spin singlet formation with coherent holes motion. This evolution has the following dramatic consequence. The BCS pairing is driven by a gain in attractive potential at the expense of kinetic energy, since the energy gap leaves out the Fermi occupation $n(k)$. With projection, $n(k)$ is already strongly erased in the non-Fermi liquid normal state, and superconductivity is instead stabilized by a gain in kinetic energy due to coherent holes motion. This picture has been verified by experiments on "undoped" samples which monitor the kinetic energy via the optical sum rule.

Why is the subject so controversial? It is because there are many states with the same energy scale but different symmetry, competing with the superconducting states. While the RVB theory leads naturally to a crossover from pseudogap to strange metal and then to Fermi liquid as one increases the doping at a temperature above the $T_C^{\max}$, the ideas presented here are no help in understanding the breakdown of Fermi liquid behavior in the strange metal. Some people ascribe the anomalous behavior to a quantum critical point which lies in the middle of the superconducting dome. The authors simply remark that the quantum critical point, if it exists, is different from any previous examples in that there is no sign of a diverging correlation length scale in any physical observable.

Lastly the authors point out that great strides have been made in the discovery of unconventional superconductors since 1986. Today, non-s-wave pairing states are almost commonplace in heavy fermions, organic superconductors and transition metal oxides. Even time-reversal symmetry is not sacrosanct. The discovery of high T_C has opened our eyes to the possibility that superconductivity is an excellent choice as the ground state of a strongly correlated system. This may be the most important message to be learned from this remarkable discovery.

15.5. Postscript

When 2002 Nobel Prize winner Laughlin proposed Gossamer superconducting theory, he pointed out that the single-band model (Hubbard model and $t-J$ model) couldn't be the starting point of the theory of high-T_C superconductivity, and triggered the recovery of RVB wave function research. As previously described, Anderson *et al.* co-proposed the "Plain Vanilla" version of RVB superconducting theory, which relied on an effective single-band model and RVB variational wave function calculation, and gave the phase diagram and included the Gossamer theory, claiming to be a complete superconducting theory. Shortly after they finished the pre-printed paper (see arXiv: cond-mat/0311467), Varma [23] (see arXiv: cond-mat/0312385) from Bell Laboratories in the United States gave a comprehensive refutation of this theory. Varma pointed out the theory had three inappropriate points: (1) it is inappropriate to adopt wave function of limited variational degree of freedom to study the Hubbard or $t-J$ models; (2) the Hubbard and $t-J$ models are not suitable for the cuprates; (3) the main calculation results are contradictory with lots of experimental results. These inconsistencies are fatal for the theory. Varma specially pointed out that the given phase diagram by the theory was wrong. In high-T_C superconductors' phase diagram, besides the superconducting transition, there are two other important points, which are the anti-ferromagnetic insulator–metal transition and the transition of $x = 0.19$ nearby. The phase diagram given in the experiments has a different nature compared with the one given in the RVB "Plain Vanilla" superconducting theory. Varama even asserted that high-temperature superconducting cuprate, which cannot contain the theory of these changes, cannot be a theory of high-temperature superconducting cuprate.

Varma's view quoted above indicates that the mechanism of high-temperature superconducting cuprate still meets no consensus.

References

[1] P. W. Anderson, *Science* **237**, 1196 (1987).
[2] M. C. Gutzwiller, *Phys. Rev. Lett.* **10**, 159 (1963).

[3] A. Paramecanti, *Phys. Rev. Lett.* **87**, 217002 (2001).
[4] F. C. Zhang, *Phys. Rev. Lett.* **90**, 207002 (2003).
[5] P. A. Lee, X. G. Wen, *Phys. Rev. Lett.* **78**, 4111 (1997).
[6] P. W. Anderson, *Mater. Res. Bull.* **8**, 153 (1973).
[7] P. W. Anderson, *Philos. Mag.* **30**, 432 (1973).
[8] H. A. Bethe, *Z. Phys.* **71**, 2051 (1931).
[9] P. W. Anderson, *Phys. Rev.* **115**, 2 (1959).
[10] J. Habbard, *Proc. Roy. Soc. London, Ser.* **A276**, 238 (1963).
[11] N. Mott, *Philos. Mag. Lett.* **64**, 211 (1991).
[12] R. J. Schrieffer, P. W. Anderson, *Phys. Today* **44**, 54 (1991).
[13] C. Gros, *Phys. Rev. B* **38**, 931 (1988).
[14] B. Edegger, *Adv. Phys.* **56**, 927 (2007).
[15] F. C. Zhang, T. M. Rice, *Phys. Rev. B* **37**, 3759 (1988).
[16] X. Campuzano, *Phys. Rev. Lett.* **83**, 3709 (1999).
[17] X. Campuzano, *Physica C* **176**, 95 (1991).
[18] R. Laughlin, arXiv: 0209269; *Philos. Mag.* **86**, 1165 (2006).
[19] H. A. Molegraaf, *Science* **295**, 2239 (2002).
[20] X. J. Zhou, *Nature* **423**, 398 (2003).
[21] M. Randaria, *Phys. Rev. B* **69**, 144509 (2004).
[22] P. D. Johnson, *Phys. Rev. Lett.* **87**, 177007 (2001).
[23] C. M. Varma, arXiv:cond.mat./0312385; *Nato Sci. Ser. Ser. II: Math. Phys. Chem.* **183**, 105 (2005).
[24] X. G. Wen, Y. Yu, *Phys. Rev. B* **72**, 045130 (2005).

16
The Two-Band Model of Electron-Doped High-T_C Superconductors

Hong-Gang Luo* and Tao Xiang*,†

*Institute of Theoretical Physics, Chinese Academy of Sciences,
Beijing 100190, China
†Institute of Physics, Chinese Academy of Sciences,
Beijing 100190, China

The parent compound of high-T_C superconductors is an anti-ferromagnetic insulator. Upon doping, the long-range anti-ferromagnetic order is suppressed and the compound becomes superconducting. The phase diagrams for the electron-doped and hole-doped are different. Based on a detailed analysis of electronic structures and experimental data, we propose that the electron-doped high-T_C superconductors are described by a two-band model. This model provides a natural account for the experimental data and suggests that the superconducting pairing of electron-doped high-T_C superconductors has the $d_{x^2-y^2}$-wave symmetry.

16.1. Introduction

In 1986, Bednorz and Müller discovered the high-temperature superconductor LaBaCuO. Since then, the research of high-temperature superconductivity has become one of the most important topics in condensed matter physics. The parent compound of high-T_C superconductors is an anti-ferromagnetic insulator. Through doping it can become a superconductor. According to the nature of the carriers introduced, the phase

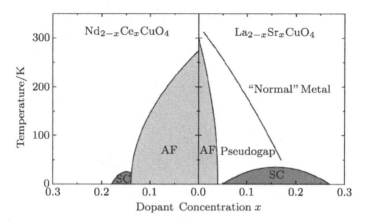

Fig. 16.1. Schematic phase diagram of high-T_C superconductors. Taken from Ref. [1].

diagram (Fig. 16.1) behaves very differently. On the hole-doped side, the anti-ferromagnetic long-range order disappears rapidly with doping, and the superconductivity occurs in a wide doping regime. However, on the electron-doped side, the anti-ferromagnetic long-range order exists in a wide doping range, and the superconductivity occurs only in a narrow doping regime. Around the optimal doping, the in-plane resistivity of the hole-doped compounds varies linearly with the temperature. But in electron-doped materials, such as $Pr_{1.85}Ce_{0.15}CuO_{4-\delta}$ (PCCO) and $Nd_{1.85}Ce_{0.15}CuO_{4-\delta}$ (NCCO) ($x = 0.15$), the resistivity varies approximately quadratically with temperature. Moreover, the pseudogap effect is observed in the hole-doped materials, but not in the electron-doped ones.

For the hole-doped cuprates, experimental measurements, including the photoemission spectroscopy, the tunneling spectroscopy, and the phase-sensitive measurements, show clearly that the superconducting pairing has the $d_{x^2-y^2}$-wave symmetry [2]. This is consistent with the theoretical prediction. On the contrary, for the electron-doped cuprates, the pairing symmetry is still an issue under debate. Some in-plane penetration depth [3–6] and tunneling measurements [7] for NCCO or PCCO suggest that the pairing is more likely to have an s-wave symmetry. However, other experimental measurements, including the penetration depth [8, 9], the phase-sensitive measurements [2, 10], the photoemission spectroscopy [11] and the grain-boundary Josephson tunneling spectroscopy for $La_{2-x}Ce_xCuO_{4-\delta}$ (LCCO) at $x = 0.105$ [12], indicate that the pairing has the d-wave symmetry. In order to clarify this issue, Skinta et al. [13] performed a systematic in-plane magnetic penetration depth measurement for the LCCO and PCCO samples. They found that in the underdoped regime the pairing symmetry is d-wave, but in the overdoped regime, the pairing symmetry is s-wave. They suggested that a transition from d- to s-wave symmetry may happen with increasing doping, and the transition point may occur around the optimal doping. A similar transition in the pairing symmetry was observed in the point contact spectroscopy for the underdoped ($x = 0.13$) and overdoped ($x = 0.17$) PCCO by Biswas et al. [14]. Khodel et al. [15] proposed a model based on the anti-ferromagnetic spin fluctuation to explain this extraordinary phenomenon. However, their theories predicted a transition from d- to p-wave phase, rather than to the experimentally observed s-wave phase.

In order to understand the origin of the difference between the n- and p-type superconductors, it is helpful to consider the difference in the electronic structure between these two kinds of compounds. The angle-resolved photoemission spectroscopy (ARPES) showed that the electronic structures of p- and n-type superconductors exhibit very different behaviors with increasing doping. In the underdoped p-type superconductors, such as $Ca_{2-x}Na_xCuO_2Cl_2$ [1] with $x = 0.1$, a small Fermi surface is formed around $(\pm\pi/2, \pm\pi/2)$, as shown in Fig. 16.2. With increasing

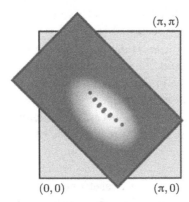

Fig. 16.2. ARPES intensity integration of 10% sodium doped $Ca_{2-x}Na_xCuO_2Cl_2$. Taken from Ref. [1].

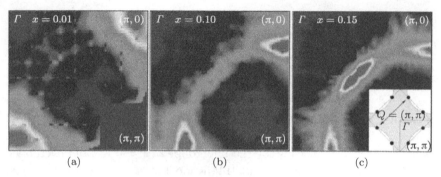

Fig. 16.3. ARPES intensity integration of three different doping NCCO samples. Taken from Ref. [16].

doping, this Fermi surface extends toward the zone edges at $(\pi,0)$ and $(0,\pi)$. However, in the electron-doped compounds, such as NCCO [16], small Fermi pockets first appear around $(\pi,0)$ and $(0,\pi)$ at low doping ($x = 0.04$). At optimal doping ($x = 0.15$), a small Fermi pocket appears around $(\pi/2,\pi/2)$, which is not connected to the Fermi surface near $(\pi,0)$, as shown in Fig. 16.3. Thus, the electron-doped materials contain two disconnected Fermi surface patches. This is entirely different from the Fermi surface shape for the p-type superconductors.

To explain the doping evolution of the Fermi surface of the n-type superconductor, Kusko et al. [17] proposed that the effective Hubbard U_{eff} is doping-dependent in the single-band Hubbard model. On the other hand, Kusunose and Rice [18] suggested that this peculiar doping dependence of the Fermi surface results from the in-gap states generated by the spectral weights transformation from high energy. However, it is difficult to understand the dramatic reduction of U_{eff} by doping as suggested by Kusko et al. The picture of the in-gap states is phenomenological and the related microscopic mechanism is not clear.

The n- and p-type superconductors are different in many respects, but they share many common features. For example, their parent compounds are all half-filled anti-ferromagnetic Mott insulators, which are believed to be the key in understanding the mechanism of high-T_C superconductors. Based on the analysis of the electronic structure of n-type superconductors and the experimental data, we propose that the physical properties of electron-doped cuprate superconductors are governed by a two-band model [26].

16.2. Experimental Evidence for the Two-Band Model

16.2.1. ARPES [16]

ARPES measures the single-particle excitation spectrum. It provides an ideal probe to detect the Fermi surface, as shown in Fig. 16.3. At low doping, the electronic spectrum weight of the n-type superconductor NCCO first appears around $(\pi, 0)$ and its equivalent symmetric points. With increasing doping, a new spectrum weight emerges from the energy below the Fermi level around $(\pi/2, \pi/2)$, which forms a small Fermi surface pocket for the superconducting samples. Kusko et al. [17] attributed these two kinds of Fermi surfaces to the upper and lower Hubbard bands. To understand the doping evolution of these Fermi surfaces, they suggested that the effective Hubbard repulsion energy U_{eff} is strongly suppressed by doping. However, physically, this strong suppression of the effective Coulomb interaction is hard to understand. Another suggestion is that the existence of the two separated Fermi pockets is due to the band folding effect associated with the anti-ferromagnetic fluctuation. However, the band gap caused by the anti-ferromagnetic fluctuation is very small, which is just of the order of the anti-ferromagnetic interaction J.

The appearance of these two separated Fermi surface pockets can in fact be described by a two-band model. In 1988, Zhang and Rice derived a low-energy effective single-band $t-J$ model from the three-band model for the p-type high-T_C superconductors at low doping. This single-band model captures certain characteristics of p-type superconductors and is commonly adopted in the study of high-T_C superconductivity. For the n-type superconductors, this single-band picture has also been used. However, it should be pointed that this simple single-band $t-J$ model is not adequate to describe the electronic structures of the n-type superconductors. Instead, a two-band $t-J$ model which should be introduced can describe the n-type superconductors [19].

16.2.2. Transport Measurements [20–23]

The existence of two bands in the n-type cuprate superconductors was in fact first revealed by the transport measurements, such as the Hall coefficient, magnetoresistance, thermopower, and Nernst effect measurements.

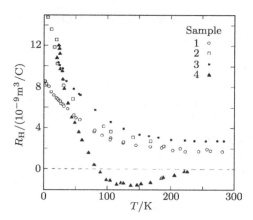

Fig. 16.4. Temperature dependence of Hall coefficients for four NCCO samples. Taken from Ref. [20].

Unlike in the p-type superconductors, there are sign changes in the Hall coefficient and thermopower with temperature. The change of the Hall angle is, on the other hand, small. It is difficult to understand these experimental data if there is only one kind of carrier.

Figure 16.4 shows the doping and temperature dependence of the Hall coefficient for NCCO. At high doping, the Hall coefficient changes sign at high temperature. This sign change suggests that besides the electron-band, which exists at low doping, there is also hole-band on the Fermi surface at high doping. In a two-band system, the Hall coefficient can be written as [20]:

$$R_H = \frac{\sigma_{H_1} + \sigma_{H_2}}{B(\sigma_1 + \sigma_2)},$$

where σ_{H_i} and σ_i are the Hall conductivity and conductivity of corresponding band, respectively, B is a constant. For the electron-band, $\sigma_H < 0$, and for the hole-band, $\sigma_H > 0$. Thus, the combination of the contributions of these two bands to the Hall coefficient can lead to a sign change in the Hall coefficient. Figure 16.5 shows the magnetoresistance for NCCO under different annealing conditions [21]. The positive magnetoresistance shown in Fig. 16.5(b) indicates that the two carriers are involved in the transport process. In particular, the sign change of the magnetoresistance is only observed in the superconducting samples. Moreover, experimentally it is also found that the hole-type carriers play an essential role in the n-type superconductivity [24].

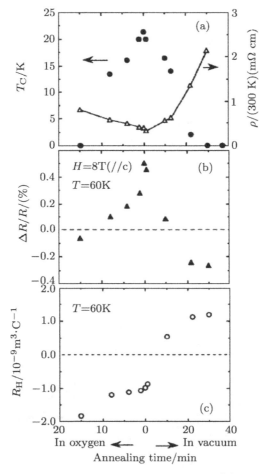

Fig. 16.5. (a) The superconducting transition temperature, (b) magnetoresistance and (c) Hall coefficient of NCCO samples under different annealing conditions. Taken from Ref. [21].

16.2.3. *Penetration Depth Experiments* [5, 6, 24]

The penetration depth is proportional to the superfluid density. The low-temperature of the superfluid density is governed by the pairing symmetry. Figure 16.6 illustrates the typical temperature dependence of the superfluid density in a superconductor with two weakly coupled bands. It generally shows a positive curvature near T_C. This positive curvature in the superfluid density near T_C is not observed in not strongly doped

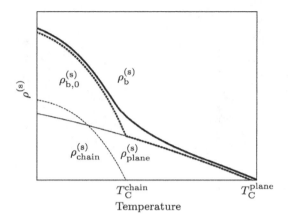

Fig. 16.6. The superfluid density of two weakly coupled bands [26].

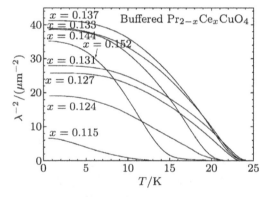

Fig. 16.7. The superfluid density of n-type PCCO with different dopings. Taken from Ref. [6].

p-type superconductors, but is ubiquitously observed [5, 6, 24] in the n-type superconductors (Fig. 16.7).

16.3. The Two-Band Model

The n-type cuprate superconductors have one hole-band and one electron-band. These two bands can be described by a reduced two-band BCS

Hamiltonian as:

$$H = \sum_{i,k\sigma} \xi_{i,k} c^+_{i,k\sigma} c_{i,k\sigma} + \sum_{i,k\sigma} V_{i,kk'} c^+_{i,k'\uparrow} c^+_{i,-k'\downarrow} c_{i,-k\uparrow} c_{i,k\downarrow}$$
$$+ \sum_{k\sigma} (V_{3,kk'} c^+_{1,k'\uparrow} c^+_{1,-k'\downarrow} c_{2,-k\uparrow} c_{2,k\downarrow} + \text{h.c.}),$$

where $c_{i,k\sigma}$ ($i = 1, 2$) is the electron annihilation operator of the ith band with the wave vector k and the spin σ. $V_{1,kk'}$ and $V_{2,kk'}$ are the pairing energies of the band 1 and the band 2, respectively. $V_{3,kk'}$ is the pairing correlation potential between the two bands. This two-band model provides a natural account for the experimental measurements, including superfluid density [27], the Raman spectra [28], the point contact spectroscopy [29] and the neutron scattering [30] of the n-type high-T_C superconductors. Based on the calculation for this model, we find that the electron-doped cuprate superconductors should also have d-wave pairing symmetry, same as in their hole-doped counterparts [31].

References

[1] A. Damascelli, Z. Hussain and Z.-X. Shen, *Rev. Mod. Phys.* **75**, 473 (2003).
[2] C. C. Tsuei, J. R. Kirtley, *Phys. Rev. Lett.* **85**, 182 (2000); *Rev. Mod. Phys.* **72**, 969 (2000).
[3] D. H. Wu et al., *Phys. Rev. Lett.* **70**, 85 (1993).
[4] L. Alff et al., *Phys. Rev. Lett.* **83**, 2644 (1999).
[5] J. A. Skinta et al., *Phys. Rev. Lett.* **88**, 207005 (2002).
[6] M. S. Kim et al., *Phys. Rev. Lett.* **91**, 087001 (2003).
[7] L. Alff et al., *Nature* **422**, 698 (2003).
[8] J. D. Kokales et al., *Phys. Rev. Lett.* **85**, 3696 (2000).
[9] R. Prozorov et al., *Phys. Rev. Lett.* **85**, 3700 (2000).
[10] B. Chesca et al., *Phys. Rev. Lett.* **90**, 057004 (2003).
[11] N. P. Armitage et al., *Phys. Rev. B* **68**, 064517 (2001).
[12] B. Chesca et al., *Phys. Rev. B* **71**, 104504 (2005).
[13] J. A. Skinta et al., *Phys. Rev. Lett.* **88**, 207003 (2002).
[14] A. Biswas et al., *Phys. Rev. Lett.* **88**, 207004 (2002).
[15] V. A. Khodel, V.M. Yakovenko, M.V. Zverev, H. Kang, *Phys. Rev. B* **69**, 144501 (2004).
[16] N. P. Armitage et al., *Phys. Rev. Lett.* **88**, 257001 (2002).
[17] C. Kusko et al., *Phys. Rev. B* **66**, R140513 (2002).
[18] H. Kusunose, T. M. Rice. *Phys. Rev. Lett.* **91**, 186407 (2003).
[19] T. Xiang, H. G. Luo, D. H. Lu, K. M. Shen, Z. X. Shen, *Phys. Rev. B* **79**, 014524 (2009); arXiv: 0807.2498 (2008).

[20] Z. Z. Wang et al., *Phys. Rev. B* **43**, 3020 (1991).
[21] W. Jiang et al., *Phys. Rev. Lett.* **73**, 1291 (1994).
[22] P. Fournier et al., *Phys. Rev. B* **56**, 14149 (1997).
[23] P. C. Li, R. L. Greene, *Phys. Rev. B* **76**, 174512 (2007).
[24] Y. Dagan, R. L. Greene, *Phys. Rev. B* **76**, 024506 (2007).
[25] A. V. Pronin et al., *Phys. Rev. B* **68**, 054511 (2003).
[26] T. Xiang, J. M. Wheatley, *Phys. Rev. Lett.* **76**, 134 (1996).
[27] H. G. Luo, T. Xiang, *Phys. Rev. Lett.* **94**, 027001 (2005).
[28] C. S. Liu, H. G. Luo, W. C. Wu, T. Xiang, *Phys. Rev. B* **73**, 174517 (2006).
[29] C. S. Liu, W. C. Wu, *Phys. Rev. B* **76**, R220504 (2007).
[30] C. S. Liu, W. C. Wu, *Phys. Rev. B* **76**, 014513 (2007).
[31] T. Xiang, *D-Wave Superconductors* (Science Press, Beijing, 2007, in Chinese).

17
Theoretical Investigations on the Spin Dynamics in High-T_C Cuprates

Jian-Xin Li

Department of Physics and National Laboratory of Solid State Microstructure, Nanjing University, Jiangsu 210093, China

High-T_C superconductivity is realized by doping the antiferromagnetic Mott insulators, so that its superconducting phase is in proximity to the antiferromagnetic phase. Therefore, the understanding of the spin dynamics in high-T_C cuprates is considered to be essential to uncover the superconducting mechanism in these materials. Here, we will first give a brief review of the experimental results and then the theoretical understandings.

17.1. Experiments of Spin Dynamics — Nuclear Magnetic Resonance and Neutron Scattering Experiment

17.1.1. *Nuclear Magnetic Resonance*

Nuclear magnetic resonance (NMR) can give local, atomic site information on spin susceptibility. The magnetic hyperfine interaction of distinct nuclei at different lattice sites couples the nuclear moments to the spins of the conduction electrons, which leads to a shift in NMR frequencies (Knight shift), and the coupling to local electron spin dynamics determines the relaxation of the field-aligned nuclear moments in NMR. The relaxation rate — so-called spin–lattice relaxation rate T_1 is given by:

$$\frac{1}{T_1 T} = \lim_{\omega \to 0} \sum_q F^2(q) \frac{\mathrm{Im}\chi(q,\omega)}{\omega},$$

where $\chi(q,\omega)$ is the dynamical spin susceptibility, ω the Zeeman splitting frequency which is about $10^{-7} J$ (J is the anti-ferromagnetic exchange coupling constant), T the temperature. $F(q)$, the hyperfine form factor which is different for different nuclear sites, acts to filter out different parts of the momentum space. Because the spin–lattice relaxation rate measures the magnetic response averaged over all momenta, NMR is a local probe. It can give spin dynamics at very low energies on different lattice sites.

17.1.2. *Neutron Scattering Experiment*

The cross-section measured in a magnetic neutron scattering experiment is given by:

$$\frac{\mathrm{d}^2\sigma}{\mathrm{d}\Omega \mathrm{d}E_\mathrm{f}} \propto \sum_{\alpha,\beta} (\delta_{\alpha\beta} - \hat{q}_\alpha \hat{q}_\beta) S^{\alpha\beta}(q,\omega),$$

where the spin structure factor is:

$$S^{\alpha\beta}(q,\omega) = \frac{1}{\pi}[1+n(\omega)]\mathrm{Im}\chi^{\alpha\beta}(q,\omega),$$

and $N(\Omega)$ is the Bose distribution function. From the above formula, we can see that among various experimental tools, the neutron scattering experiment is unique in the sense that it can give in principle the full frequency, momentum, temperature dependence of the spin structure factor and eventually, the dynamical spin susceptibility. In the following, we will focus on the experimental results obtained by neutron scattering experiments.

17.1.3. *Commensurate and Incommensurate Spin Response*

The crystal structure of an undoped (parent) high-T_C cuprate La_2CuO_4 is shown in Fig. 17.1(a). It is a layered structure in which the main electronic conductions and magnetic properties are related to the CuO_2 plane which is shown in Fig. 17.1(b), where the arrows denote the spin orientation. Due to the anti-ferromagnetic spin arrangement, the translation period for the magnetic structure is two times larger than that of the underlying crystal structure, so that a commensurate neutron scattering

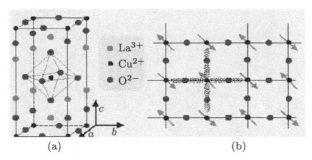

Fig. 17.1. (a) Crystal structure of La_2CuO_4. (b) CuO_4 plane where the positions of Cu and O atoms, and the spin arrangements in an antiferromegnetic order of Cu moments are indicated. Taken from Ref. [1].

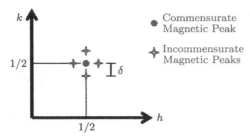

Fig. 17.2. Schematic representation of the commensurate and incommensurate magnetic peaks. The incommensurability is denoted by δ.

peak will occur at the wave vector (π, π) (the position denoted by the blue circle in Fig. 17.2). After doping, the long-range anti-ferromagnetic order will be destroyed quickly and spin fluctuations will still be observed. In this case, the commensurate peak will split, and form four incommensurate peaks (the positions of the four peaks are denoted as purple crosses in Fig. 17.2). The width of the splitting is called the incommensurability.

17.1.4. *Universality in the Spin Dynamics — The Hourglass Dispersion*

The neutron scattering experiment reveals the universality of the spin dynamics, i.e. the hourglass dispersion relation as presented in Fig. 17.3 [2]. The hourglass dispersion is characterized by the incommensurate spin excitations both in the low- and high-energy (frequency) regimes and the commensurate spin excitations at (π, π) defining a crossover point. With the increase of frequency, the incommensurability decreases in the low-energy

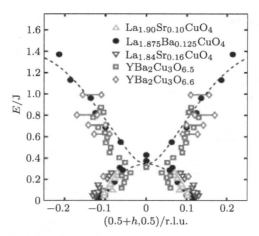

Fig. 17.3. Comparison of measured dispersion for spin excitations along the $(0.5+h, 0.5)$ direction for various high-Tc cuprates. Taken from Ref. [2].

regime, so this leads to a downward dispersion. Whereas incommensurability increases in the high-energy regime, and leads to an upward dispersion. The two branches meet at the (π,π) point at which the so-called "spin resonance" emerges. (Note that the spin resonant mode has not been identified in the La systems).

Another interesting feature is that although the low- and high-energy spin responses are incommensurate, the incommensurate peaks distribute axially in the low-energy regime while diagonally in the high-energy regime, as can be seen from Fig. 17.4 [3].

17.1.5. Comparison with the Dispersion in the Parent Compounds La_2CuO_4 and YBCO

In order to have a better understanding of the aforementioned spin dynamics in the doped systems, let us first examine the characteristics of magnetic excitations in the parent compounds. The filled and open symbols in Fig. 17.5 denote the experimental results, and the solid line represents the dispersion calculated by the spin wave theory. As can be seen from the figure, the spin wave theory gives a good description of the spin excitation spectrum in the parent compound La_2CuO_4 [5]. Similar spin excitation spectra have also been observed in the parent compound YBCO [6] (see Fig. 17.6(a)), where the magnetic response consists of both optical and acoustic modes resulting from the fact that it is a bilayer system.

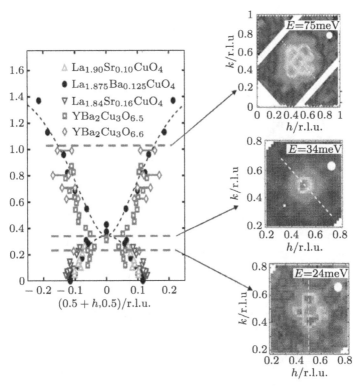

Fig. 17.4. Schematic plots intended to represent the measured dispersion for spin excitations as presented in Fig. 17.3. Right panels represent the distribution of scattering in reciprocal space at relative energies indicated by the dashed lines in left panel. Taken from Refs. [3, 4].

A comparison between the underdoped compound and the parent compound of YBCO can be found in Fig. 17.6(b) [7]. One can see that one of the effects of the hole-doping is to decrease the velocity of the high-energy spin excitations. It can be understood easily if the high-energy spin excitations can still be described by the spin wave theory in this case, since the doping weakens the effective spin coupling constant. Another effect of the doping is that it changes the low-energy feature of the dispersion significantly. After doping, the dispersion becomes downward in shape in the low frequency regime, while it shows an upward spectrum in the undoped systems. Such behavior cannot be explained by the conventional spin wave theory and, nevertheless, is crucial to our understanding of the spin excitations in the doped system.

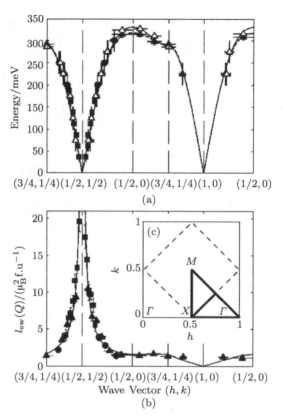

Fig. 17.5. Spectrum for the spin-wave in parent compound La_2CuO_4 along the high symmetry directions in the 2D Brillouin zone, see inset (C). (b) Wavevector dependence of the spin-wave intensity compared with predictions of linear spin-wave theory shown by the solid line. Taken from Ref. [5].

17.1.6. Doping Dependence of the Incommensurability

Figures 17.7 and 17.8 present the doping dependence of the incommensurability [8–10]. In the underdoped regime, the incommensurability is proportional to the doping concentration, i.e. the Yamada plot or Yamada relation [8]. Near the optimal doping, the incommensurability reaches its highest value and then decreases slowly with further increase of doping. The Yamada relation can be expressed by $\delta = x$ (δ is the incommensurability and x is the doping concentration), or by $\delta = 2x$ theoretically, because of the different measurement unit used (2π is usually taken as the measuring unit in experiments, while π in theoretical work).

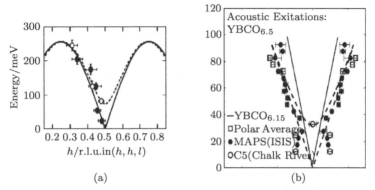

Fig. 17.6. Dispersion relations for the spin-wave excitations in the insulating $YBa_2Cu_3O_{6.15}$. Closed circles and solid line are for acoustic modes, open circles and dotted line are optical modes. Taken from Ref. [6]. (b) Dispersion of the acoustic and optic modes with respect to the (π, π) position in $YBa_2Cu_3O_{6.5}$. Taken from Ref. [7].

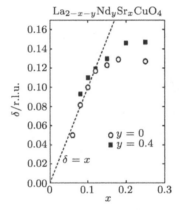

Fig. 17.7. Dependence of the magnetic incommensurability δ on doping concentration x. Taken from Ref. [2].

17.1.7. *Frequency Dependence of the Dynamic Spin Susceptibility–Spin Resonance*

Figures 17.9(a)–17.9(d) give the energy (frequency) dependence of the difference in neutron scattering intensity between the superconducting and the normal states at (π, π) on $YBa_2Cu_3O_{6.95}$, $YBa_2(Cu_{1-y}Ni_y)_3O_7$, $Tl_2Ba_2CuO_{6+\delta}$ and $Bi_2Sr_2CaCu_2O_{8+\delta}$, respectively [11]. The prominent feature in Fig. 17.9 is the appearance of a peak at certain energy. The peak

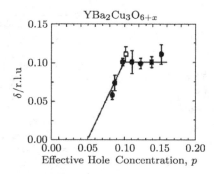

Fig. 17.8. Dependence of the magnetic incommensurability δ on effective electron concentration in $YBa_2Cu_3O_{6+x}$. Taken from Ref. [9].

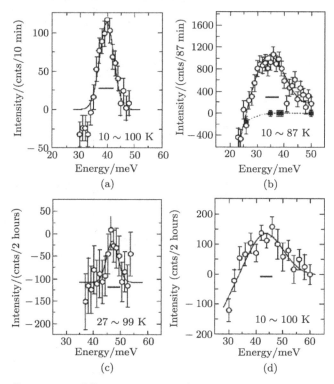

Fig. 17.9. Temperature difference spectrum of neutron scattering intensity (spin susceptibility) at the wave vector (π, π). Taken from Ref. [11].

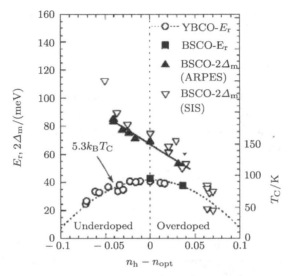

Fig. 17.10. Doping dependence of the energy of the magnetic resonance peak at (π, π) Er, and of twice the maximum of the superconducting gap, $2\Delta m$. Taken from Ref. [11].

at (π, π) in the spin excitation spectrum is called the "spin resonance", and it defines the cross-over point of the two branches in the dispersion of the spin excitations, as shown in Fig. 17.3.

It should be noted that the spin resonance only occurs in the superconducting state and in the pseudogap state for an underdoped compound. Also, the spin resonant mode has not been identified in the LaSrCuO systems to date.

The doping dependence of the spin resonance energy E_r is shown in Fig. 17.10. One can see that E_r shows an increase with doping in the underdoped regime, and then decreases in the overdoped regime, which follows roughly the doping dependence of the superconducting transition temperature T_C.

17.2. Two Theoretical Pictures for the Explanation of the Neutron Scattering Data

(1) The striped-phase scenario: this scenario is based on the observation that the superconducting state evolves from the anti-ferromagnetic insulator.

(2) The weak coupling theory: it is based on the itinerant magnetic theory with emphasis on the metallic property in the doped system.

17.2.1. *Striped-phase Scenario*

In this scenario, the doped holes organize themselves regularly into 1D charge stripes, which segregate the parent compound into a series of anti-ferromagnetic domains with the phase inverted as crossing the charge stripe, as shown in Fig. 17.11.

17.2.1.1. *Striped-phase model and Yamada plot*

In the stripe phase, the distance between the charge stripes is assumed to vary with the doping concentration. A specific illustration of the stripe structure at 1/8 doping is given in Fig. 17.12, where the red line encloses a unit cell for the charge degree of freedom and that, enclosed by the green line, denotes a unit cell for the spin degree of freedom along the direction perpendicular to the charge stripe. With this modulated magnetic structure, the anti-ferromagnetic background within each anti-ferromagnetic domain has the wave vector $Q = 2\pi(1/2, 1/2)$ and the modulated wave vector due to the charge strip is $\delta Q = 2\pi(1/8, 0)$. Hence, the observed wave vector in neutron scattering experiments should be $Q+\delta Q$, and accordingly the

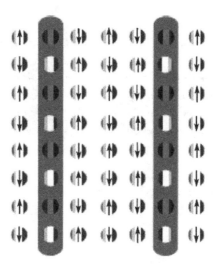

Fig. 17.11. Schematic illustration of the static stripe order for 1/8 doping.

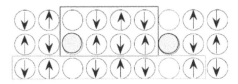

Fig. 17.12. Same with Fig. 17.11, where the red line encloses a unit cell for the charge degree of freedom and that enclosed by the green line denotes a unit cell for the spin degree of freedom along the direction perpendicular to the charge stripe.

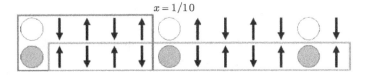

Fig. 17.13. Similar with Fig. 17.11, but for 1/10 doping.

incommensurability is $\delta = 1/8$, consistent with the Yamada relation. Therefore, the Yamada relation has been considered direct evidence of the existence of the magnetic stripe order.

Similarly, when doping concentration $x = 1/10$, the magnetic translation invariant period along the direction perpendicular to the charge stripes becomes 10 lattice constant, so $\delta = 1/10$ (see Fig. 17.13). Nevertheless, the hole density in the charge stripe will saturate when the doping concentration reaches 1/8. Further increase of doping will lead to the phase separation, where some regions still preserve the same stripe structure to that in the 1/8 doping while the excess holes are randomly distributed on other regions. Since the magnetic response is determined by the stripe order in the hole-poor phase, the modulation period will remain unchanged. This also accounts for the flat structure in the doping dependence of the incommensurability near the optimal doping (see Section 17.1.6).

17.2.1.2. *Metallic stripe phase*

The static stripe model mentioned above provides an explanation of the Yamada plot, but it gives no answers to the hourglass dispersion and the spin resonance. To account for these two phenomena in the striped-phase scenario, two groups have studied the coexistence of the stripe order and d-wave superconductivity [14], as well as the spin dynamics in the metallic stripe phase [15]. Unlike the static stripe order, the charge stripes in the metallic stripe phase fluctuate strongly so as to result in metallic behavior, and evolve into the d-wave superconducting state below T_C.

The metallic stripe phase can be described by the extended Hubbard model. In the mean-field approximation, it can be written as [14]:

$$H = -\sum_{\langle ij \rangle \sigma} (tc_{i\sigma}^+ c_{j\sigma} + \text{h.c.}) + \sum_{i\sigma} (U\langle n_{i\bar{\sigma}} \rangle - \mu) n_{i\sigma} + \sum_{\langle ij \rangle} (\Delta_{ij} c_{i\uparrow}^+ c_{j\downarrow}^+ + \text{h.c.}),$$

where U is the on-site Coulomb repulsive interaction, and Δ_{ij} is the d-wave superconducting order parameter resulting from the attractive interaction of electrons between the nearest-neighbor sites. By adjusting the chemical potential, the metallic stripe phase can be achieved.

After solving self-consistently the Bogoliubov–de Gennes equations derived from the above Hamiltonian, one can calculate the real space distribution of the charge density, the anti-ferromagnetic order as well as the superconducting order parameter, all of which are displayed in Fig. 17.14(a). As shown in the figure, the modulation period of the spin structure ($8a$) is exactly twice the modulation period of the charge distribution ($4a$), and there is a non-zero value of superconducting order parameter in this stripe structure. Figure 17.14(b) shows the momentum distribution of the single-particle spectral function, which is consistent with the ARPES experiments qualitatively.

Then, the bare spin susceptibility without the inclusion of the interaction effects can be calculated by including the normal and abnormal particle–hole contributions, as illustrated in Fig. 17.15. Due to the $8a$ period

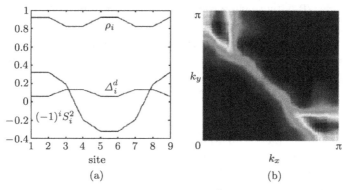

Fig. 17.14. (a) Bond centered spin density $(-1)^i s_i^z$, charge density ρ_i, and pairing potential Δ_i^d as a function of lattice site. (b) The spectral weight distribution in the k space for disordered stripes. Taken from Ref. [14].

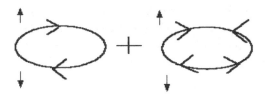

Fig. 17.15. Feynman diagram for the calculation of the bare spin susceptibility by including the normal and abnormal particle–hole contributions.

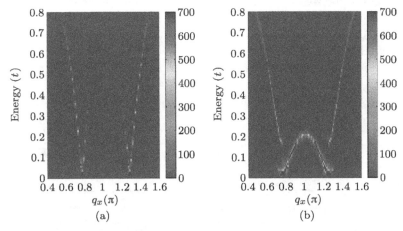

Fig. 17.16. Imaginary part of the full spin susceptibility at $q_y = \pi$ in the coexisting phase of superconducting and stripe order. (a) is for the insulating stripe and (b) the metallic stripe. Taken from Ref. [14].

of the spin modulation, the spin susceptibility should be written in the form of an 8×8 matrix, and so should the correction in the random phase approximation (RPA), i.e.

$$\chi^{+-}(q,\omega) = \chi_0^{+-}(q,\omega)[1 - U\chi_0^{+-}(q,\omega)]^{-1}.$$

The numerical results of the spin excitation spectrum are presented in Fig. 17.16 [14]. The spectrum in the stripe state without the superconducting pairing is shown in Fig. 17.16(a), where the spin wave excitation centered at (π, π) in the parent compound now splits into two branches centered at $(\pi \pm \delta\pi, \pi)$, respectively, due to the modulation of the charge stripe. Figure 17.16(b) presents the numerical result in the coexisting phase of the stripe and the d-wave superconducting phase. It shows that the including of the d-wave superconducting phase suppresses the spectral weight in the inner part of two branches, and simultaneously introduces a new branch of a downward dispersion for spin excitations below 0.2t. Thus, the overall dispersion shown in Fig. 17.16(b) reproduces the experimentally observed hourglass dispersion.

The above results indicate that the spin wave excitations in the antiferromagnetic domain contribute mainly to the high-energy excitations, while the low-energy branch comes from the particle–hole excitations in the superconducting phase even in the striped-phase model.

It is also worthwhile to point out that the spin resonance mode in this model originates from the flat portion of the dispersion around the resonant energy [15]. The flat portion in the dispersion results in an accumulation of the spectral weights of spin excitations, so that a peak will appear in the excitation spectrum (see Fig. 17.17). As for the distribution of the incommensurate peaks in momentum space, the numerical results obtained in this scenario [15] are also consistent with the experimental observations (see Fig. 17.18).

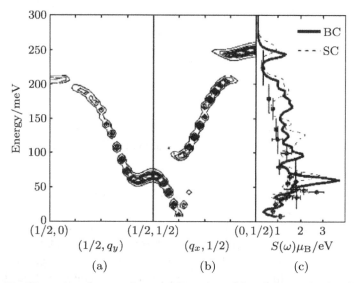

Fig. 17.17. Dispersion of magnetic excitations along (a) and perpendicular (b) to an array of stripes. Panel (c) shows $S(\omega, q)$ integrated on the magnetic Brillouin zone around (π, π) for the bond-centered (BC) and site-centered (SC) stripes together with the experimental data. Taken from Ref. [15].

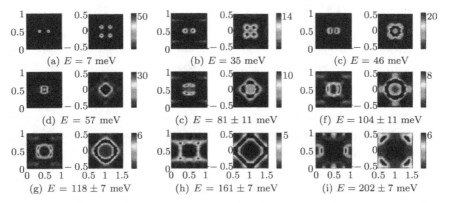

Fig. 17.18. Constant frequency scans of the imaginary part of the spin susceptibility for bond-centered stripes. Taken from Ref. [15].

17.2.2. The Weak-coupling Theory Based on the Fermiology and D-wave Superconductivity

The basic strategies of these theories are:

(1) The bare spin susceptibility that comes from particle–hole excitations is calculated with the following Feynman diagram (see Fig. 17.19):

(2) The renormalization of the spin susceptibility due to the electron–electron interactions are considered through RPA which is shown by the following Feynman diagram (see Fig. 17.20):

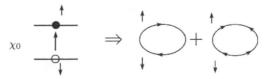

Fig. 17.19. Schematic plot for a spin-slipping particle-hole excitation and the corresponding Feynman diagrams. The line with two arrows represents the abnormal Green's function in the superconducting state.

Fig. 17.20. Feynman diagram for the renormalized spin susceptibility in the random-phase approximation (RPA).

Three critical ingredients involved in these weak-coupling calculations are: the energy-band structure ε_k of quasi-particles, the ratio Δ_0/w of the magnitude of the superconducting gap to the energy-band width, and the interacting vertex v_q in RPA. Different choices of these three ingredients can be found in various theoretical calculations. In the following, we will introduce the slave-boson mean-field approach to the $t - t' - J$ model to determine self-consistently the parameters ε_k, Δ_0, and v_q.

This approach starts from the two-dimensional $t-t'-J$ model [30]:

$$H = - \sum_{\langle ij \rangle, \sigma} t c_{i\sigma}^+ c_{j\sigma} - \sum_{\langle ij \rangle, \sigma} t' c_{i\sigma}^+ c_{j\sigma} - \text{h.c.} + J \sum_{ij} S_i S_j.$$

In the slave-boson approach, an electron operator is presented as the combination of the fermionic operator $f_{i\sigma}$ which carries the spin degree of freedom, and the slave-boson operator b_i^+, which carries the charge degree of freedom:

$$c_{i\sigma} = b_i^+ f_{i\sigma}.$$

We can choose the following mean-field order parameters:

$$\sum_\sigma \langle f_{i\sigma}^+ f_{j\sigma} \rangle = \chi_0, \quad \Delta_{ij} = \langle f_{i\uparrow} f_{j\downarrow} - f_{i\downarrow} f_{j\uparrow} \rangle = \pm \Delta_0,$$

where the signs $\pm$ are introduced to give a d-wave superconducting pairing. In the superconducting state, the bosons condense. So, we have:

$$b_i \to \langle b_i \rangle = \sqrt{\delta}.$$

Here, δ is the doping concentration.

Given the input parameters t, t', J, doping δ and temperature, one can obtain the mean-field order parameters χ_0, Δ_0 and the chemical potential μ self-consistently. The band structure ε_k is determined by χ_0 and μ. Thus, we can study systematically the evolution of the spin dynamics with the doping concentration. In RPA, the renormalized spin susceptibility is given by:

$$\chi(q,\omega) = \frac{\chi_0(q,\omega)}{1 + \alpha J \gamma_q \chi_0(q,\omega)}.$$

In this modified RPA, a phenomenological parameter α is introduced. This is because the usual RPA with $\alpha = 1$ overestimates the AF spin fluctuations. The criteria for choosing α is to set the theoretical AF instability at the doping concentration observed by experiments, such as $\delta \approx 0.02$. In the following, we show the theoretical results which are obtained by the above method.

17.2.2.1. *Incommensurate peaks along* $(\pi, \pi \pm \delta\pi)$ *direction in the momentum space*

The numerical results are presented in Figs. 17.21 and 17.22. One can find that the incommensurate peaks are along the diagonal direction when the frequency of spin fluctuations is very low. With the increase of frequency, the incommensurate peaks will move to the axial directions. In the superconducting state, the minimal energy for a particle–hole excitation is twice the superconducting gap. For a d-wave superconducting pairing, the gap has nodes along the diagonal directions, so the very low energy

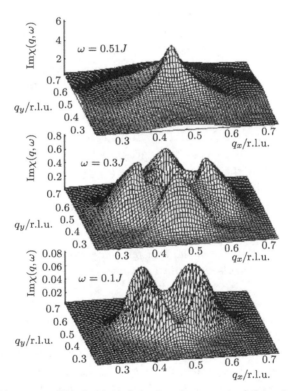

Fig. 17.21. Momentum distribution of the imaginary part of the spin susceptibility calculated via the modified RPA for the $t - t' - J$ model with doping $x = 0.12$. Taken from Ref. [16].

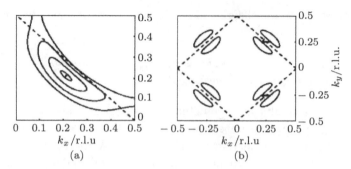

Fig. 17.22. (a) Contour plot of the superconducting quasiparticle energy in the superconducting state with doping $x = 0.12$, which is shown in the upper right 1/4 Brillouin zone (BZ). (b) Contour for $\omega = 0.3J$ and its by (π, π) shift image in the full BZ. Arrow represent the best nesting vector relative to (π, π). Taken from Ref. [16].

excitations are always along the diagonal directions. With the increase of frequency, the quasi-particles which participate in the particle–hole excitations will come from a larger momentum region. Figure 17.22(a) gives the equal-energy contour of Bogliubov quasi-particles (corresponding to $\omega = 0.1J, 0.3J, 0.51J, 0.7J$ from the smallest circle to the largest circle). The equal-energy contour contains a flat piece for $\omega = 0.3J$, allowing for a nesting contribution to the particle–hole excitations. In this case, the flat piece of the contour in the third quadrant can be moved to overlap with that in the first quadrant via a translation with the wave vector $(\pi, \pi - \delta\pi)$ ($\delta\pi$ is illustrated by the arrow in Fig. 17.22(b)). This means that there are many particle–hole excitations with the same momentum $(\pi, \pi - \delta\pi)$ which contribute to the spin response, and it results in the axial incommensurate peaks around $(\pi, \pi - \delta\pi)$.

17.2.2.2. *Spin resonance at the anti-ferromagnetic wave vector (π, π) — collective spin excitation mode*

In the weak-coupling theory, the spin resonance is explained as a collective mode of spin excitations (or spin exciton). The collective spin excitation mode occurs when the conditions $1 - |v_q|\text{Re}\,\chi_0 = 0$ and $\text{Im}\,\chi_0 \to 0$ are satisfied and thus χ diverges. The numerical results calculated based on the slave-boson mean-field approach to the t–t''–J model and the RPA approximation are shown in Fig. 17.23(a). Due to the property $\Delta_{k+Q} = -\Delta_k$ for the translation momentum Q for a d-wave superconducting pairing and

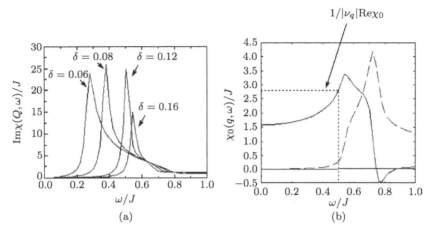

Fig. 17.23. Imaginary part of the spin susceptibility calculated via the modified RPA for the $t-t'-J$ model for different dopings δ. (b) The bare spin susceptibility as a function of ω, where the solid line denotes the real part and the dashed line the imaginary part.

the van Hove singularity near $(0, \pi)$, the imaginary part of the bare spin susceptibility has a step-like rise at the edge of the spin excitation gap which is approximately $2\Delta_0$ (see Fig. 17.23(b)). This abrupt rise of $\mathrm{Im}\chi_0$ results in the logarithmic divergence of the real part via the Kramers–Kronig relation, and this singularity shifts the collective mode energy downwards and leads it to situate in the spin gap, so no damping is expected for the mode.

17.2.2.3. *Spin excitation spectrum*

Figure 17.24 shows the calculated spin excitation spectrum using this approach [17]. In the high-energy region (above the spin-resonance energy), the incommensurate spin excitations come from the heavily damped particle–hole excitations, while the low-energy spin fluctuations arise from well-defined excitations. Figure 17.25 shows the experimental results [2]. We can find that the theoretical results exhibit an hourglass dispersion, which is in agreement with experiments.

It is noticed that the incommensurate peaks at frequencies above the spin-resonance energy rotate 45° with respect to those at low frequencies below the spin-resonance energy. This is also consistent with the experimental result.

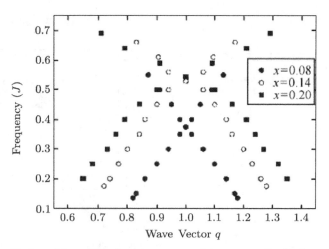

Fig. 17.24. Peak positions of the imaginary part of the renormalized spin susceptibility in the momentum space as a function of frequency calculated via the modified RPA for the $t-t'-J$ model for dopings $x = 0.08$, 0.14, and 0.20. Taken from Ref. [17].

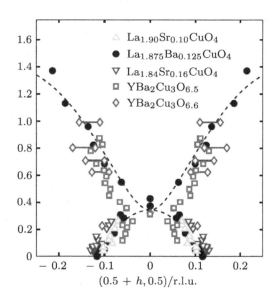

Fig. 17.25. The same as Fig. 17.3.

17.2.2.4. Yamada plot — δ (incommensurability) = $2x$ (doping concentration)

As discussed above, in the slave-boson mean-field theory of the $t-J$ model, we can calculate the doping dependence of the mean-field parameters self-consistently. This facilitates our study of the dependence of the incommensurability on the doping concentration. As shown in Fig. 17.26(a), the numerical results at low frequencies are in good agreement with the Yamada plot [17].

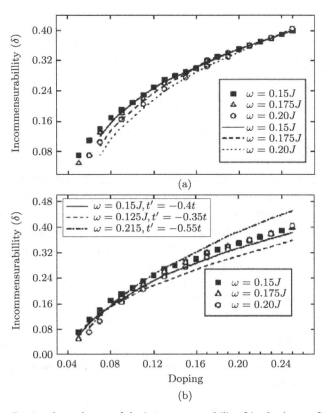

Fig. 17.26. Doping dependences of the incommensurability δ in the $(\pi, \pi + \delta\pi)$ direction for different frequencies calculated via the modified RPA for the $t - t' - J$ model. (a) A comparison for different adjusting parameters in the modified RPA. (b) A comparison for different t'. For details, please see Ref. [17].

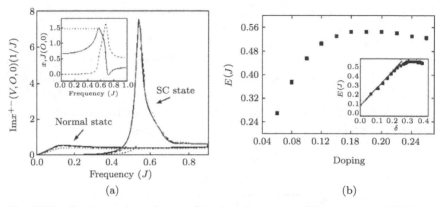

Fig. 17.27. Frequency dependences of the imaginary part of the spin susceptibility at the superconducting and normal states calculated via the modified RPA for the $t-t'-J$ model. Inset shows the bare spin susceptibility. (b) Doping dependence of the resonance energy. The inset of (b) shows the ratio between the incommensurability and resonance energy, and the solid line is a guide to the eye. Taken from Ref. [17, 31].

17.2.2.5. *Temperature and doping dependences of the spin-resonance mode energy*

Figure 17.27(a) presents the imaginary part of the spin susceptibility in the superconducting and the normal states. In the superconducting state, the spin susceptibility shows a sharp resonance peak. However, above T_C, the resonance peak disappears and the spin susceptibility shows a broad hump structure. Figure 17.27(b) presents the doping dependence of the spin-resonance mode energy. The resonance mode energy increases with the increase of doping below optimal doping, and reaches the highest value around the optimal doping. With further increase of doping, the resonance mode energy exhibits a slow decrease. This is also in good agreement with experimental results.

17.2.2.6. *Anisotropy of the spin incommensurate structure in detwinned $YBaCuO_{6.6}$*

As mentioned above, the spin fluctuations associated with a stripe phase are expected to be one-dimensional. Thus, the corresponding magnetic incommensurate peaks also exhibit the 1D structure. However, all neutron scattering experiments we have discussed up to now show that the spin fluctuations display a 2D symmetry with a fourfold pattern at the incommensurate points around the magnetic reciprocal lattice position (π, π). One explanation is that this apparent two-dimensionality results from measurements on twinned crystal where stripes alternate in direction as the planes are stacked along the c axis.

To resolve this issue, a magnetic measurement on an untwinned crystal is needed. Mook *et al.* have carried out this kind of neutron scattering measurements on a partially detwinned YBCO sample [22]. Figures 17.28(a)–17.28(b) are their experimental results. They show a strong anisotropy in spin fluctuations along the crystal a and b directions, which is suggested to be 1D in nature. Thus, it is considered to give a direct experimental support on the striped-phase model.

We have examined the possibility of the anisotropic spin fluctuation in the framework of weak-coupling theory, and find that the coupling between the CuO chain and the CuO_2 plane can result in similar anisotropy of spin fluctuations [23]. Therefore, we provide an alternative explanation of the anisotropic spin fluctuations other than the striped-phase model.

In the calculation, we consider the following two ingredients:

(1) The anisotropy between the length of a and b axis in the orthogonal lattice.

YBCO has an orthogonal lattice structure, and the asymmetry of the crystal lattice between the a and b axis is $b/a = 1.015$. This results in the asymmetry of the energy-band and superconducting gap, i.e. $t_x = (1+c)t, t_y = (1-c)t, J_x = (1+c)^2 J, J_y = (1-c)^2 J$, and $c = 0.03$. According to the LDA band calculation, the ratio between the hopping integrals along the a and b directions scales roughly as $t_x/t_y \approx (b/a)^4$.

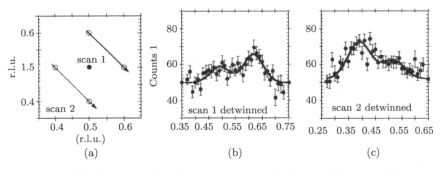

Fig. 17.28. (a) The arrows show the two scans used in the experiment, and the open circles denote the antiferromagnetic wave vectors. (b) and (c) Scans of the magnetic incommensurate scattering for the twinned and partially detwinned sample taken at 10K. Taken from Ref. [22].

(2) The coupling between the chain and plane.

We notice that there is a quasi-1D CuO chain structure in YBCO. It will couple to the CuO_2 plane and lead to another anisotropy of the electronic structure in the plane between the a and b directions. Thus, we start with a Hamiltonian which describes a system with two layers per unit cell. One layer represents a CuO_2 plane and is described by the 2D $t-t'-J$ model. The other represents a CuO chain:

$$H = H_{t-t'-J} + H_c + H_I,$$

$$H_c = -\sum_{i,\alpha,\sigma} t_c d^+_{i\sigma} d_{i+\hat{y}\sigma} - \text{h.c.},$$

$$H_I = -t_\perp \sum_{i,j,\sigma}(c^+_{i\sigma} d_{j\sigma} + \text{h.c.})$$

$$-\lambda/4 \sum_{i,\alpha}(\hat{\Delta}^+_{1,i,\alpha}\hat{\Delta}_{2,i+c/2,\alpha} + \hat{\Delta}^+_{1,i,\alpha}\hat{\Delta}_{2,i-c/2,\alpha} + \text{h.c.}).$$

The numerical results are presented in Fig. 17.29. Without the coupling between the chain and plane, the incommensurate peaks are asymmetric, arising from the asymmetry in the hopping integrals along a and b directions. However, in this case the stronger incommensurate peak is along the (π, q), i.e. the k_y direction. This is not consistent with the experimental result as discussed above. However, if the effect of the coupling between the plane and chain is considered, the stronger incommensurate peak will turn to the k_x direction, as shown in Fig. 17.29(b). Therefore, we conclude that

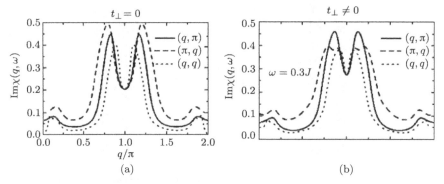

Fig. 17.29. Imaginary part of the spin susceptibilities as a function of wave vector q for a zero (a) and nonzero (b) hopping parameter between the plane and chain. Taken from Ref. [23].

the $a-b$ anisotropy of the neutron scattering experiments on untwinned YBCO can be explained when both the coupling between the plane and chain and the asymmetry of the crystal lattice between a and b axis are considered. Essentially, the 1D feature of the incommensurate neutron peak is ascribed to the effect of the Fermi surface nesting in this kind of weak-coupling calculations.

We note that spin fluctuations exist in both the k_x and k_y directions. In our case, it is the intensity and width of the peaks that are asymmetric. In this sense, the spin fluctuations are 2D in nature. This is different from the stripe model in which the incommensurate peaks are 1D. Later on, Hinkow et al. have carried out the neutron scattering measurement in the fully detwinned samples [24], and shown that the spin fluctuations exhibit the anisotropic 2D structure. It is consistent with our theoretical results.

We can understand the above theoretical results of the anisotropic spin fluctuations based on the nesting of the Fermi surface. Due to the coupling between the Cu_2O plane and chain, the equal-energy contour is not symmetric with respect to the diagonal direction (see Fig. 17.30). Therefore, that part of the nesting pieces along the k_x direction is larger than that along the k_y directions. Thus, the intensity of the incommensurate peak along the k_x direction is larger than the latter.

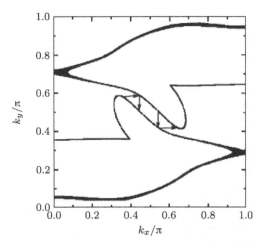

Fig. 17.30. Contour plot of the superconducting quasiparticle excitation energy $E_k = 0.3J$, and its image shifted by (π, π). The arrows denote the best nesting vectors relative to (π, π). Taken from Ref. [23].

17.2.2.7. *Spin dynamics in electron-doped high-T_C cuprates*

Neutron scattering experiment shows that the spin response in electron-doped high-T_C superconductors exhibits a commensurate peak in a large range of energy [25]. Based on the slave-boson approach to the t–t'–J model and the random phase approximation to the spin response, we have obtained the theoretical results for the electron-doped system which is consistent with experiments [26] (as shown in Fig. 17.31(b)). Here, because the Fermi surface moves close to the boundary of the magnetic Brillouin zone with electron doping, the low-energy spin response arising from the particle–hole excitations near the Fermi levels will be commensurate (see Fig. 17.31(a)).

In the meantime, we predict theoretically that a spin resonance exists in the electron-doped high-T_C superconductors (as shown in Fig. 17.32). A similar resonance mode has been observed in electron-doped compounds by neutron scattering experiments [27, 28] (see Fig. 17.33).

We also note that it has been found experimentally that the superconducting pairing symmetry in several electron-doped compounds may not be pure-wave, namely there is a higher harmonics component $\cos(6\theta)$ in addition to the dominant d-wave term. Based on the RPA approximation, Ismer *et al.* [27] have obtained similar theoretical results to ours after considering this non-monotonic superconducting pairing symmetry [29] (as shown in Fig. 17.34).

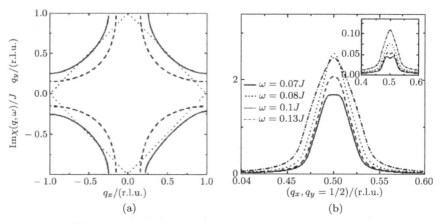

Fig. 17.31. (a) Fermi surface at doping concentration $x = 0.15$ for electron doping and the dashed line denotes that for hole doping. Taken from Ref. [26].

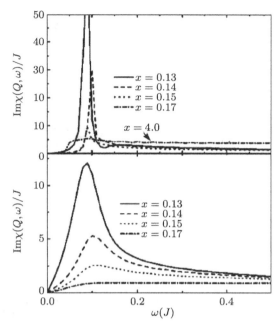

Fig. 17.32. Frequency dependence of the imaginary part at the antiferromagnetic wavevector for different dopings with the quasiparticle damping $0.004J$ (a) and $0.04J$ (b). For details, see Ref. [26].

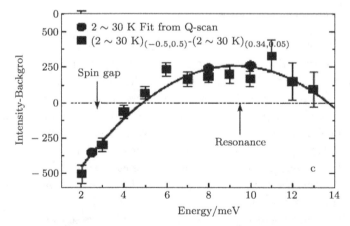

Fig. 17.33. Temperature difference plot of the neutron scattering on $Nd_{1.85}Ce_{0.15}CuO_{4-\delta}$ showing the resonance at $Er = 9.5 \pm 2$ meV. Taken from Ref. [28].

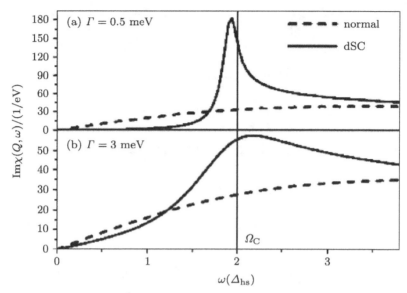

Fig. 17.34. Frequency dependence of the imaginary part of the spin susceptibility in the normal and superconducting states for electron-doped cuprates. For details, see Ref. [29].

17.3. Comparison between the Striped-Phase Scenario and the Weak-Coupling Theory

17.3.1. Universal Dispersion

Let us first look at how these two theories interpret the so-called hourglass dispersion observed in neutron scattering experiments. In fact, we may look at the hourglass dispersion separately. For the upward part above the spin resonance, it arises from the spin-wave excitations of the local magnetic moments in the striped-phase scenario, while in the weak-coupling theory it comes from the overdamped collective spin excitations. Because this branch of dispersion extends to nearly 100 meV in experiments, it is hard to consider it coming from the usual particle–hole excitations in the weak-coupling theory, instead it is most likely due to the spin-wave excitations of the local magnetic moments as described in the striped-phase scenario. As for the downward low-energy dispersion, it may be due to the itinerant magnetism which is well captured by the weak-coupling calculations. Though the downward dispersion is reproduced by both the striped-phase scenario and weak-coupling theory (see Section 17.2.1.2), one will find that the main contributions to the excitations arise from the energy-band structure and d-wave superconductivity.

17.3.2. Yamada Plot and Anisotropic Spin Fluctuations

Yamada plot shows an equality between the incommensurability and the doping concentration which follows naturally from the static stripe model, so it has been taken as demonstrating strong support for the striped-phase scenario. However, as discussed above, the weak-coupling calculations have also reproduced the Yamada plot and the anisotropic spin fluctuations.

17.3.3. Origin of the Spin Resonance

In the framework of the static stripe model (spin-only model), we have not been able to find an explanation for the spin resonance up to now. Even with a possible explanation, it is still hard to understand why the spin resonance appears in the superconducting state.

For the metallic stripe, as discussed in Section 17.2.1.2, the spin resonance is due to the accumulation of the spectral weights of spin excitations. In this sense, it does not correspond to a well-defined excitation mode, instead it is a special area in the whole dispersion.

In the weak-coupling theory, the spin resonance is ascribed to a collective spin excitation mode (or a spin exciton) . Experiments show that the spin resonance more or less (depends on the compounds where it is observed) corresponds to well-defined excitations.

17.3.4. Incommensurate Peaks along Diagonal Directions at Very Low Dopings (0.02−0.05)

In the stripe phase, the incommensurate spin peaks are ascribed naturally to the modulated stripe structure along the diagonal directions at very low dopings. In the weak-coupling calculations, no explanation for this phenomenon is found. In fact, the weak-coupling calculation does not apply to the description of the physics at very low dopings, because the system has entered into (or is very close to) the insulating phase.

17.4. Concluding Remarks

From the comparisons of the results obtained based on the two kinds of models of experiments, one can see that it is impossible to select or rule out any of the above models up to now. There is also a possibility that the two models are different limits of a more complete theory that properly includes the full nature of both the spin and charge degrees of freedom.

One of the important issues for further theoretical studies is to find a "smoking-gun" character in the spin excitations which can be used to discriminate these theoretical models, or to understand how they may be related to each other.

Acknowledgments

I thank Dr. H. M. Jiang and S. L. Yu for their help.

References

[1] J. Orenstein and A. J. Millis, *Science* **288**, 468 (2000).
[2] J. M. Tranquada, *Handbook of High-Temperature Superconductivity: Theory and Experiment*, Chapter 6 (Spnnger New York, 2007).
[3] S. M. Hayden et al., *Nature* **429**, 531 (2004).
[4] J. M. Tranquada et al., *Nature* **429**, 534 (2004).
[5] R. Coldea et al., *Phys. Rev. Lett.* **86**, 5377 (2001).
[6] S. M. Hayden et al., *Phys. Rev. B* **54**, R6905 (1996).
[7] C. Stock et al., *Phys. Rev. B* **71**, 024522 (2005).
[8] K. Yamada et al., *Phys. Rev. B* **57**, 6165 (1998).
[9] P. Dai et al., *Phys. Rev. B* **63**, 054525 (2001).
[10] N. Ichikawa et al., *Phys. Rev. Lett.* **85**, 1738 (2000).
[11] Y. Sidis et al., *Phys. Stat. Sol.* **241**, 1204 (2004).
[12] J. Zaanan and O. Gunnarson, *Phys. Rev. B* **40**, 7391 (1989).
[13] J. M. Tranquada et al., *Nature* **375**, 561 (2005).
[14] B. M. Andersen and P. Hedegard, *Phys. Rev. Lett.* **95**, 037002 (2005).
[15] G. Seibold and J. Lorenzana, *Phys. Rev. Lett.* **94**, 107006 (2005).
[16] J. Brinckmann and P. A. Lee, *Phys. Rev. Lett.* **82**, 2915 (1999).
[17] J. X. Li and C. D. Gong, *Phys. Rev. B* **66**, 014506 (2002).
[18] D. Z. Liu et al., *Phys. Rev. Lett.* **75**, 4130 (1995).
[19] N. Bulut et al., *Phys. Rev. B* **53**, 5149 (1996).
[20] J. X. Li et al., *Phys. Rev. B* **58**, 2895 (1998).
[21] M. R. Norman, *Phys. Rev. B* **63**, 092509 (2001).
[22] H. A. Mook et al., *Nature* **404**, 729 (2000).
[23] T. Zhou and J. X. Li, *Phys. Rev. B* **69**, 224514 (2004).
[24] V. Hinkov et al., *Nature* **430**, 650 (2004).
[25] K. Yamada et al., *Phys. Rev. Lett.* **90**, 137004 (2003).
[26] J. X. Li et al., *Phys. Rev. B* **68**, 224503 (2003).
[27] S. D. Wilson et al., *Nature* **442**, 59 (2006).
[28] J. Zhao et al., *Phys. Rev. Lett.* **99**, 017001 (2007).
[29] J. P. Ismer et al., *Phys. Rev. Lett.* **99**, 047005 (2007).

18
Slave-Boson Effective-Field Theory of RVB State and Its Application for the Mechanism of High-Temperature Superconductivity

Tao Li

Department of Physics, Renmin University of China, Beijing, 100872

We present a review of the slave-boson mean-field theory and the related effective gauge-field theory of the RVB state and its application in the mechanism of high-temperature superconductivity. We also introduce the recent progress in RVB theory made possible by the combination of the effective-field theory approach and the variational approach based on Gutzwiller projected wave functions.

18.1. Background

This chapter aims to review the current status of the resonating valence bond theory and the related effective-field theory approach based on slave-boson representation, which is considered one of the most important schools of thoughts on the mechanism of high-temperature superconductivity. Most of the contents in this chapter have already appeared in previous papers [1, 2]. At the same time, we will discuss the close relationship between the effective field theory approach and the Gutzwiller projected wave function study and their application to the problems of high-temperature superconductivity [3–5].

The effective-field theory approach is widely used in the study of the mechanism of high-temperature superconductivity. Some important examples in the history of high-T_C research includes anyonic superconductivity [6], mutual Chern–Simons theory based on Schwinger boson representation [8], SU(2)$\otimes$ U(1) Chern–Simons theory [7] based on slave Fermion representation, phenomenological Z_2 gauge-field theory [9] and QED$_3$ gauge-field theory [10], and the U(1) and SU(2) gauge theory [1, 2] based on slave-boson representation that will be reviewed in this paper. All these effective theories share a common characteristic in that the gauge field fluctuation plays a crucial role in the description of the ground state structure and the low-energy excitation spectrum. This is related to the strongly correlated nature of high-T_C problem.

We will start our discussion from the following well-known debating issue about the mechanism of high-T_C superconductivity: is some bosonic mode needed to induce the electron pairing in high-T_C cuprates as in the conventional superconductors? Or does the high-T_C superconductivity need some pairing glue? From conventional many-particle physics, superconductivity results from Bose condensation of Cooper pairs, which again is the result of the effective attractive interaction induced by exchanging some kind of bosonic excitation between electrons in a Fermi liquid background. While from the strong correlation perspective of high-T_C superconductivity, for example, in the so-called resonating valence bond (RVB) theory, no such glue is needed to induce electrons pairing, the electrons involved in the pairing may not be described by the Fermi liquid theory. Studies along both of these lines have contributed many useful understandings about high-T_C superconductivity. Here, we will focus on the strong correlation perspective of the high-T_C problem. There are three reasons for such a choice:

(1) The electron in the normal state of a high-T_C superconductor exhibits strong electron incoherence. This is especially obvious in the measurement of transport properties such as optical conductivity, Raman spectrum and so on. These anomalies, which are called non-Fermi liquid behaviors, tell us that we cannot interpret high-T_C superconductivity as the Cooper pairing of pre-existing quasi-particle as in Fermi liquid theory. There is simply no well-defined quasi-particle in the normal state in high T_C cuprates.
(2) In both its normal state and the superconducting state, the high-T_C superconductor exhibits unique and strong low-energy spin fluctuation.

The appearance of strong spin fluctuation in low-energy regime is a direct evidence of the strong correlation nature of the system. Although the RPA theory can also provide some understanding about the spin dynamics of the cuprates, a self-consistent and microscopic theory is impossible from the conventional many-body theory approach.

(3) High-T_C superconductors are so-called doped Mott insulators. For example, the superfluid density is approximately proportional to the density of the doped holes, which is much smaller than the prediction of the band theory. Any theory starting from weak correlation picture will have great difficulty accounting for this fact, while in the effective theory based on slave particle representation this is easily explained.

18.2. Doped Mott Insulator and RVB State

The following discussion will be based on the $t-J$ model, which is believed to describe the physics of a doped Mott insulator. The model reads:

$$H = -t \sum_{\langle i,j \rangle, \sigma} (\hat{c}^\dagger_{i,\sigma} \hat{c}_{j,\sigma} + \text{h.c}) + J \sum_{\langle i;j \rangle} (\bm{S}_i \cdot \bm{S}_j - \frac{1}{4} n_i n_j), \tag{18.1}$$

in which the electron operator $\hat{c}_{i,\sigma}$ satisfies the following all important constraint:

$$\sum_\sigma \hat{c}^\dagger_{i,\sigma} \hat{c}_{i,\sigma} \leq 1 \tag{18.2}$$

which indicates no double occupancy. This constraint is the key to the strong electron correlation and Mott physics. $S_i = \frac{1}{2} \sum_{\alpha,\beta} \hat{c}^\dagger_{i,\sigma} \bm{\sigma}_{\alpha,\beta} \hat{c}_{i,\beta}$ describes local spin, $n_i = \sum_\sigma \hat{c}^\dagger_{i,\sigma} \hat{c}_{i,\sigma}$ is the electron number. At half-filling, when each site is occupied by one and only one electron, any electron transport will necessarily induce doubly occupied site, which is prohibited by the constraint. Thus, the $t-J$ model describes an insulator at half-filling. Such an insulator, which is caused by correlation effect rather than Pauli exclusion principle, is called a Mott insulator.

At half-filling, $t-J$ model reduces to the spin-1/2 Heisenberg model and its ground state is long-range ordered. If we choose $t-J$ model to describe the high-T_C superconductor, a natural question is what is the state of these local spins when the magnetic long-range order is erased by the doped mobile holes (for example in the superconducting state). In his pioneering work, Anderson suggested the concept (and the term) of spin liquid. He argued that the local spins should stay in a liquid-like state without any conventional symmetry breaking and ordering pattern. He suggested further that such a state can be described by the so-called resonating valence bond wave function. The RVB state is a state composed of coherent superposition of different singlet pairing patterns between the pairs of local spins in the system:

$$|\text{RVB}\rangle = \sum_{\text{dimer coverings}} a[i_1, j_1, \ldots i_{N/2}, j_{N/2}] \prod_{k=1}^{N/2} (i_k, j_k), \quad (18.3)$$

in which the sum is over all possible pairing patterns (or all possible dimer covering patterns). $(i,j) = (1/\sqrt{2})|\uparrow_i\downarrow_j\rangle - |\uparrow_j\uparrow_i\rangle)$ denotes a singlet pair between site i and site j, $a[i_1, j_1, \ldots, i_{N/2}, j_{N/2}]$ is amplitude related to a given pairing pattern.

At half-filling, the RVB wave function can describe a disordered spin state without any conventional symmetry-breaking, i.e. a Mott insulator. When mobile holes are doped into such a system, the singlet spin pairs will acquire the ability of coherent motion in the charge channel and thus be transformed into real Cooper pairs that can carry supercurrent. The system thus enters the superconducting state. Such a deep connection between quantum magnetism and high-T_C superconductivity is encoded naturally in Anderson's projective construction of RVB wave function for the high-T_C cuprates. Realizing the close similarity between Cooper pair and the RVB singlet, he suggested to construct the RVB wave function by Gutzwiller projection of the BCS wave function of conventional superconductors. The BCS wave function is well known to have the form of a Bose condensation of Cooper pair:

$$|\text{BCS}\rangle = \prod_k (u_k + v_k c_{k\uparrow}^\dagger c_{-k\downarrow}^\dagger)|0\rangle$$

$$\propto \exp\left(\sum_{i,j} a(i-j) c_{i\uparrow}^\dagger c_{j\downarrow}^\dagger\right)|0\rangle, \quad (18.4)$$

in which:

$$a(i-j) = \sum_k \frac{v_k}{u_k} \exp(ik(R_i - R_j)) \quad (18.5)$$

is the real space wave function of the Cooper pair. The BCS wave function violates the no double occupancy constraint, however, we can enforce such a constraint through Gutzwiller projection procedure which results in a many-body singlet state. Especially, at half-filling, we will arrive at a singlet wave function describing the local spins.

$$|\text{RVB}\rangle = P_G|\text{BCS}\rangle = \sum_{\text{dimer coverings}} \prod_{k=1}^{N/2} a(i_k - j_k)[i_k, j_k], \quad (18.6)$$

in which the Gutzwiller projector has the form $P_G = \prod_i (1 - n_{i\uparrow} n_{i\downarrow})$. $[i, j] = (c_{i\uparrow}^\dagger c_{j\downarrow}^\dagger + c_{j\uparrow}^\dagger c_{i\downarrow}^\dagger)|0\rangle$ which is the spin singlet written in the form of Fermionic operators.

18.3. Slave-Boson Representation and the Mean-Field Theory of $t-J$ Model

Although it is interesting to describe high-T_C physics with the RVB wave function directly, it proves hard to do analytical calculations on such a wave function due to the no double occupancy constraint. We thus have to rely on approximate methods. The slave-boson mean-field theory is most frequently used and is the most successful approach in this respect. In the slave-boson representation, we rewrite the constrained electron operator $\hat{c}_{i,\sigma}$ (note that as a result of the no double occupancy constraint, $\hat{c}_{i,\sigma}$ does not respect the usual anti-commutation relation between Fermionic operators) in terms of a composite made of a Fermionic operator and a Bosonic operator:

$$\hat{c}_{i,\sigma} = f_{i,\sigma} b_i^\dagger, \quad (18.7)$$

in which $f_{i,\sigma}$ denotes a spin-$\frac{1}{2}$ neutral Fermion called spinon, $b_i^\dagger$ denotes a spinless charged Boson called holon. They satisfy the standard anti-commutation and commutation relation for Fermions and Bosons. Such a form may imply that the spin and charge quantum number of the electron are carried by two different kinds of particles. However, before we can prove that the spinon and the holon can execute independent motion, we can only take it as a pure mathematical representation. There is one obvious advantage to introducing the slave particles: the no double occupancy constraint, which is an inequality in the original electron representation, now becomes an equality:

$$\sum_\sigma f_{i,\sigma}^\dagger f_{i,\sigma} + b_i^\dagger b_i = 1. \quad (18.8)$$

It is easy to verify that both the commutation relation between the constrained electron operators and the Hilbert space they span are exactly reproduced when the above constraint is satisfied. The slave-boson representation is thus a faithful representation of the original system.

18.3.1. Half-filled System

At half-filling, the $t-J$ model reduces to the Heisenberg model. Although we know the Heisenberg model on square lattice has magnetic long-range order in its ground state, a study of the spin liquid state in this case is still of great importance, as we will see what follows. Roughly speaking, the spin liquid state can be stabilized by finite hole doping.

In the slave-boson representation, the Heisenberg model reads:

$$H = \frac{J}{4} \sum_{\langle i,j \rangle} f^\dagger_{i,\alpha} \boldsymbol{\sigma}_{\alpha,\beta} f_{i,\beta} \cdot f^\dagger_{j,\gamma} \boldsymbol{\sigma}_{\gamma,\delta} f_{j,\delta}, \quad (18.9)$$

in which $\boldsymbol{\sigma}$ denotes the Pauli matrices. The constraint becomes:

$$\sum_\sigma f^\dagger_{i,\sigma} f_{i,\sigma} = 1. \quad (18.10)$$

In the mean-field approximation, we can relax this constraint into a requirement on the mean value of the occupation number, which can be enforced by a chemical potential term. We can also decouple the interaction term with the RVB mean-field order parameters. Here, we will introduce two kinds of RVB order parameters in the singlet channel:

$$\chi_{i,j} = \sum_\sigma \langle f^\dagger_{i,\sigma} f_{j,\sigma} \rangle, \quad \Delta_{i,j} = \sum_\sigma \langle \epsilon_{\sigma,\bar{\sigma}} f_{i,\sigma} f_{j,\sigma} \rangle. \quad (18.11)$$

We note that special care should be taken to interpret χ and Δ as order parameters. Under the gauge transformation $f_{i,\sigma} \to f_{i,\sigma} e^{i\phi_i}$, $b_i \to b_i e^{i\phi_i}$, both of them are not invariant and thus cannot have direct physical meaning. In fact, the two quantities transform in the following way under the gauge transformation:

$$\chi_{i,j} \to \chi_{i,j} e^{i(\phi_i - \phi_j)}, \quad \Delta_{i,j} \to \Delta_{i,j} e^{i(\phi_i + \phi_j)}. \quad (18.12)$$

To simplify the discussion, we ignore the order parameter Δ for a moment. At the same time, we will assume that the magnitude of χ is uniform. The phase of χ, on the other hand, is in general non-uniform as a result of the presence of the gauge flux. Such gauge flux is defined as the

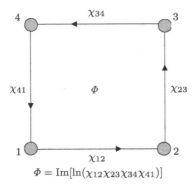

$$\Phi = \text{Im}[\ln(\chi_{12}\chi_{23}\chi_{34}\chi_{41})]$$

Fig. 18.1. U(1) gauge flux. Although the phase of χ is not gauge invariant, the phase of product of all the χ along a closed loop, Φ, is gauge invariant. Such gauge flux can be used to classify the structure of the RVB mean-field state. At the same time, the fluctuation of these gauge flux away from their mean-field value constitute the fluctuating gauge degree of freedom of the system, which plays an important role in the low-energy effective theory.

phase of the product of all the χ along a closed loop. It can be easily seen that this gauge flux is gauge invariant under the U(1) gauge transformation (Fig. 18.1). When the magnitude of χ is fixed, the gauge flux is the only quantity to classify the different mean-field states. At the same time, the fluctuation of Φ from its mean-field value constitutes an important fluctuation mode of the system in the low-energy regime. Such fluctuation is called the gauge fluctuation as it is coupled to the slave particles in the same way as a gauge field. The gauge fluctuation around the mean-field state not only determines the stability of the mean-field itself, but also may induce exotic physical effects such as the non-Fermi liquid behavior. For these reasons, the effective-field theory of the $t-J$ model built on the slave-boson representation is often called the effective gauge-field theory of the the high-T_C superconductivity.

When the flux $\Phi = 0$ or π, the mean-field state is time-reversal symmetric. In this chapter, we will concentrate on such a time-reversal symmetric state, although in the early age of the high-T_C study, anyonic mechanism for high-T_C superconductivity based on time-reversal symmetry breaking had attracted a lot of attention. The state with $\Phi = 0$ is the so-called uniform-RVB and the state with $\Phi = \pi$ is called the π flux state. These two states are of central importance for the RVB theory of high-T_C superconductivity. The dispersion of the uniform-RVB state is given by:

$$E_k = -2\chi(\cos(k_x) + \cos(k_y)). \tag{18.13}$$

This is just the dispersion of the free particle hopping on square lattice. It has large Fermi surface. The uniform RVB state is the starting point for the description of the strange metal phase and the Fermi liquid phase in the phase diagram of high-T_C superconductors. The dispersion relation of the π-flux phase is given by:

$$E_k = \pm\chi\sqrt{\cos^2(k_x) + \cos^2(k_y)}. \tag{18.14}$$

The excitation spectrum is gapless at the four momentums in the Brillouin zone $((\pm\pi/2, \pm\pi/2))$. Around these nodes, the dispersion is linear, identical to the dispersion in a d-wave superconductor. In fact, the π-flux and the more general staggered-flux that will be introduced below is the starting point for the description of the d-wave superconducting state and the pseudogap phase in the high-T_C cuprates.

18.3.2. *The Physical Meaning of the Gauge Field*

Before presenting the theory in more detail, we first provide a physical understanding of the gauge fluctuation in RVB theory. The gauge degree of freedom is non-local in nature and redundancy is required if we choose to describe it with local degrees of freedom. In quantum theory, the Aharanov–Bohm effect is one typical manifestation of such non-locality of the electromagnetic field while the redundancy is manifested in the choice of gauge. Since the gauge field acts in a similar way as the Berry phase, we can also take the electromagnetic field as some kind of Berry phase structure in space–time. Such an analog is useful for the understanding of the origin of gauge field in RVB theory.

The gauge field in RVB theory can be taken as some kind of Berry phase of the system. Usually, a non-trivial Berry phase is always related to the projection to some subspace. One typical example is Berry phase in the momentum space for band insulators (also called the Chern number). A more conventional example is provided by the change of direction when a vector is parallelly transported on a curved surface along a closed loop. Figure 18.2 shows a parallel transport on sphere and the related change of direction. What is important here is the projection of the vector onto the local tangential plane of the curved surface. It is just such a projection that renders the change of direction of a vector dependent on the loop along which the vector has traveled. In this example, the change of the direction of

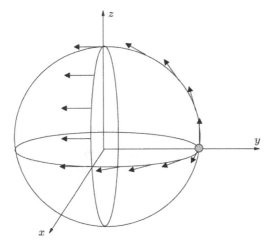

Fig. 18.2. The parallel transport on sphere and the related change of direction. Here, the parallel transport is along three segments of great circles. The shaded circle indicates the starting and end point of the closed loop. In fact, the change of direction has nothing to do with the choice of the starting point and the initial direction of the vector. It only depends on the path along which the vector is transported and is thus an intrinsic property of the sphere.

the vector is defined by the Berry phase related to the parallel transport. It plays the same role as the gauge flux in gauge-field theory.

From this example, we see the non-local nature of the gauge field (or Berry phase) and that the redundancy involved in the description is always related to the projection to some subspace. It is then easy to understand why gauge field emerges in the RVB theory. In the mean-field treatment of the RVB theory, the no double occupancy constraint is relaxed to an equality on the average occupation number. The Hilbert space is thus dramatically enlarged (now we have four states per site rather than two states per site). This will introduce huge amounts of redundant degrees of freedom. To recover the physical Hilbert space we have to project back to the subspace of no double occupancy. The gauge field emerges in just such a projection.

However, the slave particle representation is actually an identical representation of original physical system. It cannot introduce any physics that is absent in the original electron model. Does the gauge field represent some intrinsic degree of freedom of the original electron system? The answer is yes. To be clear, we will first discuss the physical meaning of the gauge flux involved in a loop made of three sites. In the triangular loop shown in

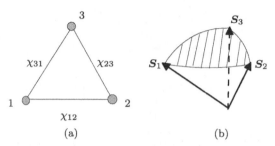

Fig. 18.3. The relation between gauge field flux and the spin chirality. (a) The triangular loop; (b) The solid angle subtended by the three spins along the loop.

Fig. 18.3, we can construct the following two gauge invariant quantities:

$$\begin{cases} P_{123} = \langle \widehat{\chi}_{12}\widehat{\chi}_{23}\widehat{\chi}_{31} \rangle = \langle f_{1,\alpha}^\dagger f_{2,\alpha} f_{2,\beta}^\dagger f_{3,\beta} f_{3,\gamma}^\dagger f_{1,\gamma} \rangle, \\ P_{132} = \langle \widehat{\chi}_{13}\widehat{\chi}_{32}\widehat{\chi}_{21} \rangle = \langle f_{1,\alpha}^\dagger f_{3,\alpha} f_{3,\beta}^\dagger f_{2,\beta} f_{2,\gamma}^\dagger f_{1,\gamma} \rangle \end{cases} \quad (18.15)$$

(in which the summation over repeated index is assumed). Then, it is easy to show that:

$$P_{123} - P_{132} = 4\mathrm{i}\langle \boldsymbol{S}_1 \cdot (\boldsymbol{S}_2 \times \boldsymbol{S}_3) \rangle. \quad (18.16)$$

Here, $\langle \boldsymbol{S}_1 \cdot (\boldsymbol{S}_2 \times \boldsymbol{S}_3) \rangle$ is quantity with clear physical meaning. It is called spin chirality and it can be interpreted as the solid angle that the three spins subtended on the sphere. Thus, if we substitute $\widehat{\chi}_{ij} = f_{i,\alpha}^\dagger f_{j,\alpha}$ with its mean-field average χ_{ij}, then the gauge flux deduced from the phase of χ_{ij} along the loop is nothing but the spin chirality, or, $\sin \Phi_{123} = 2\langle \boldsymbol{S}_1 \cdot (\boldsymbol{S}_2 \times \boldsymbol{S}_3) \rangle$. For this reason, the gauge fluctuation in RVB theory is also known by some as the fluctuation of spin chirality.

18.3.3. *The Staggered-Flux Phase and the SU(2) Gauge Structure*

In the discussion in the subsection preceding, the RVB order parameter Δ is ignored. However, such order parameter can play an important role in the physics of RVB state. For example, when the flux value ϕ is neither π or 0, the U(1) mean-field state is expected to break the time-reversal symmetry, since the sign of the flux will be reversed under time-reversal transformation (the π-flux phase is invariant as flux π is equivalent to flux $-\pi$). However, such a conclusion is not always true. In the case of the staggered-flux phase (see Fig. 18.4), the mean-field state after Gutzwiller projection is in fact a

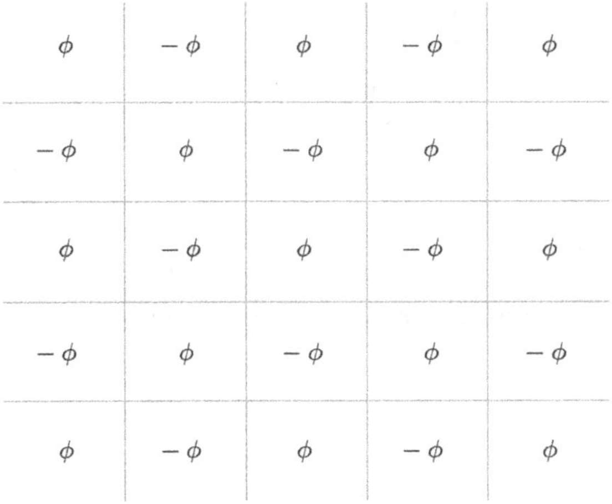

Fig. 18.4. An illustration of the staggered-flux phase.

state with a real wave function. The time-reversal symmetry is recovered in the physical state.

The reason for such a discrepancy between the mean-field state and the Gutzwiller projected state lies in the fact that the gauge symmetry of the slave particle representation at half-filling is not U(1), but the higher SU(2). At half-filling, the local spin operator is the only physical degree of freedom. However, the local spin operator is invariant not only under the U(1) gauge transformation mentioned above, but also under the following SU(2) gauge transformation:

$$\psi_i \to \psi'_i = W_i \psi_i, \qquad (18.17)$$

in which $\psi_i^T = (f_{i\uparrow}, f_{i\downarrow}^\dagger)$ is a spinor of two components, $W_i \in \mathrm{SU}(2)$ is a SU(2) matrix defined on each lattice site. It is then easy to show that with suitable choice of W_i we can transform the staggered-flux phase into the form of a d-wave pairing state:

$$\chi_{i,i+x} = \chi_{i,i+y} = \chi, \quad \Delta_{i,i+x} = -\Delta_{i,i+y} = \Delta, \qquad (18.18)$$

in which the order parameters χ and Δ are all real numbers. These two order parameters are related to the gauge flux ϕ in the staggered-flux phase through the following relation:

$$\frac{\Delta}{\chi} = \tan \phi. \qquad (18.19)$$

Since both χ and Δ are real, we see that the staggered-flux phase is time reversal symmetric.

As a result of such close relations between the staggered-flux phase and the d-wave pairing phase, the pseudogap phase with d-wave pairing is also known by some as staggered-flux phase. On the other hand, although the SU(2) gauge symmetry is exact only at half-filling, we expect that approximate SU(2) gauge symmetry will play an important role in determining the physics of the very underdoped system. This is the starting point of the SU(2) gauge theory for high-T_C superconductors that will be discussed later in this chapter.

18.3.4. *Finite Doping*

At finite doping, besides the order parameter χ and Δ, we should also introduce the mean-field order parameter b to describe the coherent motion of the charge. Here, b can be interpreted as the amplitude of the Bose condensation of holon, i.e. $b = \langle b_i \rangle$. The mean-field phase diagram constructed from these three different kinds of order parameters is shown schematically in Fig. 18.5. The phase digram is divided into five different regions according to the values of the three mean-field order parameters.

The order parameter χ, Δ, b are all zero at high temperature. There is then no coherent motion for the charge, no correlation between the local spins. Systems in such a state will behave like a group of independent

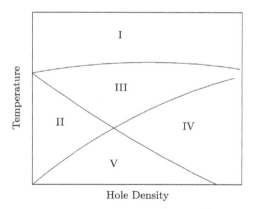

Fig. 18.5. The mean-field phase diagram of the $t-J$ in the slave-boson representation. The five regions in the phase diagram are, respectively: (I) local spin state, (II) pseudogap phase, (III) strange metal phase, (IV) Fermi liquid phase and (V) d-wave superconducting phase.

local spins. The state is thus called the local spin state. A typical feature in this state is the Curie-type temperature dependence of the magnetic susceptibility.

When χ becomes non-zero at lower temperature, four phases are possible depending on the value of the Δ and b. In the so-called underdoped region, the pairing-order parameter will appear at a temperature higher than holon condensation, while in the overdoped region the reverse is true. When $\Delta \neq 0$ and $b = 0$, the system will have non-zero spin pairing but no charge condensation, and thus will behave as a normal metal with non-zero gap. This is just the characteristic of the pseudogap phase. When $\Delta = 0$ and $b \neq 0$, the pairing gap vanishes but the Fermionic spinon will now behave like a charged quasi-particle on the background of the holon condensation. The system will thus behave as a Fermi liquid, with a Fermi surface satisfying the Luttinger theorem. When both Δ and b are zero, the quasi-particle weight will vanish but the spinon Fermi surface will still be there. Such a state will exhibit many exotic properties and will be called the strange metal phase. Finally, when both Δ and b are non-zero, the sytem will have both spin pairing gap and coherent charge motion and will behave just like a superconductor. According to the mean-field theory, the spin pairing always has the d-wave symmetry. The relation between the mean-field order parameters and the different phases mentioned above are tabulated in Table 18.1.

It should be noted that the mean-field order parameters χ, Δ and b are not gauge invariant and do not have direct physical meaning. As a result, the phase boundaries on the mean-field phase diagram, except that of the superconducting state, can only be interpreted as regions of some kind of crossover, rather than real phase transition lines. At the same time, the anti-ferromagnetic ordered state is not involved in the phase diagram and the superconducting state extends to the half-filling limit. It is still an unsolved problem in RVB theory to include the anti-ferromagnetic ordered state in a consistent manner. Another problem with the mean-field phase

Table 18.1. Mean-order parameters and the corresponding phases.

Non-zero-order parameters	Phase	Physical characteristics
None	Local spin phase	Curie behavior
χ	Strange metal phase	Non-Fermi liquid behavior
χ, Δ	Pseudogap phase	Pseudogap, Nernst effect
χ, b	Fermi liquid phase	Fermi surface, Drude behavior
χ, Δ, b	Superconducting phase	d-Wave superconductivity

diagram is that as the model under consideration is 2D, no true symmetry breaking at finite temperature is possible. There are two ways to resolve such a problem. One can either introduce small but finite coupling in the third dimension, or interpret the superconducting transition as a KT-type transition. The latter possibility has received a lot of support from experiments in recent years.

On the whole, the slave-boson mean-field theory predicts successfully the basic structure of the phase diagram. It predicts correctly the d-wave symmetry of the pairing gap. It also provides an appealing understanding of the origin of the pseudogap phase. However, to address the robustness of the mean-field predictions, one should still consider the correction effect of the fluctuation around the mean-field state. Among these fluctuations, the most important fluctuation mode in the low-energy regime is the gauge fluctuation.

18.4. Gauge Fluctuation and Its Physical Consequences

The slave-boson mean-field theory provides us with a qualitative understanding of the phase diagram of the high-T_C cuprates. However, to be consistent, one must ensure that the predictions of the mean-field theory are robust against the fluctuation around the saddle point. This is much like the situation we encounter in systems with spontaneous symmetry breaking. What is different here is that the symmetry broken in the RVB mean-field theory is not the conventional global symmetry, but the local gauge symmetry. Correspondingly, we have to analyze the stability of the mean-field state against the gauge fluctuations. The gauge fluctuation is quite different from the Goldstone mode in symmetry breaking system where they can interact with particles non-locally. At the same time, gauge fluctuation effect can also result in important corrections in quasi-particle properties, transport properties, flux quantization, spin dynamics and many other different physical properties. In the following, we will touch upon a few of these gauge fluctuation effects.

18.4.1. *U(1) Gauge Fluctuation in the Strange Metal Phase*

To be clear, we will start from the simplest case of the U(1) gauge fluctuation in the strange metal phase (in which $\chi \neq 0$, $\Delta = 0$, $b = 0$;

$\mu \neq 0$). In the strange metal phase, the low-energy gauge fluctuation takes the form of U(1) gauge field. The gauge rotation between Δ and χ becomes massive when $\mu \neq 0$ and can be neglected in the low-energy limit. The action of the effective theory in the strange metal phase then takes the form:

$$Z = \int \mathcal{D}f^\dagger \mathcal{D}f \mathcal{D}b^* \mathcal{D}b \mathcal{D}\chi \mathcal{D}\lambda \exp\left(-\int_0^\beta \mathcal{L} d\tau\right), \tag{18.20}$$

in which the integration over λ is introduced to enforce the no-double-occupancy constraint. The integration over χ is the result of the Hubbard–Stratnovich transformation in the original action. The Lagrangian in the above equation reads:

$$\begin{aligned}\mathcal{L} = &\sum_i b_i^*(\partial_\tau - i\lambda_i - \mu_\mathrm{B})b_i + \sum_{i,\sigma} f_{i,\sigma}^\dagger(\partial_\tau - i\lambda_i)f_{i,\sigma} \\ &- \tilde{J}\sum_{\langle i,j\rangle,\sigma}(\chi_{i,j}^* f_{i,\sigma}^\dagger f_{j,\sigma} + \mathrm{c.c.}) - t\sum_{\langle i,j\rangle,\sigma} f_{i,\sigma}^\dagger f_{j,\sigma} b_i b_j^* \\ &+ \tilde{J}\sum_{\langle i,j\rangle}|\chi_{i,j}|^2,\end{aligned} \tag{18.21}$$

in which $\tilde{J} = 3J/8$. This Lagrangian is obviously invariant under the following U(1) gauge transformation:

$$f_i \to f_i e^{i\phi_i}, \quad b_i \to b_i e^{i\phi_i}, \quad \chi_{ij} \to \chi_{ij} e^{i(\phi_j - \phi_i)}, \quad \lambda_i \to \lambda_i + \partial_\tau \phi_i. \tag{18.22}$$

Under the gauge transformation, the phase of χ_{ij} and the Lagrangian multiplier λ_i combine into the space and time components of the U(1) gauge field. Since the action is gauge invariant under the global U(1) gauge transformation, the gauge fluctuation composed of the phase of χ_{ij} and λ_i should be gapless in the long wavelength limit. This is the most important fluctuation mode in the low-energy effective theory. In the following, we will denote both fluctuations as a_{ij} and a_i^0. Thus, the low-energy effective theory can be written in the form of a compact U(1) gauge field coupled to both spinon and holon (the massive fluctuation of the magnitude of χ_{ij} is neglected in the low-energy effective theory):

$$\mathcal{L} = \sum_i b_i^*(\partial_\tau - ia_i^0 - \mu_B)b_i + \sum_{i,\sigma} f_{i,\sigma}^\dagger(\partial_\tau - ia_i^0)f_{i,\sigma}$$
$$- \tilde{J}\chi \sum_{\langle i,j\rangle,\sigma} (e^{ia_{ij}} f_{i,\sigma}^\dagger f_{j,\sigma} + \text{c.c.})$$
$$- t\eta \sum_{\langle i,j\rangle} (e^{ia_{ij}} b_i^* b_j + \text{c.c.}). \tag{18.23}$$

Here, the correction to the magnitude of χ_{ij} by the hopping term is included in a correction of $\tilde{J}$ and we will assume that $\chi = \eta$ holds always. It should be noted that there is no dynamical term for the gauge field itself in the above action. However, this does not imply that the gauge field can fluctuate arbitrarily. If we integrate over the spinon and the holon degree of freedom, an effective action for gauge field would emerge. The low-energy spinon and holon can then couple to this renormalized gauge fluctuation. The effective action for the gauge field reads:

$$e^{-\int_0^\beta \mathcal{L}_{\text{eff}}} = \int \mathcal{D}f^\dagger \mathcal{D}f \mathcal{D}b^* \mathcal{D}b e^{-\int_0^\beta \mathcal{L}}. \tag{18.24}$$

To avoid double counting, we should interpret the above procedure as a renormalization process: the gauge field obtains its own dynamics through its coupling to the excitation of the spinon and holon above some energy cutoff, and then the renormalized gauge field couples again with low-energy spinon and holon. Through such a procedure, we have integrated out the high-energy part of the spinon and holon degree of freedom. The resultant theory can thus only describe the long wavelength limit of the excitation spectrum.

In the long wavelength limit, the coupling between the gauge field and the spinon and holon degree of freedom can be described by the following continuous action:

$$\mathcal{L}_{\text{int}} = \int_q (j_\mu^f(q) + j_\mu^b(q))a_\mu(-q), \tag{18.25}$$

in which $q = (\boldsymbol{q},\omega), j_\mu^f(q), j_\mu^b(q)$ are the current operators of the spinon and holon. Up to the Gaussian level, the effective action of the gauge field reads:

$$S_{\text{eff}}(a) = \int_q (\Pi_{\mu,\nu}^f(q) + \Pi_{\mu,\nu}^b(q))a_\mu(q)a_\nu(-q), \tag{18.26}$$

in which:

$$\Pi^f_{\mu,\nu}(q) = \langle j^f_\mu(q) j^f_\nu(-q)\rangle, \quad \Pi^b_{\mu,\nu}(q) = \langle j^b_\mu(q) j^b_\nu(-q)\rangle \qquad (18.27)$$

are the current correlation functions of the spinon and the holon.

In the low-energy effective theory, the coupling to the U(1) gauge field will cause inelastic scattering of the spinon and the holon. As a result of the long-ranged nature of the gauge coupling, such inelastic scattering will induce non-Fermi liquid behavior. A typical consequence of such scattering by gauge field is the linear temperature dependence of the resistivity in the strange metal phase. However, the theory about such a scattering mechanism for non-Fermi liquid behavior is still not generally accepted. The interested reader is encouraged to look at the review papers listed at the beginning of this chapter.

18.4.2. Z_2 Gauge Fluctuation in the Pseudogap Phase

In the pseudogap phase ($\Delta \neq 0, b = 0$), the U(1) gauge fluctuation becomes massive as a result of the Higgs mechanism. However, as Δ carries twice the gauge charge as compared to the slave particles, gauge fluctuation at finite energy is still possible. This is the result of the Z_2 gauge invariance. Using the Hubbard–Stratnovich transformation, the effective action of the system with $\Delta \neq 0$ reads:

$$\mathcal{L} = \sum_i b_i^*(\partial_\tau - i\lambda_i - \mu_B)b_i + \sum_{i,\sigma} f_{i,\sigma}^\dagger (\partial_\tau - i\lambda_i) f_{i,\sigma}$$

$$- \frac{\tilde{J}}{2} \sum_{\langle i,j\rangle,\sigma} (\Phi_{i,\sigma}^\dagger U_{ij} \Phi_{j,\sigma} + \text{c.c.})$$

$$- t \sum_{\langle i,j\rangle,\sigma} f_{i,\sigma}^\dagger f_{j,\sigma} b_i b_j^* + \frac{\tilde{J}}{2} \sum_{\langle i,j\rangle} \text{Tr}[U_{ij}^\dagger U_{ij}], \qquad (18.28)$$

in which:

$$\Phi_{i,\uparrow} = \begin{pmatrix} f_{i,\uparrow} \\ f_{i,\downarrow}^\dagger \end{pmatrix}, \quad \Phi_{i,\downarrow} = \begin{pmatrix} f_{i,\downarrow} \\ -f_{i,\uparrow}^\dagger \end{pmatrix},$$

while the order parameter matrix becomes:

$$U_{ij} = \begin{pmatrix} -\chi_{ij}^* & \Delta_{ij} \\ \Delta_{ij}^* & \chi_{ij} \end{pmatrix}.$$

With non-zero Δ, the action is no longer invariant under the gauge transformation Eq. (18.22). The U(1) gauge fluctuation is thus massive. However, as the action is invariant under the following Z_2 gauge transformation:

$$f_i \rightarrow -f_i,$$
$$b_i \rightarrow -b_i, \qquad (18.29)$$

Z_2 gauge fluctuation (whose effect is equivalent to a π flux tube) at finite energy is still possible and the phase is known by some as the Z_2 phase. The finite energy Z_2 gauge excitation is termed as vison excitation.

18.4.3. The Ioffe–Larkin Composition Rule

As the first example for the gauge fluctuation effect, we discuss the conductivity of the system. When we introduce the slave particle, we have assumed that the charge of the electron is carried by the holon degree of freedom and that the spinon is a charge-neutral Fermion. It then seems that the charge transport property of the system is determined by the holon system alone. This is actually incorrect. Since the slave particle should satisfy the constraint Eq. (8) to be a faithful representation of the original system, the motion of the spinon and the holon is entangled with each other. The result of such entanglement is the so-called Ioffe–Larkin composition rule. According to this rule, the inverse of the conductivity of the system is given by the sum of the inverse of the spinon conductivity and the inverse of the holon conductivity, i.e.:

$$\frac{1}{\sigma(q)} = \frac{1}{\sigma^f(q)} + \frac{1}{\sigma^b(q)}. \qquad (18.30)$$

From this rule we know that the conductivity of the system is controlled by the smaller one of the spinon conductivity and the holon conductivity.

The origin of the Ioffe–Larkin composition rule can be understood as follows. Since the spinon and the holon must satisfy the constraint Eq. (18.8), the spinon current $j^f(q)$ and the holon current $j^b(q)$ must satisfy the following constraint:

$$j^f(q) + j^b(q) = 0. \qquad (18.31)$$

From Eq. (18.25), we find that such a constraint on spinon current and holon current actually amounts to the integration over the gauge fluctuation. Here, the spinon current is induced by the internal gauge field, while the holon current is coupled to both the internal gauge field and the external electromagnetic field. Up to the level of Gaussian fluctuation, the internal gauge field can be replaced by its mean value and we then have:

$$j_\mu^f(q) = \Pi_{\mu,\nu}^f(q)\bar{a}_\nu(q), \quad j_\mu^b(q) = \Pi_{\mu,\nu}^b(q)(\bar{a}_\nu(q) + A_\nu), \quad (18.32)$$

in which $\bar{a}_\nu(q)$ denotes the average of the internal gauge field, A_ν is the external electromagnetic field. From the constraint Eq. (18.31), we can solve $\bar{a}_\nu(q)$ and the result is:

$$\bar{a}_\nu(q) = -\frac{1}{\Pi^f + \Pi^b}\Pi^b A \quad (18.33)$$

(note that Π^f, Π^b is a matrix rather than a number). Thus, up to the Gaussian level, the effect of the internal gauge field is to provide a bias that is proportional to the external electromagnetic field. When this result is inserted back into Eq. (18.32) we have:

$$j = j^b = \Pi^f \frac{1}{\Pi^f + \Pi^b}\Pi^b A. \quad (18.34)$$

It is straightforward to see that the conductivity deduced from this equation follows the Ioffe–Larkin composition rule.

18.4.4. *Flux Quantization and Vortex Core State*

In type II superconductors, the magnetic field will penetrate the superconductor in the form of quantized vortex lines. In a conventional superconductor, the vortex line carries a quantized flux of $hc/2e$, since the condensed Cooper pair carries twice the charge of an electron. The superconducting order in the core of such vortex lines is suppressed. The high-T_C cuprates are typical type II superconductors. It is found experimentally that the magnetic flux is still quantized in unit of $hc/2e$. However, it is found that the core region of the vortex line is still gapped in the underdoped regime.

The observation of a vortex line with a quantized flux of $hc/2e$ and a gapped core is actually a serious challenge to the RVB theory. In the mean-field theory that we have introduced above, the holon, which carries charge e, condenses in the superconducting state. It then seems that the magnetic flux of the vortex line should be quantized in unit of hc/e, rather than $hc/2e$ as in a Bose condensate of charge $2e$. However, one should also take into account the gauge flux of the internal gauge field that the slave particle is coupled to when considering the quantization of the magnetic flux. When $\Delta \neq 0$, the gauge flux of the internal gauge field will act as Z_2 vorticity, each of which will carry a gauge flux of $hc/2e$. Such Z_2 flux can combine with the $hc/2e$ flux contributed by the external electromagnetic field and the total flux can be either zero or hc/e. This makes the $hc/2e$ flux vortex line in a charge e condensate possible.

However, even if the $hc/2e$ vortex is allowed, there remains the problem of whether the core region is gapped or not. According to the discussion above, the internal gauge field should contribute an additional Z_2 flux in a $hc/2e$ vortex. Such a Z_2 gauge flux from the internal gauge field also couples to the spinon degree of freedom. The spinon system is described by a BCS type mean-field Hamiltonian. The Z_2 gauge flux will act as a $hc/2e$ vortex in the condensate of the spinon pairs. In a conventional BCS superconductor, such vortex will have a normal core in which pairing is suppressed. It thus seems that the pairing gap should vanish in the core region of a $hc/2e$ vortex line. If this occurs, what we get would just be a conventional vortex line found in conventional superconductors. The only effect of the internal gauge field is that it seems to make the $hc/2e$ vortex possible in a condensate of charge e particles. This is obviously not consistent with the observations made in the underdoped cuprates.

One fact should be noted before further discussion. Although the spinon system is described by a mean-field Hamiltonian as a BCS superconductor, both the kinetic term χ and pairing term Δ are actually RVB order parameters describing the structure of the spin liquid. These two kinds of order parameters can transform between each other under the SU(2) gauge transformation. At half-filling, the staggered-flux phase and the d-wave pairing phase are gauge equivalent under such gauge transformation. Although exact gauge equivalence between the two is lost at finite doping, it is natural to expect that for small enough doping, the energy difference between the two should be small. In other words, the d-wave superconducting state is actually competing with a state that is very close in energy and has a similar gap structure — the staggered flux phase, at small doping concentration. We thus can construct a vortex with a core in the staggered flux phase, rather than in the gapless normal state. Such a vortex would have lower energy, since the staggered flux phase is very close in energy to the d-wave superconducting state at low doping.

To understand how the d-wave superconducting order outside the vortex core evolves into the staggered-flux order in the vortex core, we

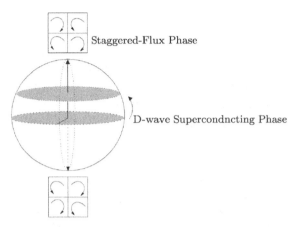

Fig. 18.6. The relation between the staggered-flux phase and the d-wave superconduting phase under SU(2) gauge transformation. The points on the equator correspond to d-wave superconducting state. Around the equator the phase of the superconducting order parameters wind by 2π. The north and south poles of the sphere correspond to states with spontaneous current circulating in clockwise and anti-clockwise manner. The evolution from the d-wave superconducting order to the staggered flux order in the vortex line is also illustrated in the figure.

show in Fig. 18.6 the relation between the two orders under SU(2) gauge transformation. The points on the equator represent states with non-zero d-wave pairing order parameter Δ, whose phase changes by 2π around the equator. The north and south poles represent states with staggered flux $\phi = \tan^{-1} \Delta/\chi$ and $\phi = -\tan^{-1} \Delta/\chi$. At half-filling, all points on the sphere are equivalent under SU(2) gauge transformation. When the system is doped with holes, such an exact SU(2) symmetry is destroyed and the different points on the sphere now correspond to different states. When the holon is Bose-condensed, the points on the equator correspond to d-wave superconducting states. Around the equator, the phase of the superconducting order parameter winds by 2π. The north and south poles of the sphere correspond to states with spontaneous current circulating in clockwise and anti-clockwise manner.

Now we consider the structure of a $hc/2e$ vortex. Far away from the core region, the system is in the d-wave superconducting state. The phase of the order parameter changes by 2π when we travel around the vortex line once. Thus, a loop around the vortex line in real space that is far away from the vortex core maps into the equator of the order parameter space. When the loop becomes smaller, the pairing order parameter is suppressed gradually. This can be achieved by moving the image in the order parameter space away from the equator. At the center of the vortex line, the superconducting order parameter will approach zero and thus the image in the order parameter space will approach the north or south pole of the sphere. This is the process of "escape to the third dimension". The vortex line, so constructed, will have a core with spontaneous circulating current and non-zero excitation gap. This is a unique prediction of the RVB theory.

18.4.5. *Quasi-particle Properties*

The quasi-particle property is the most deeply studied issue in high-T_C physics. Well-defined quasi-particle has been observed in both the superconducting state and the Fermi liquid phase of the phase diagram. The Fermi surface of the quasi-particle is found to satisfy the Luttinger theorem. It is also found that the quasi-particle weight is suppressed rapidly with the decrease of the doping concentration. In the underdoped regime, strong anisotropy in the quasi-particle weight is noticed along the Fermi surface: the quasi-particle weight is large around the gap nodes and becomes very small around the gap maximum. On the other hand, it is found that both the effective mass and the charge response coefficient of the quasi-particle are almost independent of the doping concentration. This makes a dramatic contrast with the superfluid density at zero temperature, which is approximately proportional to the density of the doped holes as a result of the strong electron correlation.

The quasi-particle properties of the high-T_C cuprates differ greatly from the prediction of the standard Fermi liquid theory. For example, the quasi-particle weight and the effective mass exhibit totally different doping dependence, and quasi-particle weight is strongly non-uniform on the Fermi surface. The slave-boson mean-field theory only provides understandings on some of them and the others remain unsolved.

In both the superconducting phase and the Fermi liquid phase, the mean-field theory predicts finite quasi-particle weight. Since the gauge fluctuation is gapped when holon condenses, the mean-field prediction should be reliable. According to the mean-field theory, each spinon has probability of x to behave as an electron in a background with holon condensate, where x is the density of the doped holes. This explains the strong doping dependence of the quasi-particle weight. Since the quasi-particle formed in such a way preserves the spinon dispersion, the doping independence of the effective mass is also explained naturally.

However, it is hard to understand the near doping independence of the quasi-particle charge from the slave-boson mean-field theory. The quasi-particle charge can be inferred from the temperature dependence of superfluid density, which can be measured directly from the penetration depth of a superconductor. As a linear response coefficient, the superfluid density also satisfies the Ioffe–Larkin composition rule discussed earlier. We thus have:

$$\frac{1}{\rho_s} = \frac{1}{\rho_s^f} + \frac{1}{\rho_s^b}, \tag{18.35}$$

in which ρ_s^f and ρ_s^b are the superfluid densities of the spinon and holon system. It is expected that ρ_s^f will behave just as that of a BCS superconductor, while ρ_s^b will behave as that of an XY model type system. We thus assume:

$$\rho_s^f(T) \approx \rho_s^f(0)(1 - \alpha T), \quad \rho_s^b(T) \approx \rho_s^b(0) \approx x, \tag{18.36}$$

in which α is determined by the spinon dispersion around its nodes and is independent of the hole density. Taking into account the fact that $\rho_s^f(0) \gg x$, we have at low temperature that:

$$\rho_s^f(T) \approx x - \frac{x^2}{\rho_s^f(0)} \alpha T. \tag{18.37}$$

Thus, the superfluid density at zero temperature is proportional to the hole density and should vanish in the half-filling limit. This is consistent with both the physical picture of a doped Mott insulator and the experimental observations. However, the theory also predicts that the coefficient of the temperature dependence of the superfluid density is proportional to the square of the hole density. This is against experimental observation, which shows that the same quantity is almost doping-independent. This dramatic contrast between the doping dependence of the superfluid density and the coefficient of its temperature dependence is one of the most fundamental problems in the physics of high-T_C cuprates.

Besides, the strong anisotropy of the quasi-particle weight on the Fermi surface in high-T_C cuprates also constitutes a big challenge to the RVB theory.

18.4.6. Spin Dynamics

The spin dynamics of the high-T_C superconductors is studied by many researchers and with many different approaches. In both its normal and superconducting state, the high-T_C superconductor exhibits strong anti-ferromagnetic spin fluctuation at low energy. In particular, in the superconducting state, a dramatic spin resonance mode will emerge at the anti-ferromagnetic ordering wave vector below the continuum of the spin fluctuation spectrum.

The strong spin fluctuation in the low-energy regime is a direct manifestation of the strongly correlated nature of the system. In the $t-J$ model, the lattice site occupied by the electron will always behave as a local spin as a result of the no double occupancy constraint, even in the superconducting state. This is naturally encoded in the RVB theory. However, in the mean-field treatment (or the effective-field theory at the Gaussian level) of the RVB state the no double occupancy constraint is almost neglected. As a result, the mean-field theory seriously underestimates the spin correlation in the RVB state. For example, the spin correlation in the superconducting state predicted by the mean-field theory has the following asymptotic behavior:

$$S(r, t=0) \sim \frac{1}{r^4}, \qquad (18.38)$$

and the spin fluctuation spectrum has the following asymptotic behavior:

$$S(q, \omega) \sim \omega^3. \qquad (18.39)$$

There is no collective mode at all. Although we can introduce phenomenological RPA correction to induce collective excitation, it is quite hard to find a self-consistent description within such a scheme.

18.4.7. The Issue of Confinement

Many issues remain open in the development of the RVB theory that we cannot touch in this short review. Here, we will present a brief introduction of the so-called confinement issue. Readers interested in more details about this issue should read the two long review papers mentioned at the beginning of this chapter.

As we know, the slave particles are coupled to a compact U(1) gauge field. However, it is well known that the $2+1$D compact U(1) lattice gauge field is confining as a result of the excitation of the monopole (or instanton), no matter what the strength of the coupling is. In the confining phase, particles with non-zero gauge charge are forbidden in the energy spectrum. For our problem, this would imply that neither spinon nor holon alone can appear in the energy spectrum at any finite energy. Thus, if the gauge field is in the confining phase, it is invalid to construct the dynamics of the system with spinon and holon. It is now unclear if the feedback effect from coupling to the finite density of the spinon and holon would change the dynamics of the gauge field qualitatively. It is argued by some that the dissipation effect from such coupling would suppress the confining effect of the instanton excitation. However, up to now, no generally accepted calculation exists on such a deconfinement mechanism.

18.5. Variational Study of the RVB Theory: Gutzwiller Projected Wave Function

From the above discussion we know the slave-boson mean-field theory and the related effective gauge-field theory can account for the basic characteristics of a doped Mott insulator. They predict successfully the basic structure of the phase diagram and the d-wave symmetry of the superconducting gap. They also provide qualitative understanding on the quasiparticle property and the vortex structure. However, we should also note that the success of the theory is itself a surprise: in all calculations, the all-important no-double-occupancy constraint of the $t-J$ model is only treated very crudely. For this reason, it is quite natural to question the validity of the calculations done in the unphysical space that violate the no-double-occupancy constraint.

However, there are still serious difficulties with accounting for the physics of the high-T_C superconductors with the slave-boson mean-field theory or the related effective gauge theory. For example, many issues concerning the quasi-particle of the cuprates such as the quasi-particle charge and anisotropy on the Fermi surface, remain open in such a scheme. More importantly, neither the anti-ferromagnetic long range, nor the low-energy spin dynamics of the cuprates can be described in a consistent way in such a scheme. Finally, no quantitative result can be expected from such a scheme as the crude treatment of the all-important no-double-occupancy constraint is far beyond any quantitative approximation.

For these reasons, a formalism which can incorporate ideas on the RVB state from the slave-boson mean-field theory and the related effective gauge theory, but at the same time account for the no double occupancy constraint in an exact manner is highly desirable. The variational theory based on Gutzwiller projected wave function provides just such a possibility. In such a scheme, the mean-field state, or state with gauge fluctuation correction is projected into the subspace of no double occupancy to arrive at a variational description of the system. As a variational approach, the study of Gutzwiller projected wave function has received much attention in recent years. One trend is to push the formalism to study not only the ground state, but also the excitation spectrum. At the same time, the Gutzwiller projected wave function is used to check the ideas proposed at the effective-field theory level. Since the projection operator is hard to treat analytically, in most cases one should rely on the Monte Carlo method to arrive at the final result.

It should be noted before further discussion that although the no double occupancy constraint is built into the Gutzwiller projected wave function, a whole account of the physical effect of the no double occupancy constraint is beyond simple projection. From the point of view of the gauge-field theory, the Gutzwiller projection only amounts to the integration of the time component of the gauge fluctuation. The effect of the fluctuation of the spatial component of the gauge field is left untouched. Thus, one way to improve the Gutzwiller projected wave function is to include the effect of the fluctuation of spatial component of the gauge field before Gutzwiller projection.

18.5.1. *Ground State*

The slave-boson mean-field theory and the effective gauge theory predict spin correlation that is much weaker than its actual value in the RVB state, as a result of the crude treatment of the no double occupancy constraint. The theory cannot provide a reasonable description of the relation between the anti-ferromagnetic order and the superconducting order. The situation improves in the variational theory based on Gutzwiller projected wave functions.

At half-filling, the spin correlation in the RVB state is much enhanced by the Gutzwiller projection and it was recently found that Gutzwiller projected mean-field state can even describe states with magnetic long-range order. For example, the spin correlation between neighboring sites in a Gutzwiller projected BCS state with $\chi = \Delta$ (or the π-flux phase) is $\langle \boldsymbol{S}_i \cdot \boldsymbol{S}_j \rangle \approx -0.32$, which is much larger than the mean-field prediction (≈ -0.08) and is already very close to the exact value of the Heisenberg model (≈ -0.3344). To describe a state with magnetic long-range order, we can also introduce magnetic-order parameter in the projected wave function as:

$$|\Psi\rangle = P_N P_G |\text{d} - \text{BCS} + \text{AF}\rangle, \qquad (18.40)$$

in which $|\text{d} - \text{BCS} + \text{AF}\rangle$ denotes mean-field state with both d-wave pairing order parameter and SDW order parameter, P_N is the projection to the subspace with N electrons. Such a wave function provides a rather good description of the ground state of the Heisenberg model. However, at the mean-field level, the RVB-order parameter is totally dominated by the SDW-order parameter and the result would be the trivial classical Neel state.

At finite doping, we can adopt the following Gutzwiller projected BCS wave function as a variational description of the superconducting state of the cuprates

$$|\Psi\rangle = P_N P_G |\text{d} - \text{BCS}\rangle. \qquad (18.41)$$

This wave function has proved to be the best wave function known for the $t-J$ model at finite doping. It behaves as a BCS superconductor with its superfluid density reduced by the correlation effect. The parameters in the wave function can be determined by the optimization of the ground-state energy. The result of such optimization agrees qualitatively with the mean-field theory, which indicates that the mean-field prediction on the structure of the phase diagram is qualitatively correct. If we also introduce the magnetic-order parameter in the wave function, we can get the whole phase diagram. For the $t-J$ model with $t/J = 3$, one finds that the antiferromagnetic order vanishes around $x = 0.1$ and the superconducting order vanishes around $x = 0.4$. These results are in qualitative agreement with observations.

18.5.2. *Quasi-particle Properties*

In the mean-field theory, the quasi-particle weight becomes non-zero when the holon Bose-condenses. The quasi-particle excitation in the Gutzwiller projected wave function formalism is constructed from projection of the mean-field excitation:

$$|\Psi_k\rangle = P_{N\pm 1} P_G \gamma_{k\uparrow}^\dagger |\text{d} - \text{BCS}\rangle, \qquad (18.42)$$

in which $\gamma_k^\dagger$ is the creation operator for the Bogliubov quasi-particle in the mean-field theory. From Monte Carlo simulation, it is found that the property of the quasi-particle described by this wave function agrees qualitatively with the mean-field prediction. For example, both the quasi-particle weight and its electromagnetic response coefficient (or quasi-particle charge) vanish in the half-filling limit, while the effective mass is almost independent of the doping. The quasi-particle weight is also found to be uniform across the whole Fermi surface. On the whole, this wave function has both the virtues and drawbacks of the mean-field theory, although closer inspection does find subtle differences in the asymptotic behavior in the half-filling limit between the two. For example, the quasi-particle weight calculated from the Gutzwiller wave function is larger than the mean-field prediction. In the half-filling limit, it vanishes as $\sim \sqrt{x}$, rather than the $\sim x$ manner predicted by mean-field theory (see Figs. 18.7 and 18.8). This implies that the Gutzwiller projection encourages the spinon and holon to recombine into quasi-particle.

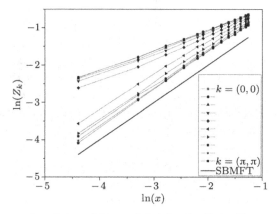

Fig. 18.7. The quasi-particle weight predicted by the Gutzwiller wave function as a function of doping. The four curves on the upper part of the figure correspond to the quasi-particle momentum inside the Fermi surface and the five curves in the lower part of the figure correspond to the quasi-particle momentum outside the Fermi surface. The solid line denotes the result of the mean-filed theory, which is uniform in momentum space. The calculation of the Gutzwiller wave function is done on a 18×18 lattice. Note that the quasi-particle weight scales differently with doping for momentum inside and outside the Fermi surface. $Z \sim \sqrt{x}$ inside the Fermi surface but $Z \sim x$ outside the Fermi surface.

To improve the description of the quasi-particle excitation, ideas of spin-charge recombination are proposed in the effective-field theory formalism. Under the correction of gauge fluctuation, it is natural to expect that the spinon and holon should be correlated with each other. Since both spinon and holon carry unit gauge charge, the coherent motion of either spinon or holon alone will be disturbed by the gauge fluctuation. However, as the composite of the two, the quasi-particle excitation is gauge neutral and thus the gauge fluctuations prefer a larger quasi-particle coherence, which is partly confirmed by the calculation on Gutzwiller projected wave function. Since the Gutzwiller projection only accounts for the fluctuation of the time component of the internal gauge field, we still need further corrections on the Gutzwiller wave function. We have proposed the following wave function to describe the spin–charge recombination effect:

$$|\Psi_k\rangle = \sum_q \phi(q) P_{N-1} P_G \gamma^\dagger_{k-q\uparrow} b^\dagger_q |\text{d} - \text{BCS}\rangle, \qquad (18.43)$$

in which $\phi(q)$ is variational parameter describing the relative motion between spinon and holon and can be determined by optimization of energy. This wave function is a generalization of the conventional Gutzwiller projected wave function and a new numerical algorithm is needed to perform simulation on it. For this purpose, we have proposed a very efficient reweighing technique to simulate wave functions composed of a large number of Slater determinants. We have applied such a new technique to the new wave function for quasi-particle and found that the quasi-particle charge now extrapolates to a finite value in the half-filling limit (see Fig. 18.8), which is consistent with experimental observation but is hard to understand in the mean-field theory and effective-field theory scheme.

18.5.3. *Spin Dynamics*

As we have seen in the previous section, the description of spin correlation in the RVB state at the mean-field level is improved by the Gutzwiller projection. The same projection procedure also predicts a more reasonable phase diagram with both anti-ferromagnetic order and the superconducting order. For this reason, it is reasonable to assume that the Gutzwiller projected BCS wave function has already built in the essential spin correlation to understand the spin dynamics of the system in the low-energy regime. For the cuprates, the most remarkable feature of the spin dynamics in the superconducting state is the so-called spin resonance mode. Such a phenomenon

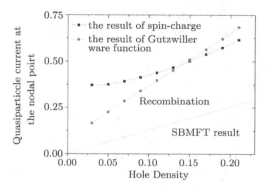

Fig. 18.8. The quasi-particle charge calculated from the new wave function with spin–charge recombination effect included as a function of doping and its comparison with the prediction of the slave-boson mean-field theory.

is studied with many different approaches and among them, the phenomenological RPA theory is the most generally accepted one. However, the description of the resonance mode with the RPA theory involves many phenomenological parameters to be determined by fitting. At the same time, there is intrinsic difficulty with such an approach on the explanation of the doping dependence of the mode energy. According to the RPA theory, the resonance energy should scale with the superconducting gap, which however follows an opposite trend with the resonance energy as a function of doping.

We thus construct the following variational wave function for the spin resonance mode:

$$\left|\Psi_{Q=(\pi,\pi)}\right\rangle = \sum_q \phi(q) P_N P_G \gamma^\dagger_{Q-q\uparrow} \gamma^\dagger_{q\uparrow} \left|d-BCS\right\rangle, \tag{18.44}$$

in which $\phi(q)$ is a variational parameter that describes the relative motion between the two excited spinons. This wave function can actually be taken as the wave function for a spin exciton in the projected Hilbert space. Unlike the conventional RPA theory, there is no need to introduce phenomenological coupling constant here. The variational parameter $\phi(q)$ can be determined by optimization of the mode energy. The theory thus contains no tuning parameter and can be viewed as *ab initio*.

The wave function above reduces to the wave function in the single mode approximation if we choose $\phi(q) = u_{Q-q} v_q$ (u_q, v_q is the coefficient of the Bogliubov transformation):

$$|\Psi_Q\rangle = S_Q^+ P_N P_G |d-BCS\rangle, \tag{18.45}$$

in which S_Q^+ is the spin density operator. The mode energy in the single mode approximation can be obtained through the calculation of double commutator. For the $t-J$ model it is given by:

$$\Delta E_Q = \frac{-\langle G|H_t|G\rangle - \frac{8}{3}\langle G|H_J|G\rangle}{\langle G|S_Q^+ S_{-Q}^-|G\rangle}, \quad (18.46)$$

in which $|G\rangle$ is the ground state. It is thus clear that the mode energy should vanish in the thermodynamic limit whenever the spin structure factor $\langle G|S_Q^+ S_{-Q}^-|G\rangle$ at momentum $Q = (\pi, \pi)$ scales with the number of lattice site with a power $\alpha > 1$ (assuming that $\langle G|S_Q^+ S_{-Q}^-|G\rangle \approx N_s^\alpha$, in which N_s is the number of lattice sites). When the system is long-range ordered, we have $\alpha = 2$ and $\Delta E_Q = 0$, as required by the Goldstone theorem. Since the mode energy of our wave function is, by construction, lower than that of the prediction of the single mode approximation, it will also satisfy the Goldstone theorem. The mode energy will approach zero when $\alpha > 1$, even if the ground state is still spin disordered.

The spin correlation in the mean-field theory is rather weak and the power calculated is $\alpha = 1$. Thus, even at half-filling, the mode energy is non-zero. However, after the Gutzwiller projection, the power is enhanced to $\alpha = 4/3 > 1$ and mode energy is then zero at half-filling.

In Fig. 18.9, we plot the energy of the spin resonance as a function of doping. Although there is still quantitative difference with the experimental

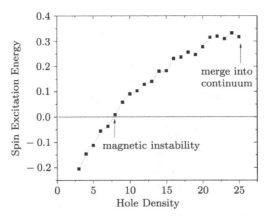

Fig. 18.9. The resonance energy calculated from the wave function Eq. (18.44) as a function of doping.

observation, it provides a rather good qualitative account of the experimental results. Below $x = 0.075$, the mode energy becomes negative, which indicates the magnetic instability of the system. Above $x = 0.25$, the spin resonance mode merges into the continuum spectrum and loses spectral weight.

18.6. Conclusions

We present a review of the slave-boson mean-field theories and related effective gauge-field theories of the $t-J$ model. Such a theoretical scheme is seen to provide a rather good overall description of the strong correlation nature of the high-T_C superconductors as doped Mott insulators. It provides systematic and enlightening understandings on the physical behavior of the high-T_C cuprates throughout the phase diagram. In particular, it predicts successfully the pseudogap phenomenon and the d-wave symmetry of the pairing gap.

However, many important issues remain open in the slave-boson framework. The spin dynamics of the cuprates, the relation between superconductivity and anti-ferromagnetism, the anomalous quasi-particle properties are three examples in this respect. At the same time, it is still not realistic to calculate the properties of the cuprates in this scheme quantitatively. All these difficulties are caused by the highly non-trivial no double occupancy constraint, which is only treated very crudely in the mean-field theory and the effective gauge theory approach.

The variational theory based on Gutzwiller projected wave function provides a possible way to develop the RVB theory further. In this scheme, the description of the spin dynamics is much improved as compared to the mean-field theory approach or effective gauge theory approach. It confirms the predictions of the mean-field theory and the effective gauge-field theory on the structure of the superconducting state. More importantly, it also provides the unique opportunity to test the theoretical suggestions proposed on the ground of mean-field theory and effective gauge-field theory. However, it should also be noted that the Gutzwiller projection alone is not enough to account for all the physical consequences of the no-double-occupancy constraint. To go beyond the Gutzwiller projected wave function description, one should still consider gauge fluctuation correction beyond the Gaussian level. In this chapter, we use the problem of quasi-particle charge and the problem of spin resonance as two examples to illustrate how to go beyond the Gutzwiller wave function description. In both examples we get satisfactory results. It is predictable that with the mutual interaction between the effective theory approach and the variational theory approach, we can make even more important progress on the mechanism of the high-temperature superconductivity.

Acknowledgments

The author acknowledges the great contributions that Prof. Ru-Shan Han has made to the study of high temperature superconductivity in China. The author is also grateful for the collaboration of Fan Yang in the study reported in this chapter and the support of NSFC grants.

References

[1] P. A. Lee, N. Nagaosa and X. G. Wen, *Rev. Mod. Phys.* **78**, 17 (2006).
[2] P. A. Lee, Cond-mat/0708.2115 (2007).
[3] H. Y. Yang, F. Yang, Y. J. Jiang and T. Li, *J. Phys. C* **19**, 016217 (2007).
[4] T. Li and F. Yang *Phys. Rev. B* **81**, 214509, (2010).
[5] F. Yang and T. Li, *Phys. Rev. B* **83**, 064524 (2011).
[6] V. Kalmeyer and R. B. Laughlin, *Phys. Rev. Lett.* **59**, 2095 (1987).
[7] P. A. Marchetti, Z. B. Su and L. Yu, *Phys. Rev. B* **58**, 5808 (1998).
[8] Z. Y. Weng, D. N. Sheng, Y. C. Chen and C. S. Ting, *Phys. Rev. B* **55**, 3894 (1997).
[9] T. Senthil and M. Fisher, *Phys. Rev. B* **62**, 7850 (2000).
[10] M. Franz and Z. Tesanovic, *Phys. Rev. Lett.* **87**, 257003 (2001).

19
Superconductivity of Cuprates — A Phenomenon of Strong Correlation of Electrons

Wei Guo* and Ru-Shan Han*,†

*School of Physics, Peking University, Beijing 100871, China
†China Center of Advanced Science and Technology,
Beijing 100080, China

The strong correlation of electrons such as the magnetic ordering in transition metal oxides and the Kondo resonance in dilute magnetic compounds rising from the local exchange interactions were well-understood many-body effects in the past decades. The superconducting cuprate discovered in 1986 inspired a fascinating idea: the magnetic origin of superfluid. We speculate that the fluctuating spins on the oxygen sites in the CuO_2 layer created by the holes may pair into a resonating valence bond state with the quantum number $S = 1, S_z = 0$ via local exchange interactions, the same force responsible for magnetic ordering in the low-doped cuprate. Impurity doping drives the cuprate from an anti-ferromagnetic insulator to a superconductor, then a Fermi liquid, resulting in a rich phase diagram.

The superconducting cuprates discovered by Müller and Bednorz in 1986 are essentially different from the conventional BCS superconductors. They are strongly correlated systems characterized by the coexistence of superconductivity and magnetism [1–8]. Like many of the transition-metal-oxides, the undoped cuprate is an anti-ferromagnetic insulator. When the cuprate is doped with impurity it becomes a superconductor with a coherent length of about 10 Å. The short coherent length is an indication of the local pairing. Is the same interaction responsible for both the magnetic ordering and the electron pairing? A classic example for strong correlation between itinerant charge carriers and local spins is the Kondo resonance where the conduction electron spins and a local spin form a resonating singlet ground state via s–d exchange interaction. In the case of the high T_C problem, we focus on the issue of the formation of the resonating valence bond state, a key concept to understanding superconductivity in cuprates.

19.1. Electronic Structure of the Cuprate and Magnetism

The structure unit responsible for superconductivity in a cuprate is the CuO_2 layer, in which the copper atoms at each corner of a single square are separated by an oxygen atom (Fig. 19.1). The Cu site is singly occupied by a d electron with $S = 1/2$ since the large on-site Coulomb interaction U splits the half-filled d band into two sub-bands, i.e. the Hubbard bands. The O site is non-magnetic where the p orbital is doubly occupied, the Hartree-Fork approximation applies, and the filled p-band sits between two Hubbard bands of d electrons [9]. The CuO_2 layer is insulating for an undoped cuprate.

The basic model describing the interactions and the electronic states of the CuO_2 layer in a cuprate is the two-band model:

$$H = \sum_{l\sigma} \varepsilon_p p_{l\sigma}^\dagger p_{l\sigma} \sum_{i\sigma} \varepsilon_d d_{i\sigma}^\dagger d_{i\sigma} + U d_{i\uparrow}^\dagger d_{i\uparrow} d_{i\downarrow}^\dagger d_{i\downarrow}$$
$$+ \sum_{i,l\in\{i\},\sigma} V_{il} \left(d_{i\sigma} p_{l\sigma}^\dagger + d_{i\sigma}^\dagger p_{l\sigma} \right), \tag{19.1}$$

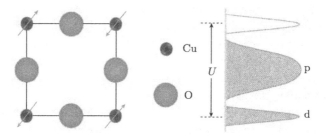

Fig. 19.1. The magnetic and the electronic structures of the CuO_2 layer.

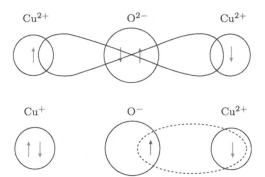

Fig. 19.2. Kramers superexchange. The ground state (above) and the intermediate states (below) relevant in the perturbation calculations. For the excited states, an O spin jumps to the Cu site on one side, while the other O spin interacts with the Cu spin on the other side via direct exchange.

where $d_{i\sigma}^\dagger, d_{i\sigma}, p_{l\sigma}^\dagger, p_{l\sigma}$ are creation and annihilation operators for d electrons at Cu sites i, and p electrons at O sites l, $l \in \{i\}$ are the nearest neighbors of the Cu site i. U is the on-site Coulomb interaction for d electrons. For undoped cuprates, one can derive the effective spin interaction of Cu-site spins by using perturbation theory, known as Kramers superexchange theory where the p electrons play the role of mediation for the magnetic interaction between Cu-site spins [10] (Fig. 19.2):

$$H = -K \sum_{i,l \in \{i\}} \mathbf{S}_i \cdot \mathbf{S}_j, \qquad (19.2)$$

where $K < 0$ is the anti-ferromagnetic coupling constant:

$$K \propto \rho^2 J_{\mathrm{d}},$$

and ρ is the transition integral p $\to$ d transition:

$$\rho = \int \mathrm{d}^3 r \phi_{\mathrm{d}}(r) \hat{V} \phi_{\mathrm{p}}(r).$$

The d orbital ϕ_{d} in a Cu^{2+} ion is a localized state with the $\mathrm{d}_{x^2-y^2}$ symmetry, whereas the p state is described by a Wannier wavefunction ϕ_{p} in 2D. The coupling between p and d electrons arises from the overlap of ϕ_{d} and ϕ_{p}, and the magnetization of Cu spins in the CuO_2 layer is observed in neutron scattering experiments [11].

19.2. The Magnetic Origin of Superconductivity in Cuprates

Anderson proposed in 1987 that the spin $S = 1/2$ lattice described by Eq. (19.2) has a type of ground state where the local spins have no preferential directions, the neighboring Cu spins are paired into the singlets, the so-called resonating valence bond state, and condense to the superconducting state when the doped CuO_2 layer becomes metallic [12].

In fact, the doped cuprate has a two-band structure. The holes introduced in the CuO_2 layer reside on O sites. In this case, whether the degrees of freedom of the p electrons are integrable remains a question. In order to derive an effective single-band model for the superconducting cuprates, Zhang and Rice proposed that the hybridization between p and d electrons binds a hole on each square of O atoms to the central Cu^{2+} ion to form a local singlet [13]. Zhang–Rice started from the two-band model (Eq. (19.1)) where $p_i^\dagger$ and p_i are creation and annihilation operators for holes. The four O hole states around a Cu ion were written in a symmetric form:

$$\sum_\delta t_\delta p_{i+\delta,\sigma} = 2t_0 P_{i\sigma},$$

where t_0 is the amplitude of the hybridization. $P_{i\sigma}$ are not orthogonal:

$$[P_{i\sigma}, P_{j\sigma'}^\dagger]_+ = \delta_{\sigma\sigma'}\left(\delta_{ij} - \frac{1}{4}\delta_{<ij>}\right).$$

We define Wannier functions:

$$\phi_{i\sigma} = \frac{1}{\sqrt{N}}\sum_k P_{k\sigma}\exp(i\mathbf{k}\cdot\mathbf{R}_i),$$

$$P_{k\sigma} = \frac{1}{\sqrt{N}}\beta_k\sum_i P_{i\sigma}\exp(-i\mathbf{k}\cdot\mathbf{R}_i),$$

where β_k is a normalization factor. $\phi_{i\sigma}$ is orthogonal:

$$[\phi_{i\sigma}, \phi_{j\sigma'}^\dagger]_+ = \delta_{\sigma\sigma'}\delta_{ij}.$$

$P_{i\sigma}$ can be expanded as:

$$P_{i\sigma} = \sum_j \lambda(R_i - R_j)\phi_{j\sigma} = \lambda_0\phi_{i\sigma} + \lambda_1\sum_{<1>}\phi_{<1>\sigma} + \lambda_2\sum_{<2>}\phi_{<2>\sigma} + \cdots,$$

(19.3)

where $\lambda_0 \approx 0.96$, $\langle 1 \rangle$ are the nearest-neighbor sites around site i with $\lambda_1 \approx 0.14$, $\langle 2 \rangle$ are the next nearest, we have:

$$H = \sum_{i,\sigma} \varepsilon_d d_{i\sigma}^\dagger d_{i\sigma} + U \sum_i n_{i\uparrow} n_{i\downarrow} + \lambda_0^2 \sum_{i,\sigma} \phi_{i\sigma}^+ \phi_{i\sigma} + H', \tag{19.4}$$

where H' is the hybridization interaction:

$$H' = 2\lambda_0 t_0 \sum_{i,\sigma} (d_{i\sigma}^+ \phi_{i\sigma} + \text{h.c.}) + 2\lambda_n t_0 \sum_{i,\sigma,n=1}^\infty (d_{i\sigma}^+ \phi_{<n>\sigma} + \text{h.c.}).$$

H' is taken as perturbation term, and the hybridization binds an O hole to a Cu at site i. Two states are relevant:

$$|\alpha\rangle = d_{i\uparrow}^\dagger \phi_{i\downarrow}^\dagger |B\rangle, \quad |\beta\rangle = d_{i\downarrow}^\dagger \phi_{i\uparrow}^\dagger |B\rangle, \tag{19.5}$$

where:

$$|B\rangle = \sum_{l=1}^M a_l |B_l\rangle, \quad |B_l\rangle = \prod_{j\neq i} d_{j\sigma_l(j)}^\dagger |0\rangle.$$

$|B_l\rangle$ is the spin configuration of the whole Cu spin-lattice excluding site i. By using second-order perturbation theory to calculate the binding energy:

$$E^\pm = -8t(1 \mp \lambda_0^2) < 0, \quad t = \frac{t_0^2}{\lambda_0^2 \varepsilon_p - \varepsilon_d},$$

where E^+ and E^- are the binding energies of spin triplet and singlet. The singlet is the ground state. The singlet effectively screens a Cu spin, and moves through the lattice of Cu^{2+} ions as a "hole" (Fig. 19.3).

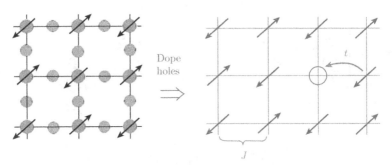

Fig. 19.3. Zhang–Rice's theory is an attempt to integrate the degrees of freedom for p electrons, reducing the two-band model to an effective single-band model.

By integrating the degrees of freedom of p electrons, Zhang and Rice obtained an effective single-band Hamiltonian:

$$H_{t\text{-}J} = -t \sum_{i,j,\sigma}(c_{i\sigma}^\dagger c_{j\sigma} + \text{h.c.}) - J\sum_{i,j} \boldsymbol{S}_i \cdot \boldsymbol{S}_j \qquad (19.6)$$

to describe the low-excitation states of the superconducting cuprate. Despite the simple appearance and the partial success in calculating for the low-excitation states for the cuprates, the single-band model failed to explain either the crossover from the anti-ferromagnetic to the superconducting states as shown in the phase diagram, or coexistence of superconductivity and magnetism in the superconducting region.

19.3. Effect of the Local Spin Polarization

In the low-doping limit the correlation of the local Cu spins in a cuprate is not negligible. We consider the competition between the local Cu spin polarization and Zhang–Rice's singlet to show that the Zhang–Rice singlet is decoupled by the arbitrarily weak spin polarization [14]. By introducing the local-spin polarization

$$\Delta = \varepsilon_{d\uparrow} - \varepsilon_{d\downarrow}$$

to the two-band model (19.1), we adopt the diagonalization method. The number of states involved in second perturbation calculation is 28. We have a 28×28 matrix for H:

$$H - E_B = \begin{pmatrix} \lambda_0^2\varepsilon_p+\varepsilon_d & \Delta & 0 & 0 & BV_1 & CV_1 & CV_1 \\ \Delta & \lambda_0^2\varepsilon_p+\varepsilon_d & -A & A & BV_2 & CV_2 & CV_2 \\ 0 & -A & U+2\varepsilon_d & 0 & 0 & 0 & 0 \\ 0 & A & 0 & 2\lambda_0^2\varepsilon_p & 0 & 0 & 0 \\ BV_1^T & BV_2^T & 0 & 0 & (2\lambda_0^2\varepsilon_p)I_8 & 0 & 0 \\ CV_1^T & CV_2^T & 0 & 0 & 0 & (U+2\varepsilon_d)I_8 & 0 \\ CV_1^T & CV_2^T & 0 & 0 & 0 & 0 & (2\lambda_0^2\varepsilon_p)I_8 \end{pmatrix},$$
$$(19.7)$$

where

$$A = \frac{4}{\sqrt{2}}\lambda_0 t_0, \quad B = \frac{2}{\sqrt{2}}\lambda_1 t_0, \quad C = \lambda_1 t_0,$$

$$D_1 = \lambda_0^2 \varepsilon_p + \varepsilon_d, \quad D_2 = U + 2\varepsilon_d, \quad D_3 = \lambda_0^2 \varepsilon_p.$$

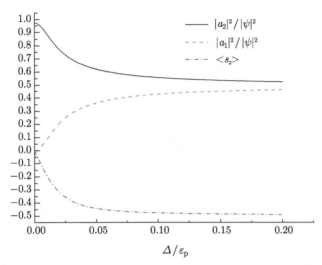

Fig. 19.4. The numerical computation for amplitudes of the triplet ($|a_1|^2$) and the singlet states ($|a_2|^2$) as functions of spin polarization Δ, where we take $t/\varepsilon_p = 0.05, U/\varepsilon_p = 3$. The average spin on Cu site $\langle s_z \rangle = a_1 a_2$. Coefficient $a_2 = -a_1 \approx -1/\sqrt{2}$ at large Δ, the eigenstate is a spin-polarized state.

I_8 is an 8×8 unit matrix, while V_1 and V_2 are 8×1 vectors:

$$V_1 = (1\ 1\ 1\ 1\ 1\ 1\ 1\ 1), \quad V_2 = (1\ 1\ 1\ 1\ -1\ -1\ -1\ -1).$$

The eigenvector of H is defined as:

$$\psi = (a_1,\ a_2, \ldots, a_{28}), \tag{19.8}$$

where a_1 is the coefficient for the triplet state, a_2 is for the singlet. a_3 to a_{28} are for the base vectors listed in Eq. (19.6), from which we have:

$$(\lambda_0^2 \varepsilon_p + \varepsilon_d) a_1 + \Delta a_2 + B \sum_{n=5}^{12} a_n + C \sum_{n=13}^{28} a_n = E a_1,$$

$$B a_1 + B a_2 + 2\lambda_0^2 \varepsilon_p a_5 = E a_5,$$

$$B a_1 - B a_2 + 2\lambda_0^2 \varepsilon_p a_9 = E a_9.$$

If there exists a singlet solution, it requires that $a_1 = 0$. It follows:

$$\sum_{n=5}^{12} a_n = 0, \quad \sum_{n=13}^{28} a_n = 0.$$

So:

$$\Delta \cdot a_2 = 0. \qquad (19.9)$$

If Δ is non-zero, a_2 the coefficient for the singlet state equals to zero, thus the singlet state is unstable.

19.4. The Resonating Valence Bond State in the Superconducting Cuprates

In this section, we show that the RVB state can be generated in the two-band structure where holes are the charge carriers and Cu spins are localized provided there is a magnetic background. The holes introduced in CuO_2 layer not only induce metallic-insulating transition, but also suppress the on-site superexchange, with the quasi-free spin flips on Cu sites breaking the long-range anti-ferromagnetic order. In low-doped cuprate, the coupling between Cu spins is Ising-like [15]:

$$H_m = -K \sum_{i,j} S_i^z S_j^z,$$

with a weakened coupling constant K:

$$K = K_0(1-v),$$

where K_0 is the superexchange for undoped cuprates, v is the average of the hole occupancy of the O sites. The hole states are described by the Wannier functions. To illustrate the formation of the RVB state, we consider the direct exchange interaction between two holes and a Cu spin at the neighbor site in a spin cluster (Fig. 19.5),

$$H_{int} = -J\boldsymbol{S} \cdot (\boldsymbol{\sigma}_1 + \boldsymbol{\sigma}_2), \quad J < 0,$$

where J is the coupling constant of the direct exchange. $\boldsymbol{S}$ is the Cu spin. The polarized $\boldsymbol{S}$ has two positions: $|0\rangle$ the ground state and $|1\rangle$ the excited state. σ_i ($i = 1, 2$) are O spins. By using second-order perturbation theory, we derive an effective spin coupling between O spins [16]:

$$H_{eff} = \frac{\langle 0|\hat{H}_{int}|1\rangle\langle 1|\hat{H}_{int}|0\rangle}{E_0 - E_1},$$

which is:

$$H_{eff} = -\frac{1}{4}\lambda J(\sigma_1^+ + \sigma_2^+)(\sigma_1^- + \sigma_2^-), \qquad (19.10)$$

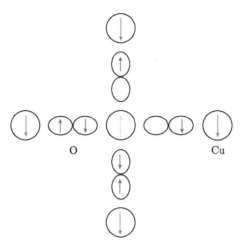

Fig. 19.5. The uncompensated spins created by holes residing on O sites in a spin cluster interact with the Cu spin at the center.

where $\lambda = J/4K$ varies with the doped hole concentration, and $\sigma^{\pm} = \sigma_x \pm i\sigma_y$. H_{eff} has an eigenstate:

$$\chi_{+} = \frac{1}{\sqrt{2}}(|\uparrow\downarrow\rangle + |\downarrow\uparrow\rangle),$$

with the binding energy:

$$E_b = -\frac{1}{2}\lambda J\left(1 - \frac{1}{2}\lambda^2\right). \tag{19.11}$$

The spin pair has the quantum numbers of $S = 1$, $S_z = 0$, so the possibility of s-wave pairing is excluded. For the local pairing, the symmetry of the electron pair is restricted by the symmetry of the crystal lattice of the CuO_2 layer, therefore, the electron pair is forced to have d symmetry [17]. We write an anti-symmetric d wavefunction in terms of scattered partial waves in $l = 2$ channel:

$$\psi = Y_2^{+1}(\theta_1, \varphi_1)Y_2^{-1}(\theta_2, \varphi_2) - Y_2^{-1}(\theta_1, \varphi_1)Y_2^{+1}(\theta_2, \varphi_2).$$

the total quantum number $m = m_1 + m_2 = 0$. Let $\theta = \theta_1 - \theta_2$, $\varphi = \varphi_1 - \varphi_2$:

$$\psi \propto \cos 2\theta \sin \varphi,$$

where $-\pi \leq \theta \leq \pi$. Since the wavefunction ψ is limited in a 2D layer, we take a fixed value for φ, $\varphi \to \pi/2$, or $-\pi/2$ (Fig. 19.6). The projection of ψ

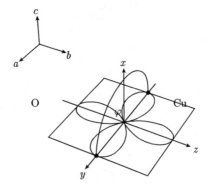

Fig. 19.6. The projection of the anti-symmetric 3D wavefunction in the $a - b$ plane, where φ is defined in the $x - y$ plane.

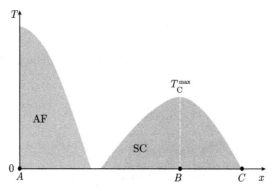

Fig. 19.7. The phase diagram of the cuprates. A, B and C are three critical doping levels. A: the undoped cuprate is an anti-ferromagnetic insulator where the p electron plays the role of mediator for the superexchange interaction for Cu spins. B: near optimal doping, two O spins generated by the doped holes form an RVB pair via the effective spin coupling mediated by a Cu spin. C: the d-band electrons are delocalized and hybridized with the p band electrons forming a Fermi liquid.

in the a-b plane has the same pattern as $d_{x^2-y^2}$, but it is consistent with the triplet pairing. The theory of spin pairing on O sites naturally explains the phase diagram (Fig. 19.7). Near the optimum doping, the superconducting transition temperatures of the cuprates obey an empirical universal law:

$$\frac{T}{T_m} = 1 - \kappa(x - x_0)^2.$$

From Eq. (19.11), we can derive the expression for κ:

$$\kappa = \left(\frac{4K_0}{J}\right)^2,$$

where K_0 is the superexchange constant of the undoped cuprate, J is the direct exchange between the hole spins and a neighbor Cu spin. The upper limit of the superconducting transition temperature is determined by K_0:

$$T_{\max} = \frac{1}{6.81 k_B}\lambda_{\max} K_0.$$

If $K_0 \approx 0.1$ eV, $T_{\max} \sim 150$ K.

Based on the effective spin coupling Eq. (19.10), we propose an effective Hamiltonian for the superconducting cuprates:

$$H = \sum_{i,j} t_{ij} p_i^\dagger p_j - g\sum_{i,\delta,\sigma} p_{i\sigma}^\dagger p_{i+\delta,-\sigma}^\dagger p_{i+\delta,\sigma} p_{i,-\sigma} - K\sum_{i,j} S_i^z S_j^z, \quad (19.12)$$

where $p_i^\dagger$ and p_i are creation and annihilation operators of holes. The effective interaction Eq. (19.12) describes the local pair pairing in the presence of a magnetic background of Cu spins. The local pairs form a RVB state and condense, and the cuprate becomes superconducting. g depends on the doping level. When the p band is filled, $g = 0$. Near the optimal doping:

$$g = \frac{J}{4K}.$$

Equation (19.12) gives a correct phase diagram.

19.5. Summary

Based on Kramers' theory, we have shown the key role that the p electrons play in superconductivity in a doped cuprate. In fact, Kramers' theory which describes the indirect interaction in the transition metal oxides where the p electrons on O sites play the role of mediation had never been tested in experiment by taking the p electrons away until the superconducting cuprate was discovered. The phase diagram shows that the anti-ferromagnetic order of Cu spins is sensitive to the impurity doping. The most important generation for Kramers' theory is the emergence of the RVB state when holes are introduced in p orbital in a doped cuprate. The results of NMR experiments indicate the pairing occurs on O sites [18]. The d electrons, which are emphasized in the single-band model, merely provide magnetic background and play the role of mediation for p electrons' pairing. Two types of the interactions, the direct exchange J between p and d electron spins and the superexchange K between d electron spins are essential to magnetic ordering and superconductivity in the cuprates. The upper limit of the superconducting transition temperature is determined by the strength of the superexchange of the undoped cuprate. Impurity doping alters the nature of the ground state. By reducing the strength of the superexchange, the cuprate varies from an anti-ferromagnetic insulator to a superconductor, then to a Fermi liquid state when d electrons become delocalized.

References

[1] B. Lake, *Nature* **415**, 299 (2002).
[2] A. Aharony, *Phys. Rev. Lett.* **60**, 1330 (1988).
[3] P. C. Dai et al., *Phys. Rev. B* **63**, 54525 (2001).
[4] T. Thio et al., *Phys. Rev. B* **38**, 905 (1988).
[5] M. Matsuda et al., *Phys. Rev. B* **66**, 174508 (2002).
[6] M. Takigawa, A. P. Reyes, P. C. Hammel, J. D. Thompson, R. H. Heffner, Z. Fisk and K. C. Ott, *Phys. Rev. B* **43**, 247 (1991).
[7] A. Kaminski, S. Rosenkranz, H. M. Fretwell, J. C. Campuzano, Z. Li, H. Raffy, W. G. Cullen, H. You, C. G. Olson, C. M. Varma and H. Hochst, *Nature* **416**, 610 (2002).
[8] W. Guo, L. S. Duan and R. S. Han, *Int. J. Mod. Phys. B* **19**, 63 (2005).
[9] S. Uchida, T. Ido, H. Takagi, T. Arima, Y. Tokura and S. Tajima, *Phys. Rev. B* **43**, 7942 (1991).
[10] Kramers, *Physica* **1**, 187 (1934).
[11] D. V. Vaknin et al., *Phys. Rev. Lett.* **58**, 280 (1987).
[12] P. W. Anderson, *Science* **235**, 1196 (1987).
[13] F. C. Zhang and T. M. Rice, *Phys. Rev. B* **37**, 3759 (1988).
[14] H. Li and W. Guo, *J. Supercond. Novel Magnet.* **23**, 679 (2010).
[15] A. N. Lavrov et al., *Phys. Rev. Lett.* **87**, 017007 (2001).
[16] W. Guo, X. G. Yin and R. S. Han, *Int. J. Mod. Phys. B* **21**, 3112 (2007).
[17] C. C. Tsuei and J. R. Kirtley, *Rev. Mod. Phys.* **72**, 969 (2000).
[18] G. Q. Zheng, Y. Kitaoka, K. Asayama, K. Hamada, H. Yamauchi and S. Tanaka, *Physica C* **260**, 197 (1996).

20
Magnetic Excitations in High-Temperature Superconductors: Search for Universal Features in Different Classes of Copper Oxides

Peng-Cheng Dai[*,†] and Shi-Liang Li[†]

[*]*Department of Physics and Astronomy, University of Tennessee Tennessee 37996-1200 USA*
[†]*Oak Ridge National Laboratory, Tennessee 37831-6393, USA*
[‡]*Institute of Physics, Chinese Academy of Sciences, Beijing 100190, China*

We review current progress in the neutron scattering studies of magnetic excitations in high-transition temperature (high-T_C) copper oxide superconductors. Since its discovery 20 years ago, understanding the microscopic origin of superconductivity in these copper oxides has been the "holy-grail" in condensed matter physics. In contrast to conventional superconductors, where the interaction that pairs the electrons to form the superconducting state is mediated by lattice vibrations (phonons), it is generally believed that magnetic excitations might play a fundamental role in the superconducting mechanism of copper oxides because superconductivity occurs when mobile "electrons" or "holes" are doped into the anti-ferromagnetic parent compounds. In this chapter, we summarize the key results on the magnetic excitations obtained by inelastic neutron scattering over the past 20 years. We discuss the status of the field and point out possible future directions.

20.1. Introduction

Not long ago, condensed matter physicists held up the banner of Bloch theorem, generally believing that if we can understand a single unit cell, then we can understand the nature of the crystal, because the translational symmetry ensures that every cell is identical. However, this assumption of uniform ground state (for most condensed matter physicists, they accept this point almost without hesitation), in fact, can only be strictly set up in the case that the solid can be described by single electron model. The latter can be summarized by Landau's Fermi liquid theory (FLT). In the last 50 years, FLT provided a far-reaching perspective for researching the nature of the material's metallic states. According to FLT, in an ordinary metal, the electronic system with interaction can be seen as the quantum gas, composed of non-interacting "quasi-particles". This kind of quasi-particles are like free electrons from every point of view, but they just do not have the same effective mass. FLT, Band theory and BCS theory based on the FLT can describe most insulators, semiconductors, metals, and superconductors well. However, the high-T_C superconducting copper oxides found in 1986 completely changed the understanding we had established about superconducting and even strongly correlated electron materials, because many of their natural characteristics are completely different from the FLT predictions.

Different from the conventional superconductors, all high-T_C superconductors can be acquired by doping "hole" and "electron" to anti-ferromagnetic parent compounds, but the former is metal. In the undoped parent compound, the anti-ferromagnetic structure is very simple, just the lattice unit cell in the copper oxide plane doubling, as shown in Fig. 20.1(a) [1]. When mobile "holes" or "electrons" are doped into the copper-oxide plane, the long-range anti-ferromagnetic order is gradually destroyed and is replaced by metal and superconducting phase. Although the static anti-ferromagnetic order has been eliminated, the neutron scattering experiments have clearly shown that in samples with all doping concentrations there are short-range anti-ferromagnetic spin fluctuations (magnetic excitation). Like phonons, spin fluctuations may also be used as a medium, which makes the electron pairs and forms the superconducting state [2, 3]. Therefore, when the system transits from the anti-ferromagnetic insulators to overdoped superconductors, determining the dependence relationship of the energy $\hbar\omega$ of spin fluctuations and the wave

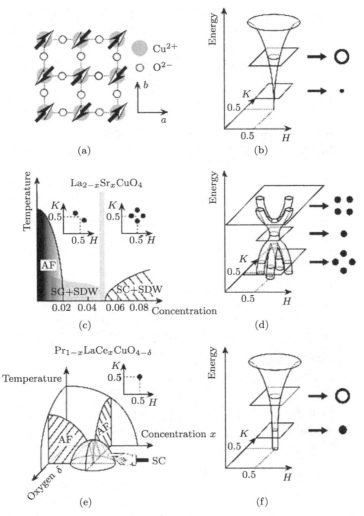

Fig. 20.1. A brief summary of the status quo in this area: (a) the parent compound of hole-doped superconductors have the spin structure with line, but the electron-doped do not; (b) the spin-wave dispersion relation from the anti-ferromagnetic order of Cu^{2+} can be well described by the nearest Heisenberg Hamiltonian; (c) in the $La_{2-x}Sr_xCuO_4$ (LSCO) system, the evolution of spin density wave (SDW) with hole density increasing, when the doping concentration is $0.02 \leq x \leq 0.05$, there are two static incommensurate SDW peaks in the spin-glass phase of LSCO, $(0.5+\delta, 0.5-\delta)$ and $(0.5-\delta, 0.5+\delta)$. When $x \gg 0.05$ in the system in the superconducting state, two pairs of incommensurate peaks simultaneously turn up at $(0.5 \pm \delta, 0.5)$ and $(0.5, 0.5 \pm \delta)$; (d) spin excitation dispersion relation of hole-doped LSCO and YBCO systems [28]; (e) the 3D phase diagram of e-doped superconductors, this chapter focuses on the oxygen concentration δ variation in the $Pr_{0.88}LaCe_{0.12}CuO_4$ samples; (f) spin excitation dispersion relation of e-doped materials.

vector Q becomes very important. The purpose of this chapter is to summarize the latest magnetic excitation results using neutron scattering in the high-T_C superconductors of different systems. If the spin fluctuations are indeed very important for high-temperature superconducting mechanism, then there will be a common feature for high-temperature superconductors of different systems.

Finally, we briefly introduce neutron scattering. Neutron is a subatomic particle, having 1.0087 unit atomic mass, without charge, but with 1/2 spin. Due to the wavelength and energy of thermal neutrons, thermal and cold neutrons can be compared to the atomic spacing in solids and the lattice vibrations. Neutron scattering plays a unique and irreplaceable role in determining the static and dynamic lattice of high-T_C superconductors and their magnetic properties for the following reasons: (1) neutrons are scattered by nuclei, so the scattering intensities of light atoms (such as oxygen) and heavy atoms are similar; (2) the neutron magnetic moment (spin 1/2) and interaction between local magnetic ions in the solid and unpaired itinerant-electron; (3) neutrons are a weak detection tool, so neutron scattering cross-section depends only on the static and dynamic correlation functions, without the need to amend the impact of the detector itself; (4) neutrons can penetrate the entire sample, so the surface defect does not affect the measurement results.

20.2. The Evolution of Static Anti-ferromagnetic Order when Hole- or Electron-Doped to the Copper–Oxygen Plane

The elastic neutron scattering can measure the static anti-ferromagnetic order, so it can be used to determine the magnetic structure of hole- and electron-doped materials [4]. While the spin waves in the undoped parent compound and spin fluctuations in superconducting materials can be determined by inelastic neutron scattering in principle, the latter needs large, high-quality single crystals, so most of the spin excitation measurements are limited to $La_{1-x}Sr_xCuO_4$ (LSCO) and $YBa_2Cu_3O_{6+x}$ (YBCO) copper-oxide systems, because it is relatively easy to obtain large single crystals in these two systems [5]. For LSCO samples with only one layer copper oxide plane in each single cell, the neutron scattering experiments have found that the superconductivity and static incommensurate spin density wave

(SDW) coexist, in the samples ($x \approx 1/8$) [6]. Here, the SDW is quasi-2D, i.e. long-range ordered in the oxygen-copper plane, only short-range correlation along c axis [6]. To the square lattice of LSCO index, the position of anti-ferromagnetic order in the reciprocal space is located in copper oxide layer (0.5, 0.5), while the position of four incommensurate SDW peaks are $(0.5 \pm \delta, 0.5)$ and $(0.5, 0.5 \pm \delta)$ [5]. At the optimally doped region, LSCO has the highest T_C, the static SDW order disappears and the spin energy gap turns up, and above the spin-gap energy the incommensurate spin fluctuations replace it [7]. On the other hand, in the low doping insulating ($x = 0.03, 0.04, 0.05$) and superconducting ($x = 0.06$) samples, the experiments show that at the boundary of non-superconducting state to superconducting state transition, the SDW features have changed [8]. For non-superconducting $La_{1-x}Sr_xCuO_4$ ($x = 0.03, 0.04, 0.05$), there exist two incommensurate SDW peaks. Compared to the incommensurate peak of superconducting samples, their direction has rotated 45°. When increasing the doping concentration x to change the samples into the superconducting state, the locations of the two incommensurate SDW peaks become four incommensurate peaks at $(0.5 \pm \delta, 0.5)$ and $(0.5, 0.5 \pm \delta)$. In addition, at low-doped samples, the incommensurate value δ depends linearly on doping x, while $x \geq 1/8$, the samples tend to saturation [7].

For the hole-doped YBCO, the situation is very different. Because it's very difficult to make the oxygen atoms uniform in $x \sim 0.3$ YBCO samples, the static anti-ferromagnetic evolution is unclear at the boundary of insulator–metal phase transition. However, recently the μ spin rotation (μSR) experiments indicate that in very clean underdoped YBCO ($x = 0.375$) single crystals, the superconducting phase and anti-ferromagnetic phase coexist [9]. The neutron scattering experiments of similar single crystals indicate the anti-ferromagnetic order is quasi-static, and its behavior is very alike to spin-glass [10].

For the typical e-doped superconductors $Nd_{2-x}Ce_xCuO_4$ (NCCO) [11, 12], the previous neutron scattering experiments show that when the superconducting state is established, the static 3D anti-ferromagnetic order is strongly suppressed [13]. However, even for the optimally doped NCCO ($T_C = 25K$) the static anti-ferromagnetic order still exists [25]. But for the same e-doped $Pr_{1-x}La_xCe_xCuO_{4-\delta}$ (PLCCO), μSR [15] and neutron diffraction results [16, 17] show that the static anti-ferromagnetic order will not exist when the PLCCO is annealed to the optimal superconductivity, as shown in Fig. 20.1(e). In the middle doping regions, there exist 2D commensurate SDW, like the incommensurate SDW in hole-doped materials. Therefore, there is an intermediate process in the copper oxide from the anti-ferromagnetic to the superconducting transition, and during this period, the material forms a microscopic (or mesoscopic) uneven state.

20.3. The Evolution of Spontaneous Excitation as a Function of Hole Concentration

Despite the parent compounds of high-T_C superconductors having antiferromagnetic spin structure, such as La_2CuO_4 and $YBa_2Cu_3O_4$, shortly after they were found, they have been solved [1, 5]. However, the complete dispersion of spin-wave excitation in La_2CuO_4 was finally determined after the ISIS's time of flight neutron spectrometer completion in Rutherford Appleton Laboratory. Figure 20.2(a) summarizes the spin-wave dispersion relation of La_2CuO_4 [18]. Using Heisenberg Hamiltonian, Radu et al. inferred the nearest-neighbor exchange $J = 111.8 \pm 4$ meV, and the second-neighbor exchange $J' = 11.4 \pm 3$ meV. This means that the magnetic exchange interactions are mainly nearest-neighbor anti-ferromagnetic interactions. Figure 20.2(b) provides the relationship between local magnetic susceptibility $\chi''(\omega)$ and energy, where $\chi''(\omega) = \int \chi''(Q,\omega) d^3Q / \int d^3Q$ [18, 19]. Results as the conventional spin wave, localized magnetic susceptibility, detected in most of the energy area is essentially constant, independent of energy. The results indicate that the anti-ferromagnetic order of high-T_C superconducting parent compounds can be well described by the Heisenberg theory. Therefore, now the question is when holes or electrons dope into the copper-oxygen layer, how will the spin wave change?

For LSCO samples, the related elastic measurements have shown the existence of four incommensurate peaks, and that the incommensurate peaks are similar to first-order phase transition, when the superconducting state turns up at $x \geq 0.06$ [8]. Inelastic neutron scattering experiments show that the incommensurate value δ of spin fluctuations is only dependent on the doping level, independent of the way of introducing the charge (Fig. 20.3). As a function of hole concentration, the incommensurate value increases linearly with hole concentration doping up to $x \geq 1/8$, which saturates at about $\delta \approx 0.125$ [7]. As a function of energy, the incommensurate value decreases as the energy increases, and gathers together to a point, and the latter is called resonance peak. Above the resonance peak,

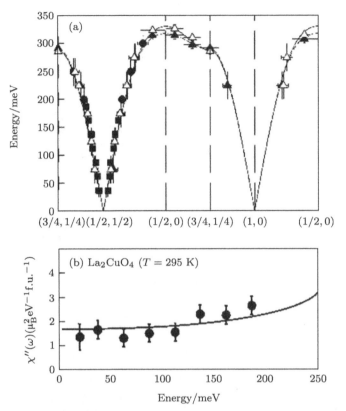

Fig. 20.2. (a) Undoped La_2CuO_4 spin-wave dispersion curve, solid and dashed lines are simulation results, by the use of Heisenberg Hamiltonian; (b) the local magnetic susceptibility of La_2CuO_4, its size is determined by the integral of the observed spin wave vector. (a) taken from Ref. [18]. The lower panel is from [19].

the re-emergence of incommensurate peaks spread out, making the spin excitation spectrum look like an hourglass (Fig. 20.1(d)) [21, 22].

For YBCO, which have two copper-oxide planes in each unit cell, the situation is more complex. Earlier, the measurement of near optimally doped YBCO was mainly the resonance peak, i.e. the sharp magnetic excitation at the centre of the 2D reciprocal space wave vector (1/2, 1/2) of copper-oxide plane [23, 24]. It is noteworthy that the wave vector corresponds to anti-ferromagnetic Bragg peak position of the undoped parent compound (Figs. 20.1(a) and 20.1(b)). However, recently, the neutron scattering experiments by the new system find that the spin excitation spectra of YBCO and LSCO are strikingly similar [25–27]. For example,

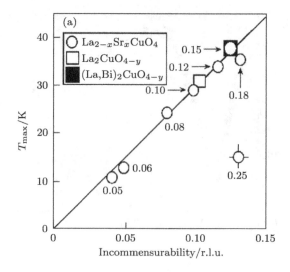

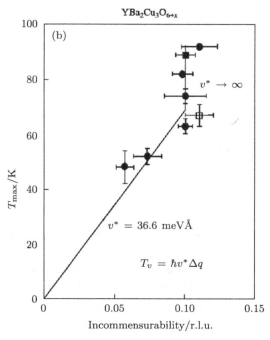

Fig. 20.3. (a) The incommensurate value as a function of T_C in single-layer high-T_C superconducting copper oxide LSCO system. Taken from Ref. [7]. (b) Similar data in the bilayer high-T_C superconductor YBCO, noting that δ becomes saturated to $\delta = 0.1$ at $T_C > 60$ K [25].

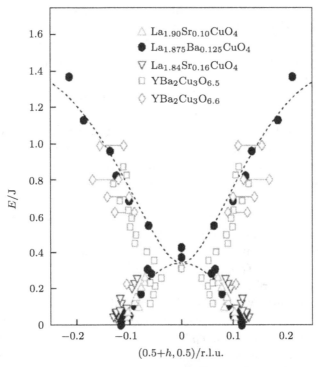

Fig. 20.4. The dispersion relation comparison of different hole-doped high-temperature superconducting copper oxides (where the energy has been scaled by the exchange energy of parent insulator, detail seen in Ref. [28]). Taken from Ref. [28].

the spin fluctuations below the resonance peak are nearly at the same incommensurate position as of LSCO [26]. Different from LSCO, the incommensurate value of YBCO dependence on doping saturates at $x \geq 0.5$ about $\delta \approx 0.1$ [25]. This difference can be explained by the existence of a large superconducting spin gap in YBCO, because the latter would suppress the low-energy spin excitation [28].

In a recent review article, Tranquada [28] discussed researching the status quo in the area of the hole-type materials. Figure 20.4 summarizes the current research about the dispersion relation of all hole-type single layer LSCO and double-layer YBCO materials, in which the vertical axis is the ratio of energy and exchange coupling energy J. The key conclusion Fig. 20.4 shows is that the dispersion relation of spin excitation in hole-type copper oxides is universal and independent of the details and materials.

One of the hole-type copper oxides $La_{0.875}Ba_{0.125}CuO_4$ has static stripe order [21]. In the stripe order, the holes are isolated into a uniform pattern (stripes), which obtains the anti-ferromagnetic region. According to the point, Tranquada thought that the spin excitation of hole-type copper oxide was derived from the dynamic form of stripes. In other words, the excitation is the result of dynamic stripes, so the stripes should be directly related to high-T_C superconducting mechanism.

While this stripes image may explain some of spin excitation features, they may also be derived from quasi-particle interactions of crossover nesting Fermi surface. Because there is sufficient evidence to show the existence of quasi-particle and Fermi surface in doped copper-oxide superconductors, the spin excitation dependence on the doping and energy may reflect the change of Fermi surface topology. For example, the hourglass-type spin excitation detail of YBCO has been predicted by nesting model before the neutron scattering experiments [29]. The above-mentioned resonance peak and incommensurate spin fluctuations are very different from the dispersion relations of the parent's spin-wave excitation. In addition, the evolution of local susceptibility (wave vector integral) after introducing hole into the copper-oxygen plane is also very interesting. As shown in Fig. 20.2(b), undoped local magnetic susceptibility of La_2CuO_4 is mostly independent of energy when the energy is smaller than 200 meV. The undoped YBCO ($x \leq 0.2$) has similar situation. However, for YBCO, the spin-wave excitation has acoustic and optical two-channel, because each unit cell has two layer copper ions. The acoustic spin wave corresponds to the same motion of spin in two copper-oxygen planes, but the optical spin wave corresponds to the opposite movement of spin. For undoped YBCO ($x \leq 0.2$), the optical gap is about 70 meV [30]. For underdoped YBCO ($x = 0.6, T_C = 63$ K), the local magnetic susceptibility of acoustic channel shows, as the superconducting state turns up, the main characteristics are resonance peak at about 34 meV and spin gap at about 20 meV (Fig. 20.5) [20]. But in the optical channel, the 70 meV optical gap in the insulator drops down to 40 meV. Obviously, compared with the local magnetic susceptibility of insulating YBCO, hole-doping simultaneously suppresses the low- and high-energy spin fluctuations, making the magnetic susceptibility of middle energy region greater than insulator.

For optimally doped single layer LSCO samples, in Fig. 20.6 the local magnetic susceptibility $\chi''(\omega)$ as a function of temperature gives the same basic image. In the normal states, there is no spin gap in $\chi''(\omega)$, but compared with spin wave of undoped samples (see Fig. 20.2), the low-energy and

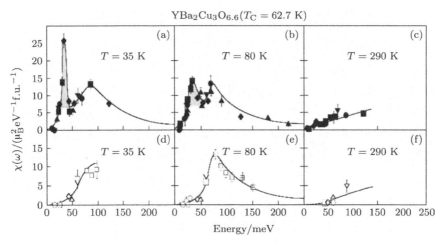

Fig. 20.5. The local magnetic susceptibility of acoustic and optical channels in the under-doped YBCO ($x = 0.6$) as a function of temperature [20].

high energy are simultaneously suppressed. However, the $\chi''(\omega)$ of middle energy region is larger than undoped spin wave. When the system enters the superconducting state at low temperature, a ~6 meV spin gap opens, and pushes density of states to the 9–30 meV energy region above spin gap (Fig. 20.6). If assuming the spin fluctuations, observed by neutron, are directly related to the density of state, the results are strikingly similar to the traditional BCS picture. Although the phenomenon of the intensity increase below T_C does not appear in anti-ferromagnetic Bragg point $(1/2, 1/2)$, if it appears in the incommensurate position, it is still considered that the phenomenon and double-resonance peak in YBCO system together give us the same signal [28]. Obviously, in order to create the common features of different superconducting copper-oxygen, we should observe other materials, measured by lots of neutron scattering experiments. Therefore, the next section of chapter this article will focus on what we have done in the past few years, the research results of electronic high-temperature superconducting materials.

20.4. The Evolution of the Spin Excitation in the Electron-Type High-T_C Superconductors

E-type high-temperature superconductors were discovered soon after the h-type materials were found [11, 12]. A unique feature of these materials

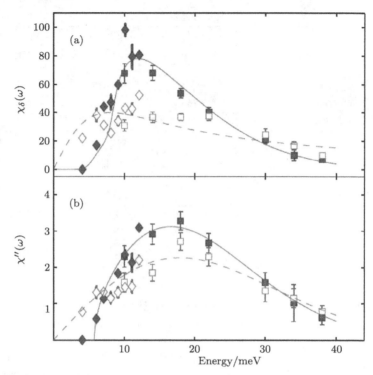

Fig. 20.6. (a) The incommensurate peak $\chi''_\delta(\omega)$ by fitting the peak of $\chi''(Q,\omega)$ at the incommensurate position and (b) local susceptibility $\chi''(\omega)$ as a function of temperature in the optimally-doped LSCO. Taken from Ref. [22].

is that the newly grown single crystals are insulators of long-range anti-ferromagnetic order. Only after they are annealed to remove excess oxygen, they can become superconductors. Although the specific microscopic mechanism of superconductivity caused by the annealing process is unclear, this necessary process itself shows that the actual doping concentration is controlled by Ce and oxygen concentration. Thus, the phase diagram of electron-type materials should be 3D. Most previously reported phase diagrams show the dependence of Ce-doping [15], while we focus on understanding the process as a function of annealing and deoxygenization from the anti-ferromagnetic transition to superconductivity. For $Pr_{0.88}LaCe_{0.12}O_{4-\delta}$ (PLCCO), decreasing oxygen concentration can suppress static anti-ferromagnetic order but enhance superconductivity.

Figure 20.1(e) indicates the phase diagram cross-section which we have measured in the last several years [16, 17].

For the magnetic excitation, Yamada and collaborators' [14] previous work about e-type NCCO shows that the scattering is commensurate, and (1/2, 1/2) is the center. And they found superconductivity would introduce a very small (about several meV) spin gap. But in our research about PLCCO, we decided on a systematic method to understand the evolution of spin excitation as a function of annealing process: we were the first ones to use the ISIS's MAPS time-of-flight spectrometer to detect all the magnetic excitation spectrum of underdoped PLCCO ($T_C = 21$ K). Lots of position-sensitive detectors allow direct measurements of spin excitation as a function of wave vector in different energies. Images from the magnetic excitation can be cut to determine the dispersion curve (Fig. 20.7). E-doped effect makes the commensurate fluctuation at (1/2, 1/2) broadening along the direction of wave vector in the low-energy region ($\hbar\omega \leq 80$ meV), but in the high-energy region $\hbar\omega \geq 100$ meV, similar to spin-wave, the excitation intensity is suppressed. This obviously leads to the redistribution of local dynamic spin susceptibility $\chi''(\omega)$ as a function of energy, resulting in a new energy scale at about 2 meV (Fig. 20.7). The new energy scale of e-doped PLCCO is significantly less than the optimally-doped LSCO [22] and underdoped YBCO ($x = 0.6$) [20].

Although the above-mentioned data provide a consistent picture of spin excitation in e-type superconductors, whether the h-type superconductors have the same characteristics remains unclear. In fact, upon measuring the whole excitation spectrum of under-doped PLCCO [32], the results seem to suggest that electron-doped materials and hole-doped materials may be fundamentally different, because the former have no observable incommensurate spin fluctuations. However, in the latest study on the optimally-doped PLCCO ($T_C = 24$ K) spin excitation, we found the resonance mode in electron-doped materials [33]. Surprisingly, the resonance energy of PLCCO is just on the universal curve of all systems, which shows that the excitation is independent of the type of carrier and the main features of all high-T_C copper-oxide superconductors. Meanwhile, we found a resonance peak in a low-energy excitation, which has commensurate dispersion relationship (Fig. 20.8), indicating that the resonance peak energy and the observed inward diverging incommensurate excitation in hole-doped copper oxide superconductors have not been essentially linked (Fig. 20.1).

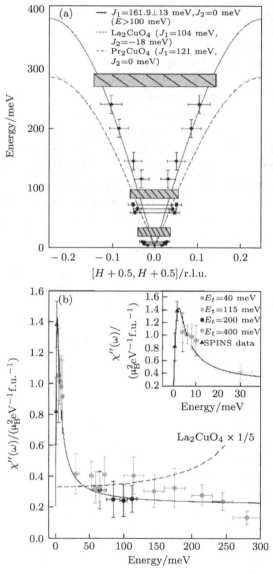

Fig. 20.7. (a) The spin excitation dispersion relation of PLCCO samples at 7 K. Solid line, dashed and dotted lines are dispersion curves of linear spin wave fitting under different exchange energies. (b) The local magnetic susceptibility $\chi''(\omega)$ of PLCCO samples as a function of energy. Dashed line is $\chi''(\omega) \times 1/5$ of La_2CuO_4 [32].

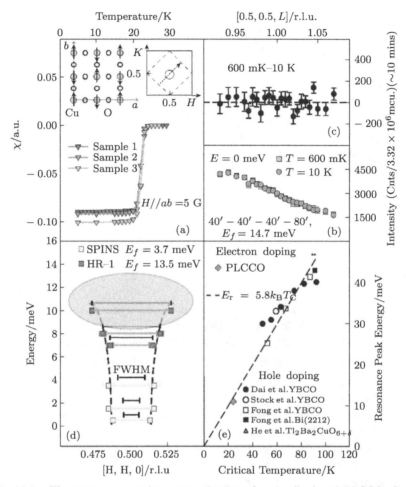

Fig. 20.8. The neutron scattering summarization of optimally-doped PLCCO. Commensurate spin fluctuations' dispersion has commensurate resonance peak at about 11 meV. The energy scale of hole- and electron-copper-oxide superconductors' resonance energy is consistent with $E_r = 5.8\kappa_B T_C$ [33].

20.5. Summary and Conclusions

This chapter briefly summarizes the latest developments of high-T_C copper-oxide superconductors by using neutron scattering. The parent compound can be well understood under the framework of 1/2 spin Mott insulator, which has long-range anti-ferromagnetic order. In the hole-doped materials, including double-layer system YBCO and single layer copper oxide LSCO, the low-energy magnetic fluctuations split at the wave vector $Q = (1/2, 1/2)$, changing to be along the incommensurate position $Q = (0.5 \pm \delta, 0.5 \pm \delta)$. When the energy increases, these low-energy fluctuations gather together at commensurate wave vector $Q = (1/2, 1/2)$. When the energy continues to increase, they outward re-diverge, but rotate outward from their original location (Fig. 20.1(d)). For YBCO, the incommensurate fluctuations converge to energy of commensurate position, and the spectrum weight increases significantly below the T_C, leading to a local magnetic pattern. This mode is known as resonant excitation, it is the strong coupling with superconductivity and in optimally-doped systems, which only appears below the T_C. The resonance pattern is also found in another two hole-doped materials, and the characteristic Er is closely related to systematic superconductivity by the universal relation $E^r = 5.8\kappa_B T_C$. This implies that the resonance peaks are associated with the mechanism of e-e pairs in high-T_C superconducting materials.

Since in the magnetic excitation spectrum of hole-doped copper oxide superconductors, the resonance mode and incommensurate peak coexist, a question naturally arises: do these characteristics also exist in spin excitation of electron-doped copper oxide superconductors? To answer this question, we detect the spin dynamics of optimally-doped PLCCO ($T_C = 24$ K). Using the cold neutron three-axis spectrometer, we studied the low-energy excitation of this system, and found spin fluctuations in at least 0.5 meV with still no gap, and maintained the commensurate position. This is different from the incommensurate phase with spin gap in optimally-doped hole-type copper oxide superconductors (Fig. 20.1). Although the commensurate characteristic of low-energy spin excitation in electron-doped copper oxide superconductors excluded the possibility that the incommensurate peak in hole-type copper oxide superconductors was universal, whether there existed a resonance mode in electron-type copper oxide superconductors is still debatable. We used the thermal neutron three-axis spectrometer, found a local magnetic mode, its intensity increased when the system was cooled to below T_C. The mode took ~11 meV as the center, and precisely by appeared at position of commensurate wave vector $Q = (1/2, 1/2)$. Temperature scans verified that the mode only appeared below T_C, so it is exactly the same with observed mode in hole-type copper oxide superconductors [33].

Therefore, this resonance energy of electron-doped system can compare with the same resonance peak in hole-type copper oxide superconductors, as shown in Fig. 20.8. Surprisingly, the Er of PLCCO is fully consistent with universal curve, so it shows that this excitation is the most basic in all high-T_C copper oxide superconductors, independent of the type of carriers. Also, the systematic commensurate low-energy excitation can propagate to resonance mode (Fig. 20.8). This discovery indicates that resonance energy in hole-type system and observed inward converging incommensurate excitation have essential links. Thus, in the last two decades, the magnetic excitation research of different materials by neutron scattering has provided an important result, i.e. the resonance mode is a basic characteristic of superconducting copper oxide plane. If we can understand the microscopic origin of the resonance peak, then perhaps we can finally determine what is the bonding "glue" in the high-T_C superconductors.

Acknowledgments

We are grateful to graduates Stephen Wilson and Hye JungKang, because their thesis work made this review possible. We also thank the following professors, because their discussion was helpful: Hai-Hu Wen, Shou-Cheng Zhang, Dong-Hai Lee and Zhen-Yu Weng. The support units are US NSF Grant No. DMR20453804, US DOE BES Grant No. DE2FG02-05ER46202. The support unit in ORNL is US DOE Grant No. DE2AC05200OR22725. This work is also supported by the Chinese Academy of Sciences "International Team on Superconductivity and Novel Electronic Materials" Program (ITSCNEM).

References

[1] D. Vaknin et al., *Phys. Rev. Lett.* **58**, 2802 (1987).
[2] D. J. Scalapino, *Phys. Reports* **250**, 330 (1995).
[3] A. Chubukov, D. Pines, J. Schmalian, *The Physics of Superconductors*, Vol. I, *Conventional and High-T_C Superconductors* (Springer, Berlin, 2003).
[4] J. W. Lynn, S. Skanthakumar, *Handbook on the Physics and Chemistry of Rare Earths* (Elsevier, Amsterdam, 2001).
[5] M. A. Kastner, R. J. Birgeneau, G. Shirane et al., *Rev. Mod. Phys.* **70**, 897 (1998).
[6] H. Kimura, K. Hirota, H. Matsushita, K. Yamada et al., *Phys. Rev. B* **59**, 6517 (1999).

[7] K. Yamada, C. H. Lee, K. Kurahashi et al., Phys. Rev. B **57**, 6165 (1998).
[8] S. Wakimoto, G. Shirane, Y. Endoh et al., Phys. Rev. B **60**, R769 (1999).
[9] R. I. Miller et al., Phys. Rev. B **73**, 144509 (2006).
[10] C. Stock et al., Phys. Rev. B **73**, 100504 (2006).
[11] Y. Tokura, H. Takagi, S. Uchida, Nature (London) **337**, 345 (1989).
[12] H. Takagi, S. Uchida, Y. Tokura, Phys. Rev. Lett. **62**, 1197 (1989).
[13] K. Yamada, K. Kurahashi, Y. Endoh et al., J. Phys. Chem. Solids **60**, 1025 (1999).
[14] K. Yamada, K. Kurahashi, T. Uefuji et al., Phys. Rev. Lett. **90**, 137004 (2003).
[15] M. Fujita, T. Kubo, S. Kuroshima et al., Phys. Rev. B **67**, 014514 (2003).
[16] P. Dai, H. J. Kang, H. A. Mook et al., Phys. Rev. B **71**, 100502 (2005).
[17] H. J. Kang, P. Dai, H. A. Mook et al., Phys. Rev. B **71**, 214512 (2005).
[18] R. Coldea, S. M. Hayden, G. Aeppli et al., Phys. Rev. Lett. **86**, 5377 (2001).
[19] S. M. Hayden et al., Phys. Rev. Lett. **76**, 1344 (1996).
[20] P. Dai et al., Science **284**, 1344 (1999).
[21] J. M. Tranquada et al., Nature **429**, 534 (2004).
[22] N. B. Christensen et al., Phys. Rev. Lett. **93**, 147002 (2004).
[23] J. Rossat-Mignod et al., Physica C **185**, 86 (1991).
[24] H. A. Mook et al., Phys. Rev. Lett. **70**, 3490 (1993).
[25] P. Dai, H. A. Mook, R. D. Hunt et al., Phys. Rev. B **63**, 054525 (2001).
[26] S. M. Hayden, H. A. Mook, P. Dai et al., Nature **429**, 531 (2004).
[27] C. Stock, Phys. Rev. B **69**, 014502 (2004).
[28] J. M. Tranquada, Hand Book of High-Temperature Superconductivity: Theory and Experiment, Chapter 6 (Springer New York, 2007).
[29] M. R. Norman, Phys. Rev. B **63**, 092509 (2001).
[30] S. M. Hayden, G. Aeppli, T. G. Perring et al., Phys. Rev. B **54**, R6905 (1996).
[31] M. Fujita et al., Phys. Rev. B **67**, 014514 (2003).
[32] S. D. Wilson et al., Phys. Rev. Lett. **96**, 157001 (2006).
[33] S. D. Wilson et al., Nature **442**, 59 (2006).

A Letter to the High-Temperature Superconducting Colleagues

Dear madam/sir:

Indisputably there have been obvious notable achievements in high-temperature superconductivity in the last two decades. But we have not made a consensus on the mechanism so far. In experiments, we have accumulated quite comprehensive and in-depth work from all aspects, different perspectives, by almost all experimental methods, including improved and new devices for careful examination. So it is time to grasp the main features in high-temperature superconductors from a global perspective. As far as I can see, many researchers have not done that. In theory, there are numerous theoretical models, including many high-profile ones. For each theoretical model, there are many researchers who have invested tremendous enthusiasm and energy, gotten a lot of achievements, and also encountered puzzling problems, but still have not been generally recognized. Fragmentation of a country by rivaling warlords has occurred. Hence, we need a serious and more objective assessment based on experiment.

It took 46 years from the discovery of the phenomenon of superconductivity in 1911 to establish the BCS theory. It is quite challenging to explain the superconducting mechanism, which is one of the basic subjects in condensed matter physics. In order to promote further study, after discussing with some colleagues, the idea of this seminar burst upon me. Different from the general seminar, young experts of the first line will be invited to do mini reviews. We hope the participants will gain overall harvest, especially the younger ones, and face the facts with an open-minded heart and form a view of further work (including preparation, characterization and theoretical simulation). It is our purpose to have adequate time for discussion. We co-launch the seminar for this purpose. "The Assessment Seminar on High-Temperature Superconducting Mechanism Research" will be jointly held by China Center of Advanced Science and Technology (CCAST); Institute of Physics, Chinese Academy of Sciences(CAS); National Laboratory for Superconductivity, Institute of Physics, CAS; School of Physics, Peking University; Center of Advanced Study, Tsinghua University, on March 1–5 2008. Experts of experimental and theoretical aspects on the first front have been invited to give reviews and attempt to conclude the main consensuses, differences, solvable and unsolvable aspects using influential theoretical models in the last 20 years in superconductivity from a global perspective to provide the striving direction and inspire a new enthusiasm and motivation for high-temperature superconductivity research. It requires the scientist to introduce the raw experimental data and the problems objectively, to point out the trend of the theoretical study impartially and discuss the real situation and provide recommendations more deeply.

Last, but not least, I appreciate your valuable advice.

Best wishes,

Ru-Shan Han
December 2007

Conference Summary

The meeting was very successful, and received the participants' good evaluation. It was considerate that the meeting was most timely, wonderful and fruitful. The meeting was welcomed and supported by more than expected, behaving as follows: every invited reporter was pleased to accept the invitation and had made very serious preparations, and the content was very exciting; each report summarized the achievements in the last two decades, and also listed the controversial issues; the overwhelming response, 150–160 people attended, the vast majority held on to the last, and warmly participated in the discussion. The meeting was voted a great success with good organization, in-depth discussion, involved range being very wide, fruitful and would be a major boost for future work. This result is not seen in similar professional meetings of domestic organizations. The conference was successful not only in inviting foreign well-known Chinese scholars Peng-Cheng Dai and Guo-Qing Zheng, who are the outstanding representatives of their fields, but also in the fact that most reporters were young domestic experts, active in the first line, who gave very wonderful reviews, and particularly luckily, brought forth remarkable achievements obtained since large equipment has been built and supported by the state investment in recent years were put into operation, such as low-temperature thermal conductivity, electronic Raman, angle-resolved photoelectron spectroscopy, precision specific heat, and so on. In theory, the reporters gave a concise introduction about the internationally influential theoretical models. During the meeting, some new ideas and work had been widely valued. It is hoped that such a conference will be held again in the near future. Thanks for the sponsorship provided by CCAST and Institute of Physics, Chinese Academy of Science.

Ru-Shan Han
March 5, 2008

CPSIA information can be obtained
at www.ICGtesting.com
Printed in the USA
BVHW041034200620
581572BV00006B/11